I0064878

An Introduction to Materials Science and Engineering

An Introduction to Materials Science and Engineering

Emilio McMahon

New York

Published by NY Research Press
118-35 Queens Blvd., Suite 400,
Forest Hills, NY 11375, USA
www.nyresearchpress.com

An Introduction to Materials Science and Engineering
Emilio McMahon

International Standard Book Number: 978-1-63238-697-7 (Hardback)

Cataloging-in-Publication Data

An introduction to materials science and engineering / Emilio McMahon.
p. cm.
Includes bibliographical references and index.
ISBN 978-1-63238-697-7
1. Materials science. 2. Materials. 3. Engineering. I. McMahon, Emilio.
TA403 .I58 2019
620.11--dc23

Contents

Preface

The field of materials science and engineering is concerned with the design and discovery of new materials. Some of the important areas of study in this field include the structure of materials such as atomic, nano, macro and microstructure, chemical bonding, materials properties, kinetics and thermodynamics of materials, synthesis and processing, besides others. Advances in this field facilitate theoretical and practical advancements in nanotechnology, metallurgy, semiconductor physics and biomaterials. Besides these, the industrial applications of materials science and engineering are immense. Some of them include development in crystal growth, ion implantation and thin-film deposition in the design and manufacture of ceramics, polymers, etc. This book is compiled in such a manner, that it will provide in-depth knowledge about the principles and applications of materials science and engineering. It unfolds the innovative aspects of this field, which will be crucial for the holistic understanding of the subject matter. This textbook will serve as a reference to a broad spectrum of readers.

A detailed account of the significant topics covered in this book is provided below:

Chapter 1- Materials science and engineering is a field of study concerned with the design and discovery of new materials. It incorporates the principles of physics, engineering and chemistry. This chapter has been carefully written to provide an introduction to materials science and materials engineering.

Chapter 2- An atomic bond is an attractive force between atoms or ions, which enables the formation of chemical compounds. It can be an ionic bond, covalent bond or a metallic bond. The topics elucidated in this chapter cover the essentials of atomic bonding such as Kössel-Lewis approach to chemical bonding, hydrogen bonding, strong chemical bond, etc.

Chapter 3- Materials exhibit diverse properties such as mechanical, chemical, thermal, magnetic, optical and electrical properties. These properties determine the usability of materials in engineering applications. An elaborate study of the varied properties of materials has been provided in this chapter.

Chapter 4- Crystalline materials have a number of crystalline defects, which have a significant influence on its properties. The diffusion process by which the random thermally activated movement of atoms results in a net transfer of atoms in a solid is known as atomic diffusion. This chapter discusses the crystalline defects and diffusion through an elucidation of topics such as point defects, line defects, planar defects and amorphous solids.

Chapter 5- A metal is any material that is a good conductor of heat and electricity. It is malleable and ductile. Some important materials are iron, aluminium, copper, etc. which have been discussed in this chapter.

Chapter 6- A polymer is a large molecule that is composed of repeating subunits. The study of the structure of a polymer is important for an understanding of polymers. This chapter closely examines the properties of polymers through the study of the physical, chemical and thermal properties, polymer crystallization, etc.

Chapter 7- A ceramic is a non-metallic material that is comprised of an inorganic compound of metal, non-metal or metalloid held together through ionic or covalent bonds. A material formed from two or more materials having widely differing physical and chemical properties, such that the combined material has unique properties from its constituents is known as a composite. The topics included in this chapter cover the diverse aspects of ceramic and composite materials such as their properties, fabrication, processing, etc.

Chapter 8- A phase diagram is a representation of the physical conditions of pressure, volume or temperature at which distinct thermodynamic phases coexist and occur at equilibrium. This is an important chapter, which analyzes in detail about phase diagrams, its various types, phase equilibria, etc.

It gives me an immense pleasure to thank our entire team for their efforts. Finally in the end, I would like to thank my family and colleagues who have been a great source of inspiration and support.

Emilio McMahon

1

Introduction to Materials Science and Engineering

Materials science and engineering is a field of study concerned with the design and discovery of new materials. It incorporates the principles of physics, engineering and chemistry. This chapter has been carefully written to provide an introduction to materials science and materials engineering.

Materials

A material is any substance or mixture of substances that occupy a volume, and has a mass. A substance usually refers to pure compound. Materials are so important in the development of civilization that we associate Ages with them. In the origin of human life on Earth, the Stone Age, people used only natural materials, like stone, clay, skins, and wood. When people found copper and how to make it harder by alloying. The use of iron and steel, a stronger material that gave advantage in wars. The next big step was the discovery of a cheap process to make steel, which enabled the railroads and the building of the modern infrastructure of the industrial world.

The tree-diagram below shows an overview of a variety of materials that might be encountered:

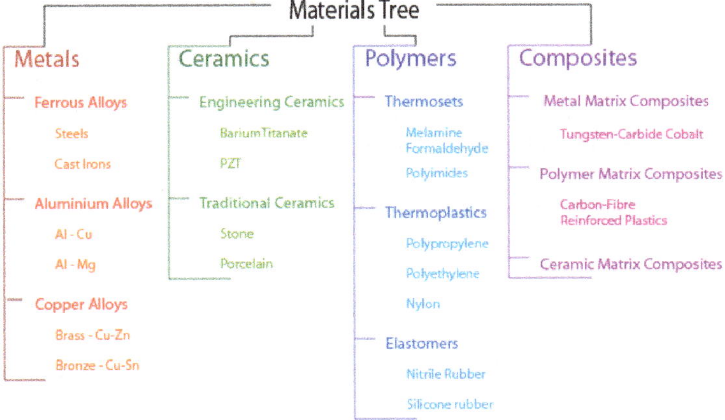

Metals

There are many metals which you are familiar with - copper pipes and wire, aluminium saucepans and cast iron stoves. Metals may be mixed with other elements especially other metals to produce alloys which will have improved properties. Heat treatment can also be used to change the properties of alloys e.g. hardening and tempering of high carbon steel.

All metals are good conductors of heat and electricity. Copper is a particularly good conductor but is not very strong, it is also fairly dense. Aluminium is a good conductor has a low density and when alloyed has a high tensile strength. Some alloys such as pewter and zinc alloys have a low melting point and can be easily formed by casting or moulding but they have a low tensile strength.

Metals in common use are corrosion resistant except iron and steel which rust quickly. Corrosion resistance is achieved by electroplating to add a layer of corrosion resistant material such as chromium or zinc, painting, plastic coating, and coating with an oil or grease. The alloy stainless steel is very rust resistant.

When choosing a metal for a particular job the properties must be carefully considered. For example aluminium could be used for overhead power lines as its lower density and good tensile strength offset its slightly lower electrical conductivity.

Polymers

Polymers are made from long chain molecules which may have cross linking bonds affecting flexibility/stiffness.

There are three groups of polymer:

- Thermoplastics: which may be reformed with heat. e.g. PVC, HIPS, nylon, polycarbonate, PET, acrylic.
- Thermosetting plastics: which once moulded or formed cannot be reformed by heat. e.g. Melamine(MF), epoxy resin, Urea formaldehyde (UF).
- Elastomers: rubbers long chain elastic molecules. e.g. neoprene, natural rubber. Used for car tyres and elastic bands.

Applications:

Nylon is used for bearings and the cases for power tools also used for fishing line and ropes. Nylon is very strong and wear resistant it is also slippery without the need for lubrication. Originally used as a silk substitute - stockings and climbing ropes.

PVC is used for casings for electrical consumer items and is also used in its flexible form as the insulating sheath on electrical cable and flex.

Melamine is used as the protective layer on work surfaces and laminated flooring.

UF is used to make electrical components when a good insulator is needed such as plug tops and switch buttons.

Acrylic is used for safety shields but is not as tough as polycarbonate. Polycarbonate is used for the lenses in safety eye protection e.g. goggles.

Ceramics

This class of material includes plates and cups, bricks, earthenware pots, engineering ceramics, glasses [glasses are non-crystalline and not normally classed as ceramics], and refractory (furnace) materials. Ceramics are made by heating together materials such as silica, chalk and clays. Other chemicals may be included to act as flux and to change colour etc.

Engineering Ceramics Include:

- Silicon carbide
- Zirconia
- Silicon nitride
- Diamond
- Cubic boron nitride
- Tungsten Carbide
- Properties:

Engineering ceramics are ideally suited for high performance applications where a combination of properties such as wear resistance, hardness, stiffness and corrosion resistance are important. In addition to these properties, engineering ceramics have relatively high mechanical strength at high temperatures. They are good electrical insulators, They often have a close thermal expansion coefficient to metals (they can be bonded to metals - e.g. carbide tipped tools).

Ceramics have been regarded as hard but brittle, however modern ceramics have been developed which are viable alternatives to metals and their alloys in many applications - engineering ceramic parts and components are more durable and have longer life-spans under given operational conditions. Ceramic cutting tools, for instance, require less sharpening or replacement due to wear, and will last at least 60 to 100 times longer than steel blades.

Engineering ceramics are chemically resistant to most acids, alkalis and organic solvents and can withstand high temperatures. Metals weaken rapidly at temperatures above 816 degrees C while engineering ceramics retain a good degree of their mechanical properties at much higher temperatures.

Applications:

Mechanical components include wear plates and thermal barriers, bearings for high speed and high stiffness spindles, bushes, gears.

Process components include pump shafts, seats, bearing surfaces, gears and even complete pump bodies, valve guides and seats.

Ceramics are used for cutting tools including razor blades for film and tape cutting to 300mm diameter circular slitters for the paper industry.

Ceramic turbine blades are used in most turbochargers providing lighter units than the steel alternatives allowing improved performance at higher temperatures.

Composites

Composites are mixtures of materials which give improved properties. One of the materials is the matrix or binding chemical and the other is the reinforcer. A good example is GRP - glass reinforced polyester (plastic) resin. where the glass fibres increase the strength of the polyester resin. Carbon fibre reinforced epoxy resin is stronger and lighter than steel.

Concrete is a composite (the cement is the matrix and the gravel and steel rods are the reinforcer) as are bricks made from clay reinforced with straw.

Natural composites include wood, shell and bone.

Applications:

Car bodies - especially sports cars, F1 racing cars, boat hulls, lightweight struts and supports in bridge building and the construction industry.

Aerospace - use of carbon fibre composites as well as high tech ceramic parts has revolutionised this industry.

Properties of Materials

Materials are most of all the objects and therefore materials have its own properties. In general different properties of materials are enlisted below. Since materials have these properties it makes the materials useful and purposeful to use. To understand the properties of material explore the article.

The material can be termed as the mixture of materials to compose a thing.

Properties of Materials

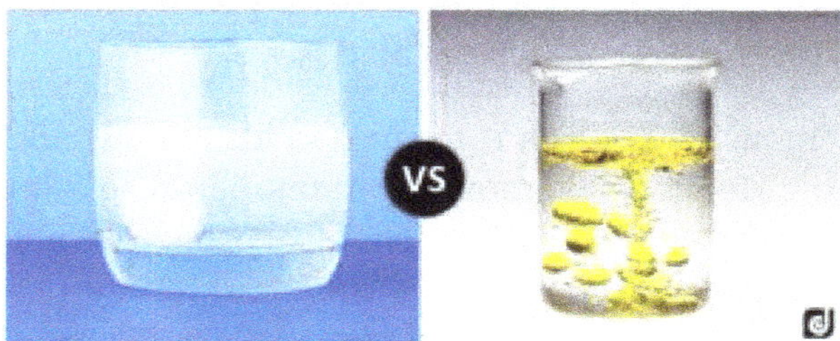

Soluble vs. Insoluble

- Transparency/Opaque: The amount of light material allow passing through it is transparency of the material. The maximum amount of light to pass through the material, therefore, they are transparent materials. Examples: Plastic, Air, and Glass.
- Translucent materials are, which that only allow light to pass through them partially. Examples: Oiled paper, Coloured syrup, and some sheer materials. An opaque object is the materials are which don't allow any light to pass through them. Examples: Wood, Cardboard, and Metals.
- The appearance of the Material: The look, feel, texture in addition to lustre, colour and quality defines the property of appearance of the material.
- Soluble/Insoluble: The nature of the material to completely dissolve in water and therefore termed as soluble material. Materials are completely dissolvable in water, therefore, they

are soluble materials. Examples: Lemon juice, Sugar and Salt. Materials don't completely dissolve in water, they form a thin film over the surface of the water, therefore, they are iron rods and copper wires. Examples: Oil, Kerosene and Sawdust.

Some More Properties of Materials

- Float/Sink: weight of the material helps to define this property. The material is lightweight it will float over the surface, therefore, they are floating material. Examples: Insoluble materials like Sawdust, Oil, Plastic and Wood. The material is heavyweight it will sink in water, therefore, they are sinking materials. Examples: Stone and Metals.
- Heat and electricity conductivity: Materials that allow heat and electricity pass through the material and hence called good conductors of heat and electricity. Good conductor of heat and electricity are which allow the head and electricity pass through them. Examples: Iron rods and copper wires. Bad conductors are the materials which don't allow heat and electricity pass through them. Examples: hydrochloric acid.
- State of a material: Compactness and the consistency of the materials describe the state of a material. There are 3 states in which a material can be classified and they are as solid, liquid and gaseous. Solid materials are the most compact and the particles are densely packed.

Examples: Wood, Paper and Glass. Liquid materials are less compact and particles are moderately densely packed. Examples: Water, Oil and Kerosene. Gaseous materials are least compact and particles are loosely packed. Examples: Air.

Nanomaterials

Nanoscale materials are defined as a set of substances where at least one dimension is less than approximately 100 nanometers. A nanometer is one millionth of a millimeter - approximately 100,000 times smaller than the diameter of a human hair. Nanomaterials are of interest because at this scale unique optical, magnetic, electrical, and other properties emerge. These emergent properties have the potential for great impacts in electronics, medicine, and other fields.

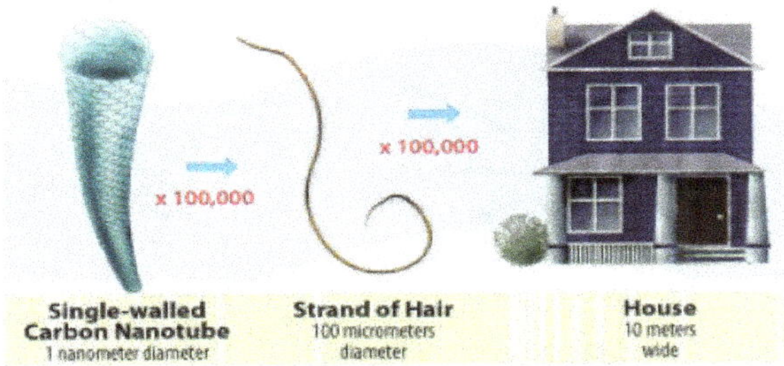

Figure: Nanomaterial (For Example:Carbon nanotube)

Classification of Nanomaterials

Nanomaterials have extremely small size which having at least one dimension 100 nm or less. Nanomaterials can be nanoscale in one dimension (eg. Surface films), two dimensions (eg. strands or fibres), or three dimensions (eg. particles). They can exist in single, fused, aggregated or agglomerated forms with spherical, tubular, and irregular shapes. Common types of nanomaterials include nanotubes, dendrimers, quantum dots and fullerenes. Nanomaterials have applications in the field of nano technology, and displays different physical chemical characteristics from normal chemicals (i.e., silver nano, carbon nanotube, fullerene, photocatalyst, carbon nano, silica). According to Siegel, Nanostructured materials are classified as Zero dimensional, one dimensional, two dimensional, three dimensional nanostructures.

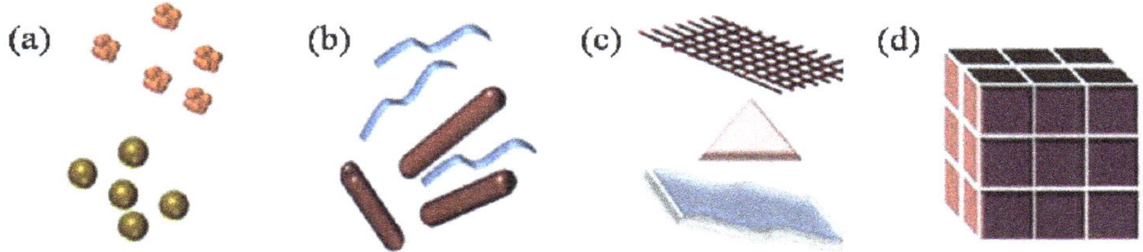

Figure: Classification of Nanomaterials (a) 0D spheres and clusters; (b) 1D nanofibers, nanowires, and nanorods; (c) 2D nanofilms, nanoplates, and networks; (d) 3D nanomaterials.

Nanomaterials are materials which are characterized by an ultra-fine grain size (< 50 nm) or by a dimensionality limited to 50 nm. Nanomaterials can be created with various modulation dimensionalities as defined by Richard W. Siegel: zero (atomic clusters, filaments and cluster assemblies), one (multilayers), two (ultrafine-grained over layers or buried layers), and three (nanophase materials consisting of equiaxed nanometer sized grains) as shown in the above figure.

Nanomaterial Synthesis and Processing

We are dealing with very fine structures: a nanometer is a billionth of a meter. This indeed allows us to think in both the 'bottom up' or the 'top down' approaches to synthesize nanomaterials, i.e. either to assemble atoms together or to dis-assemble (break, or dissociate) bulk solids into finer pieces until they are constituted of only a few atoms. This domain is a pure example of interdisciplinary work encompassing physics, chemistry, and engineering up to medicine.

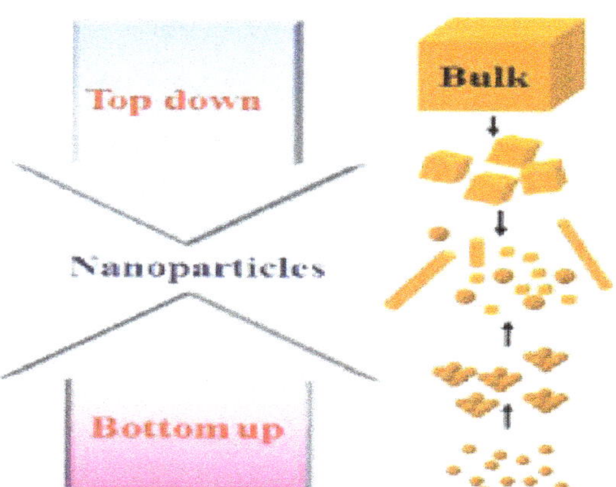

Figure: Schematic illustration of the preparative methods of nanoparticles.

Methods for Creating Nanostructures

There are many different ways of creating nanostructures: of course, macromolecules or nanoparticles or buckyballs or nanotubes and so on can be synthesized artificially for certain specific materials. They can also be arranged by methods based on equilibrium or near-equilibrium thermodynamics such as methods of self-organization and self-assembly (sometimes also called bio-mimetic processes). Using these methods, synthesized materials can be arranged into useful shapes so that finally the material can be applied to a certain application.

Mechanical Grinding

Mechanical attrition is a typical example of 'top down' method of synthesis of nanomaterials, where the material is prepared not by cluster assembly but by the structural decomposition of coarser-grained structures as the result of severe plastic deformation. This has become a popular method to make nanocrystalline materials because of its simplicity, the relatively inexpensive equipment needed, and the applicability to essentially the synthesis of all classes of materials. The major advantage often quoted is the possibility for easily scaling up to tonnage quantities of material for various applications. Similarly, the serious problems that are usually cited are:

1. Contamination from milling media and/or atmosphere, and
2. To consolidate the powder product without coarsening the nanocrystalline microstructure.

In fact, the contamination problem is often given as a reason to dismiss the method, at least for some materials. Here we will review the mechanisms presently believed responsible for formation of nanocrystalline structures by mechanical attrition of single phase powders, mechanical alloying of dissimilar powders, and mechanical crystallisation of amorphous materials. The two important problems of contamination and powder consolidation will be briefly considered.

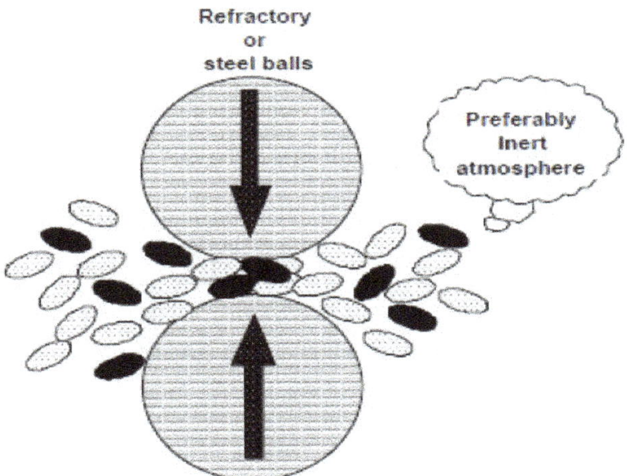

Figurre: Schematic representation of the principle of mechanical milling.

Mechanical milling is typically achieved using high energy shaker, planetary ball, or tumbler mills. The energy transferred to the powder from refractory or steel balls depends on the rotational (vibrational) speed, size and number of the balls, ratio of the ball to powder mass, the time of milling and the milling atmosphere. Nanoparticles are produced by the shear action during grinding.

Milling in cryogenic liquids can greatly increase the brittleness of the powders influencing the fracture process. As with any process that produces fine particles, an adequate step to prevent oxidation is necessary. Hence this process is very restrictive for the production of non-oxide materials since then it requires that the milling take place in an inert atmosphere and that the powder particles be handled in an appropriate vacuum system or glove box. This method of synthesis is suitable for producing amorphous or nano crystalline alloy particles, elemental or compound powders. If the mechanical milling imparts sufficient energy to the constituent powders a homogeneous alloy can be formed. Based on the energy of the milling process and thermodynamic properties of the constituents the alloy can be rendered amorphous by this processing.

Wet Chemical Synthesis of Nanomaterials

In principle we can classify the wet chemical synthesis of nanomaterials into two broad groups:

1. The top down method: where single crystals are etched in an aqueous solution for producing nanomaterials, For example, the synthesis of porous silicon by electrochemical etching.
2. The bottom up method: consisting of sol-gel method, precipitation etc. where materials containing the desired precursors are mixed in a controlled fashion to form a colloidal solution.

Gas Phase Synthesis of Nanomaterials

The gas-phase synthesis methods are of increasing interest because they allow elegant way to control process parameters in order to be able to produce size, shape and chemical composition controlled nanostructures. In conventional chemical vapour deposition (CVD) synthesis, gaseous products either are allowed to react homogeneously or heterogeneously depending on a particular application.

1. In homogeneous CVD, particles form in the gas phase and diffuse towards a cold surface due to thermophoretic forces, and can either be scrapped of from the cold surface to give nano-powders, or deposited onto a substrate to yield what is called 'particulate films'.
2. In heterogeneous CVD, the solid is formed on the substrate surface, which catalyses the reaction and a dense film is formed.

In order to form nanomaterials several modified CVD methods have been developed. Gas phase processes have inherent advantages, some of which are noted here:

- An excellent control of size, shape, crystallinity and chemical composition.
- Highly pure materials can be obtained.
- Multicomonent systems are relatively easy to form.
- Easy control of the reaction mechanisms.

Most of the synthesis routes are based on the production of small clusters that can aggregate to form nano particles (condensation). Condensation occurs only when the vapour is supersaturated and in these processes homogeneous nucleation in the gas phase is utilised to form particles. This can be achieved both by physical and chemical methods.

Furnace

The simplest fashion to produce nanoparticles is by heating the desired material in a heat-resistant crucible containing the desired material. This method is appropriate only for materials that have a high vapor pressure at the heated temperatures that can be as high as 2000°C. Energy is normally introduced into the precursor by arc heating, electron-beam heating or Joule heating. The atoms are evaporated into an atmosphere, which is either inert (e.g. He) or reactive (so as to form a compound). To carry out reactive synthesis, materials with very low vapor pressure have to be fed into the furnace in the form of a suitable precursor such as organometallics, which decompose in the furnace to produce a condensable material. The hot atoms of the evaporated matter lose energy by collision with the atoms of the cold gas and undergo condensation into small clusters via homogeneous nucleation. In case a compound is being synthesized, these precursors react in the gas phase and form a compound with the material that is separately injected in the reaction chamber. The clusters would continue to grow if they remain in the supersaturated region. To control their size, they need to be rapidly removed from the supersaturated environment by a carrier gas. The cluster size and its distribution are controlled by only three parameters:

1) The rate of evaporation (energy input),
2) The rate of condensation (energy removal), and
3) The rate of gas flow (cluster removal).

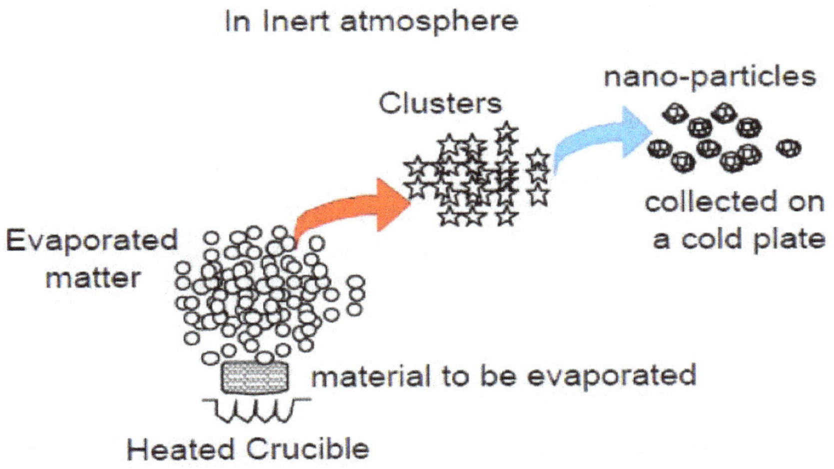

Figure: Schematic representation of gas phase process of synthesis of single phase nanomaterials from a heated crucible.

Because of its inherent simplicity, it is possible to scale up this process from laboratory (mg/day) to industrial scales (tons/day).

Flame Assisted Ultrasonic Spray Pyrolysis

In this process, precusrsors are nebulized and then unwanted components are burnt in a flame to get the required material, eg. ZrO_2 has been obtained by this method from a precursor of Zr $(CH_3 CH_2 CH_2O)_4$. Flame hydrolysis that is a variant of this process is used for the manufacture of fused silica. In the process, silicon tetrachloride is heated in an oxy-hydrogen flame to give highly dispersed silica. The resulting white amorphous powder consists of spherical particles with sizes in the range 7-40 nm. The combustion flame synthesis, in which the burning of a gas mixture, e.g. acetylene and oxygen or hydrogen and oxygen, supplies the energy to initiate the pyrolysis of precursor compounds, is widely used for the industrial production of powders in large quantities, such as carbon black, fumed silica and titanium dioxide. However, since the gas pressure during the reaction is high, highly agglomerated powders are produced which is disadvantageous for subsequent processing. The basic idea of low pressure combustion flame synthesis is to extend the pressure range to the pressures used in gas phase synthesis and thus to reduce or avoid the agglomeration. Low pressure flames have been extensively used by aerosol scientists to study particle formation in the flame.

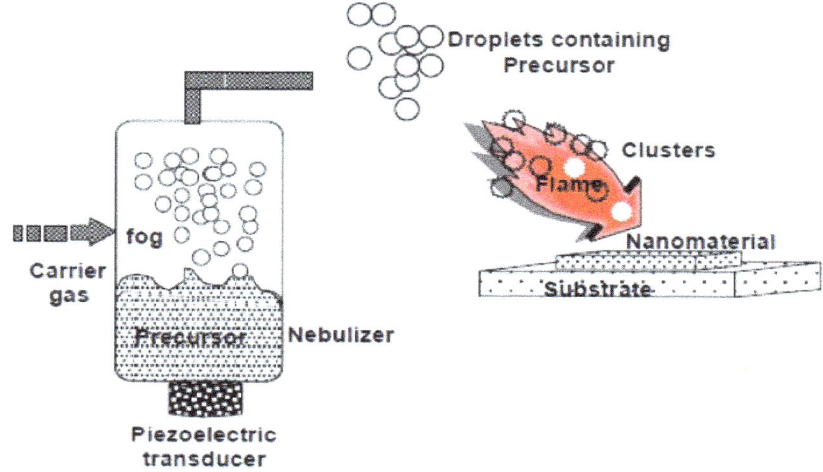

Fig.: Flame assisted ultrasonic spray pyrolysis

A key for the formation of nanoparticles with narrow size distributions is the exact control of the flame in order to obtain a flat flame front. Under these conditions the thermal history, i.e. time and temperature, of each particle formed is identical and narrow distributions result. However, due to the oxidative atmosphere in the flame, this synthesis process is limited to the formation of oxides in the reactor zone.

Gas Condensation Processing (GPC)

In this technique, a metallic or inorganic material, e.g. a suboxide, is vaporized using thermal evaporation sources such as crucibles, electron beam evaporation devices or sputtering sources in an atmosphere of 1-50 mbar He (or another inert gas like Ar, Ne, Kr). Cluster form in the vicinity of the source by homogenous nucleation in the gas phase and grow by coalescence and incorporation of atoms from the gas phase. The cluster or particle size depends critically on the residence time of the particles in the growth system and can be influenced by the gas pressure, the kind of inert gas, i.e. He, Ar or Kr, and on the evaporation rate/vapor pressure of the evaporating material. With increasing gas pressure, vapor pressure and mass of the inert gas used the average particle size of the nanoparticles increases. Lognormal size distributions have been found experimentally and have been explained theoretically by the growth mechanisms of the particles. Even in more complex processes such as the low pressure combustion flame synthesis where a number of chemical reactions are involved the size distributions are determined to be lognormal.

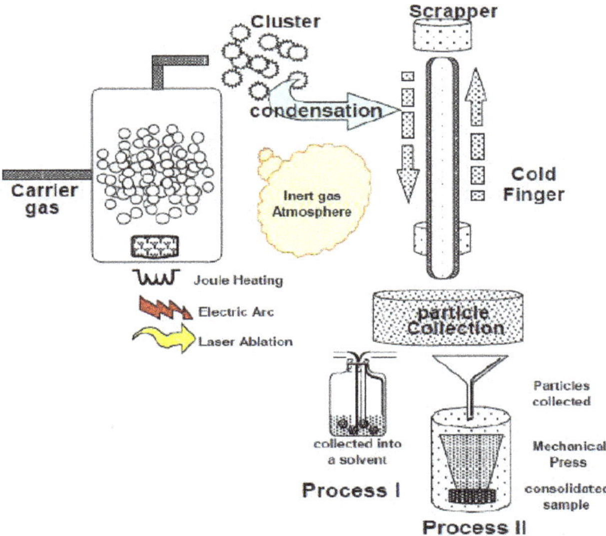

Figure: Schematic representation of typical set-up for gas condensation synthesis of nanomaterials followed by consolidation in a mechanical press or collection in an appropriate solvent media.

Originally, a rotating cylindrical device cooled with liquid nitrogen was employed for the particle collection: the nanoparticles in the size range from 2-50 nm are extracted from the gas flow by thermophoretic forces and deposited loosely on the surface of the collection device as a powder of low density and no agglomeration. Subsequenly, the nanoparticles are removed from the surface of the cylinder by means of a scraper in the form of a metallic plate. In addition to this cold finger device several techniques known from aerosol science have now been implemented for the use in gas condensation systems such as corona discharge, etc. These methods allow for the continuous operation of the collection device and are better suited for larger scale synthesis of nanopowders. However, these methods can only be used in a system designed for gas flow, i.e. a dynamic vacuum is generated by means of both continuous pumping and gas inlet via mass flow controller. A major advantage over convectional gas flow is the improved control of the particle sizes. It has been found that the particle size distributions in gas flow systems, which are also lognormal, are shifted towards smaller average values with an appreciable reduction of the standard deviation of the distribution. Depending on the flow rate of the He-gas, particle sizes are reduced by 80% and standard deviations by 18%.

The synthesis of nanocrystalline pure metals is relatively straightforward as long as evaporation can be done from refractory metal crucibles (W, Ta or Mo). If metals with high melting points or metals which react with the crucibles, are to be prepared, sputtering, i.e. for W and Zr, or laser or electron beam evaporation has to be used. Synthesis of alloys or intermetallic compounds by thermal evaporation can only be done in the exceptional cases that the vapour pressures of the elements are similar. As an alternative, sputtering from an alloy or mixed target can be employed. Composite materials such as Cu/Bi or W/Ga have been synthesised by simultaneous evaporation from two separate crucibles onto a rotating collection device. It has been found that excellent intermixing on the scale of the particle size can be obtained.

However, control of the composition of the elements has been difficult and reproducibility is poor. Nanocrystalline oxide powders are formed by controlled postoxidation of primary nanoparticles of a pure metal (e.g. Ti to TiO_2) or a suboxide (e.g. ZrO to ZrO_2). Although the gas condensation method including the variations have been widely employed to prepared a variety of metallic and ceramic materials, quantities have so far been limited to a laboratory scale. The quantities of metals are below 1 g/day, while quantities of oxides can be as high as 20 g/day for simple oxides such as CeO_2 or ZrO_2. These quantities are sufficient for materials testing but not for industrial production. However, it should be mentioned that the scale-up of the gas condensation method for industrial production of nanocrystalline oxides by a company called nanophase technologies has been successful.

Chemical Vapor Condensation (CVC)

As shown schematically in Figure, the evaporative source used in GPC is replaced by a hot wall reactor in the Chemical Vapor Condensation or the CVC process. Depending on the processing parameters nucleation of nanoparticles is observed during chemical vapor deposition (CVC) of thin films and poses a major problem in obtaining good film qualities. The original idea of the novel CVC process which is schematically shown where, it was intended to adjust the parameter field during the synthesis in order to suppress film formation and enhance homogeneous nucleation of particles in the gas flow. It is readily found that the residence time of the precursor in the reactor determines if films or particles are formed. In a certain range of residence time both particle and film formation can be obtained.

Adjusting the residence time of the precursor molecules by changing the gas flow rate, the pressure difference between the precursor delivery system and the main chamber occurs. Then the temperature of the hot wall reactor results in the fertile production of nanosized particles of metals and ceramics instead of thin films as in CVD processing. In the simplest form a metal organic precursor is introduced into the hot zone of the reactor using mass flow controller. Besides the increased quantities in this continuous process compared to GPC has been demonstrated that a wider range of ceramics including nitrides and carbides can be synthesized. Additionally, more complex oxides such as BaT_iO_3 or composite structures can be formed as well. Appropriate precursor compounds can be readily found in the CVD literature. The extension to production of nanoparticles requires the determination of a modified parameter field in order to promote particle formation instead of film formation. In addition to the formation of single phase nanoparticles by CVC of a single precursor the reactor allows the synthesis of:

1. Mixtures of nanoparticles of two phases or doped nanoparticles by supplying two precursors at the front end of the reactor, and
2. Coated nanoparticles, i.e., n-ZrO_2 coated with n-Al_2O_3 or vice versa, by supplying a second precursor at a second stage of the reactor. In this case nanoparticles which have been formed by homogeneous nucleation are coated by heterogeneous nucleation in asecond stage of the reactor.

Because CVC processing is continuous, the production capabilities are much larger than in GPC processing. Quantities in excess of 20 g/hr have been readily produced with a small scale laboratory reactor. A further expansion can be envisaged by simply enlarging the diameter of the hot wall reactor and the mass flow through the reactor.

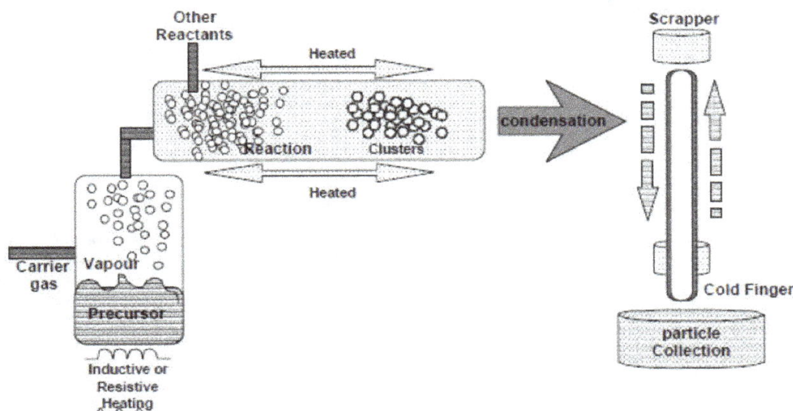

Figure: A schematic of a typical CVC reactor

Sputtered Plasma Processing

In this method is yet again a variation of the gas-condensation method excepting the fact that the source material is a sputtering target and this target is sputtered using rare gases and the constituents are allowed to agglomerate to produce nanomaterial. Both dc (direct current) and rf (radio-frequency) sputtering has been used to synthesize nanoparticles. Again reactive sputtering or multitarget sputtering has been used to make alloys and/or oxides, carbides, nitrides of materials. This method is specifically suitable for the preparation of ultrapure and non-agglomerated nanoparticles of metal.

Microwave Plasma Processing

This technique is similar to the previously discussed CVC method but employs plasma instead of high temperature for decomposition of the metal organic precursors. The method uses microwave plasma in a 50 mm diameter reaction vessel made of quartz placed in a cavity connected to a microwave generator. A precursor such as a chloride compound is introduced into the front end of the reactor. Generally, the microwave cavity is designed as a single mode cavity using the TE10 mode in a WR975 waveguide with a frequency of 0.915 GHz. The major advantage of the plasma assisted pyrolysis in contrast to the thermal activation is the low temperature reaction which reduces the tendency for agglomeration of the primary particles. This is also true in the case of plasma-CVD processes. Additionally, it has been shown that by introducing another precursor into a second reaction zone of the tubular reactor, e.g. by splitting the microwave guide tubes, the primary particles can be coated with a second phase. For example, it has been demonstrated that ZrO_2 nanoparticles can be coated by Al_2O_3. In this case the inner ZrO_2 core is crystalline, while the Al_2O_3 coating is amorphous. The reaction sequence can be reversed with the result that an amorphous Al_2O_3 core is coated with crystalline ZrO_2. While the formation of the primary particles occurs by homogeneous nucleation, it can be easily estimated using gas reaction kinetics that the coating on the primary particles grows heterogeneously and that homogeneous nucleation of nanoparticles originating from the second compound has a very low probability. A schematic representation of the particle growth in plasma's is given below:

Particle precipitation aided DVD

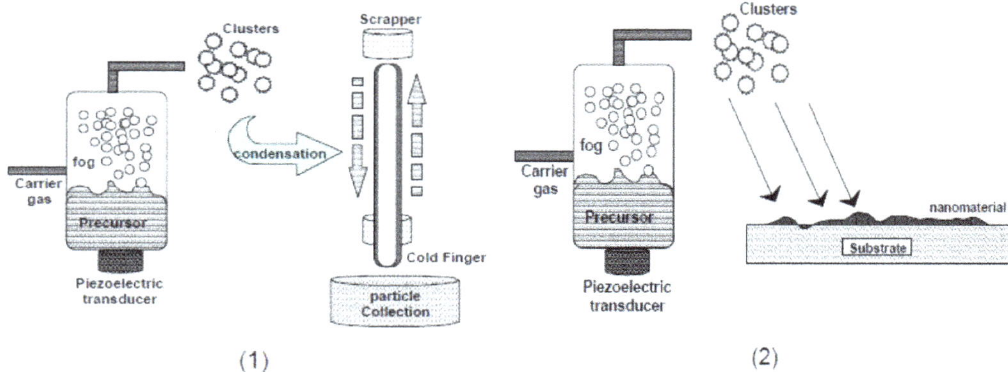

Figure: Schematic representation of (1) nanoparticle, and (2) particulate film formation.

In another variation of this process, colloidal clusters of materials are used to prepare nanoparticles. The CVD reaction conditions are so set that particles form by condensation in the gas phase and collect onto a substrate, which is kept under a different condition that allows heterogeneous nucleation. By this method both nanoparticles and particulate films can be prepared. An example of this method has been used to form nanomaterials eg. SnO_2, by a method called pyrosol deposition process, where clusters of tin hydroxide are transformed into small aerosol droplets, following which they are reacted onto a heated glass substrate.

Laser Ablation

Laser ablation has been extensively used for the preparation of nanoparticles and particulate films. In this process a laser beam is used as the primary excitation source of ablation for generating clusters directly from a solid sample in a wide variety of applications. The small dimensions of the particles and the possibility to form thick films make this method quite an efficient tool for the production of ceramic particles and coatings and also an ablation source for analytical applications such as the coupling to induced coupled plasma emission spectrometry, ICP, the formation of the nanoparticles has been explained following a liquefaction process which generates an aerosol, followed by the cooling/solidification of the droplets which results in the formation of fog. The general dynamics of both the aerosol and the fog favors the aggregation process and micrometer-sized fractal-like particles are formed. The laser spark atomizer can be used to produce highly mesoporous thick films and the porosity can be modified by the carrier gas flow rate. ZrO_2 and SnO_2 nanoparticulate thick films were also synthesized successfully using this process with quite identical microstructure. Synthesis of other materials such as lithium manganate, silicon and carbon has also been carried out by this technique.

Properties of Nanomaterials

Nanomaterials have the structural features in between of those of atoms and the bulk materials. While most micro structured materials have similar properties to the corresponding bulk materials, the properties of materials with nanometer dimensions are significantly different from those of atoms and bulks materials. This is mainly due to the nanometer size of the materials which render them: (i) large fraction of surface atoms; (ii) high surface energy; (iii) spatial confinement; (iv) reduced imperfections, which do not exist in the corresponding bulk materials. Due to their small dimensions, nano materials have extremely large surface area to volume ratio, which makes a large to be the surface or interfacial atoms, resulting in more "surface" dependent material properties. Especially when the sizes of nano materials are comparable to length, the entire material will be affected by the surface properties of nanomaterials. This in turn may enhance or modify the properties of the bulk materials. For example, metallic nanoparticles can be used as very active catalysts. Chemical sensors from nanoparticles and nanowires enhanced the sensitivity and sensor selectivity. The nanometer feature sizes of nanomaterials also have spatial confinement effect on the materials, which bring the quantum effects.

The energy band structure and charge carrier density in the materials can be modified quite differently from their bulk and in turn will modify the electronic and optical properties of the materials. For example, lasers and light emitting diodes (LED) from both of the quantum dots and quantum wires are very promising in the future optoelections. High density information storage using quantum dot devices is also a fast developing area. Reduced imperfections are also an important factor in determination of the properties of the nanomaterials. Nanosturctures and Nanomaterials favors of a self-purification process in that the impurities and intrinsic material defects will move to near the surface upon thermal annealing. This increased materials perfection affects the properties of nanomaterials. For example, the chemical stability for certain nanomaterials may be enhanced, the mechanical properties of nanomaterials will be better than the bulk materials. The superior mechanical properties of carbon nanotubes are well known. Due to their nanometer size, nanomaterials are already known to have many novel properties. Many novel applications of the nanomaterials rose from these novel properties have also been proposed.

Optical Properties

One of the most fascinating and useful aspects of nanomaterials is their optical properties. Applications based on optical properties of nanomaterials include optical detector, laser, sensor, imaging, phosphor, display, solar cell, photocatalysis, photoelectrochemistry and biomedicine.

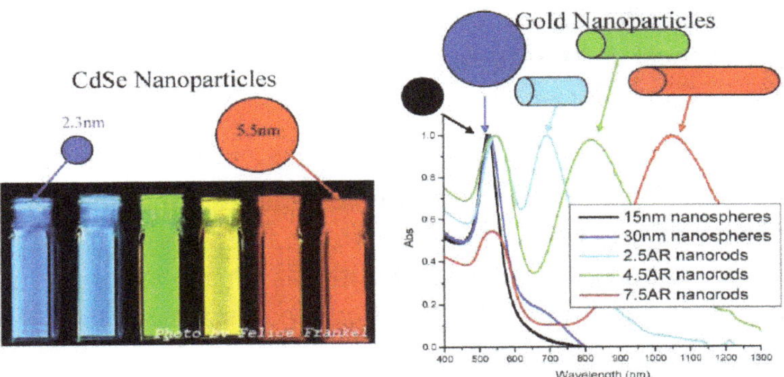

Figure: Fluorescence emission of (CdSe) ZnS quantum dots of various sizes and absorption spectra of various sizes and shapes of gold nanoparticles.

The optical properties of nanomaterials depend on parameters such as feature size, shape, surface characteristics, and other variables including doping and interaction with the surrounding environment or other nanostructures. Likewise, shape can have dramatic influence on optical properties of metal nanostructures. Figure exemplifies the difference in the optical properties of metal and semiconductor nanoparticles. With the CdSe semiconductor nanoparticles, a simple change in size alters the optical properties of the nanoparticles. When metal nanoparticles are enlarged, their optical properties change only slightly as observed for the different samples of gold nanospheres. However, when an anisotropy is added to the nanoparticle, such as growth of nanorods, the optical properties of the nanoparticles change dramatically.

Electrical Properties

Electrical Properties of Nanoparticles' discuss about fundamentals of electrical conductivity in nanotubes and nanorods, carbon nanotubes, photoconductivity of nanorods, electrical conductivity of nanocomposites. One interesting method which can be used to demonstrate the steps in conductance is the mechanical thinning of a nanowire and measurement of the electrical current at a constant applied voltage. The important point here is that, with decreasing diameter of the wire, the number of electron wave modes contributing to the electrical conductivity is becoming increasingly smaller by well-defined quantized steps.

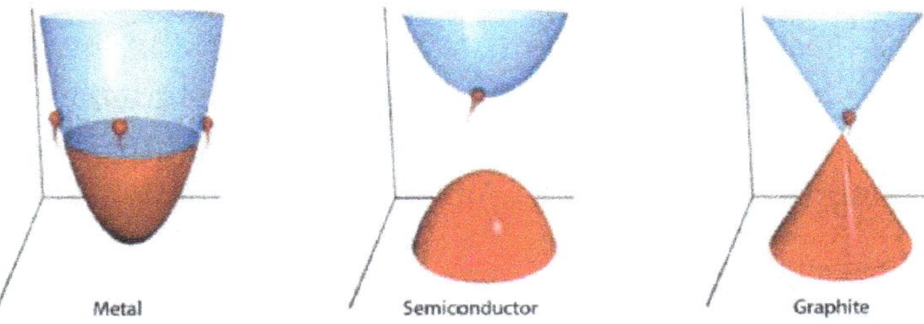

Figure: Electrical behavior of naotubes.

In electrically conducting carbon nanotubes, only one electron wave mode is observed which transport the electrical current. As the lengths and orientations of the carbon nanotubes are different, they touch the surface of the mercury at different times, which provides two sets of information: (i) the influence of carbon nanotube length on the resistance; and (ii) the resistances of the different nanotubes. As the nanotubes have different lengths, then with increasing protrusion of the fiber bundle an increasing number of carbon nanotubes will touch the surface of the mercury droplet and contribute to the electrical current transport.

Mechanical Properties

"Mechanical Properties of Nanoparticles" deals with bulk metallic and ceramic materials, influence of porosity, influence of grain size, super plasticity, filled polymer composites, particle-filled polymers, polymer-based nanocomposites filled with platelets, carbon nanotube-based composites. However, two materials, neither of which is produced by pressing and sintering, have attracted much greater interest as they will undoubtedly achieve industrial importance.

These materials are polymers which contain nanoparticles or nanotubes to improve their mechanical behaviors, and severely plastic-deformed metals, which exhibit astonishing properties. However, because of their larger grain size, the latter are generally not accepted as nanomaterials. Experimental studies on the mechanical properties of bulk nanomaterials are generally impaired by major experimental problems in producing specimens with exactly defined grain sizes and porosities. Therefore, model calculations and molecular dynamic studies are of major importance for an understanding of the mechanical properties of these materials.

Filling polymers with nanoparticles or nanorods and nanotubes, respectively, leads to significant improvements in their mechanical properties. Such improvements depend heavily on the type of the filler and the way in which the filling is conducted. The latter point is of special importance, as any specific advantages of a nanoparticulate filler may be lost if the filler forms aggregates, thereby mimicking the large particles. Particulate-filled polymer-based nanocomposites exhibit a broad range of failure strengths and strains. This depends on the shape of the filler, particles or platelets, and on the degree of agglomeration. In this class of material, polymers filled with silicate platelets exhibit the best mechanical properties and are of the greatest economic relevance. The larger the particles of the filler or agglomerates, the poorer are the properties obtained. Although, potentially, the best composites are those filled with nanofibers or nanotubes, experience teaches that sometimes such composites have the least ductility. On the other hand, by using carbon nanotubes it is possible to produce composite fibers with extremely high strength and strain at rupture. Among the most exciting nanocomposites are the polymer-ceramic nanocomposites, where the ceramic phase is platelet-shaped. This type of composite is preferred in nature, and is found in the structure of bones, where it consists of crystallized mineral platelets of a few nanometers thickness that are bound together with collagen as the matrix. Composites consisting of a polymer matrix and defoliated phyllosilicates exhibit excellent mechanical and thermal properties.

Magnetic Properties

Bulk gold and Pt are non-magnetic, but at the nano size they are magnetic. Surface atoms are not only different to bulk atoms, but they can also be modified by interaction with other chemical species, that is, by capping the nanoparticles. This phenomenon opens the possibility to modify the physical properties of the nanoparticles by capping them with appropriate molecules. Actually, it should be possible that non-ferromagnetic bulk materials exhibit ferromagnetic-like behavior when prepared in nano range. One can obtain magnetic nanoparticles of Pd, Pt and the surprising case of Au (that is diamagnetic in bulk) from non-magnetic bulk materials. In the case of Pt and Pd, the ferromagnetism arises from the structural changes associated with size effects.

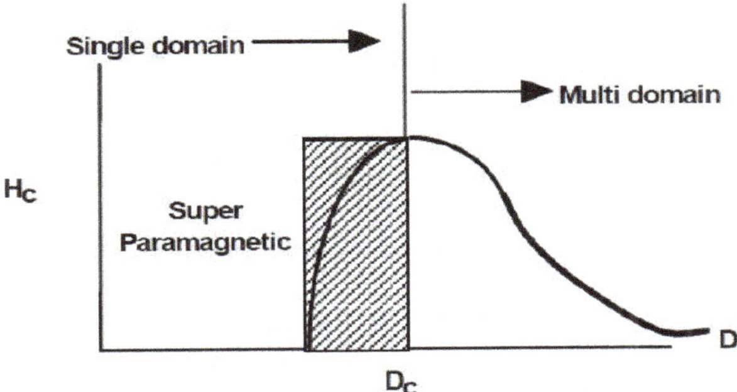

Figure: Magnetic properties of nanostructured materials.

However, gold nanoparticles become ferromagnetic when they are capped with appropriate molecules: the charge localized at the particle surface gives rise to ferromagnetic-like behavior. Surface and the core of Au nanoparticles with 2 nm in diameter show ferromagnetic and paramagnetic character, respectively. The large spin-orbit coupling of these noble metals can yield to a large anisotropy and therefore exhibit high ordering temperatures. More surprisingly, permanent magnetism was observed up to room temperature for thiol-capped Au nanoparticles. For nanoparticles with sizes below 2 nm the localized carriers are in the 5d band. Bulk Au has an extremely low density of states and becomes diamagnetic, as is also the case for bare Au nanoparticles. This observation suggested that modification of the d band structure by chemical bonding can induce ferromagnetic like character in metallic clusters.

Selected Application of nanomaterials

Nanomaterials having wide range of applications in the field of electronics, fuel cells, batteries, agriculture, food industry, and medicines, etc. It is evident that nanomaterials split their conventional counterparts because of their superior chemical, physical, and mechanical properties and of their exceptional formability.

Fuel Cells

A fuel cell is an electrochemical energy conversion device that converts the chemical energy from fuel (on the anode side) and oxidant (on the cathode side) directly into electricity. The heart of fuel cell is the electrodes. The performance of a fuel cell electrode can be optimized in two ways; by improving the physical structure and by using more active electro catalyst. A good structure of electrode must provide ample surface area, provide maximum contact of catalyst, reactant gas and electrolyte, facilitate gas transport and provide good electronic conductance. In this fashion the structure should be able to minimize losses.

Carbon Nanotubes-Microbial Fuel Cell

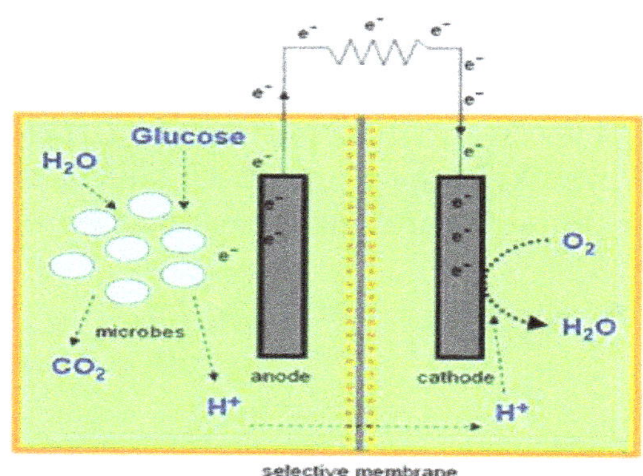

Figure: Schematic representation of microbial fuel cell.

Microbial fuel cell is a device in which bacteria consume water-soluble waste such as sugar, starch and alcohols and produces electricity plus clean water. This technology will make it possible to generate electricity while treating domestic or industrial wastewater. Microbial fuel cell can turn different carbohydrates and complex substrates present in wastewaters into a source of electricity. The efficient electron transfer between the microorganism and the anode of the microbial fuel cell plays a major role in the performance of the fuel cell. The organic molecules present in the wastewater posses a certain amount of chemical energy, which is released when converting them to simpler molecules like CO_2. The microbial fuel cell is thus a device that converts the chemical energy present in water-soluble waste into electrical energy by the catalytic reaction of microorganisms.

Carbon nanotubes (CNTs) have chemical stability, good mechanical properties and high surface area, making them ideal for the design of sensors and provide very high surface area due to its structural network. Since carbon nanotubes are also suitable supports for cell growth, electrodes of microbial fuel cells can be built using of CNT. Due to three-dimensional architectures and enlarged electrode surface area for the entry of growth medium, bacteria can grow and proliferate and get immobilized. Multi walled CNT scaffolds could offer self-supported structure with large surface area through which hydrogen producing bacteria (e.g., E. coli) can eventually grow and proliferate. Also CNTs and MWCNTs have been reported to be biocompatible for different eukaryotic cells. The efficient proliferation of hydrogen producing bacteria throughout an electron conducting scaffold of CNT can form the basis for the potential application as electrodes in MFCs leading to efficient performance.

Catalysis

Higher surface area available with the nanomaterial counterparts, nano-catalysts tend to have exceptional surface activity. For example, reaction rate at nano-aluminum can go so high, that it is utilized as a solid-fuel in rocket propulsion, whereas the bulk aluminum is widely used in utensils. Nano-aluminum becomes highly reactive and supplies the required thrust to send off pay loads in space. Similarly, catalysts assisting or retarding the reaction rates are dependent on the surface activity, and can very well be utilized in manipulating the rate-controlling step.

Phosphors for High-Definition TV

The resolution of a television, or a monitor, depends greatly on the size of the pixel. These pixels are essentially made of materials called "phosphors," which glow when struck by a stream of electrons inside the cathode ray tube (CRT). The resolution improves with a reduction in the size of the pixel, or the phosphors. Nanocrystalline zinc selenide, zinc sulfide, cadmium sulfide, and lead telluride synthesized by the sol-gel techniques are candidates for improving the resolution of monitors. The use of nanophosphors is envisioned to reduce the cost of these displays so as to render high-definition televisions (HDTVs) and personal computers affordable to be purchase.

Next-Generation Computer Chips

The microelectronics industry has been emphasizing miniaturization, whereby the circuits, such as transistors, resistors, and capacitors, are reduced in size. By achieving a significant reduction in their size, the microprocessors, which contain these components, can run much faster, thereby enabling computations at far greater speeds. However, there are several technological impediments to these advancements, including lack of the ultrafine precursors to manufacture these components; poor dissipation of tremendous amount of heat generated by these microprocessors due to faster speeds; short mean time to failures (poor reliability), etc. Nanomaterials help the industry break these barriers down by providing the manufacturers with nanocrystalline starting materials, ultra-high purity materials, materials with better thermal conductivity, and longer-lasting, durable interconnections (connections between various components in the microprocessors).

Metamaterial

Metamaterial is an artificially structured material that exhibits extraordinary electromagnetic properties not available or not easily obtainable in nature. Since the early 2000s, metamaterials have emerged as a rapidly growing interdisciplinary area, involving physics, electrical engineering, materials science, optics, and nano science. The properties of metamaterials are tailored by manipulating their internal physical structure. This makes them remarkably different from natural materials, whose properties are mainly determined by their chemical constituents and bonds. The primary reason for the intensive interest in metamaterials is their unusual effect on light propagating through them.

Metamaterials consist of periodically or randomly distributed artificial structures that have a size and spacing much smaller than the wavelengths of incoming electromagnetic radiation. Consequently, the microscopic details of these individual structures cannot be resolved by the wave. For example, it is difficult to view the fine features of metamaterials that operate at optical wavelengths with visible light, and shorter-wavelength electromagnetic radiation, such as an X-ray, is needed to image and scan them. Researchers can approximate the assemblage of inhomogeneous individual structures as a continuous substance and define their effective material properties at the macroscopic level. Essentially, each artificial structure functions as an atom or a molecule functions in normal materials. However, when subjected to regulated interactions with electromagnetic radiation, the structures give rise to entirely extraordinary properties. (Some naturally occurring materials such as opal and vanadium oxide do exhibit unusual properties when they interact with electromagnetic radiation and have been called "natural metamaterials." However, metamaterials are most often known as artificially occurring materials.)

An example of such extraordinary properties can be seen in electric permittivity (ε) and magnetic permeability (μ), two fundamental parameters that characterize the electromagnetic properties of a medium. These two parameters can be modified, respectively, in structures known as metallic wire arrays and split-ring resonators (SRRs), proposed by English physicist John Pendry in the 1990s and now widely adopted. By adjusting the spacing and size of the elements in metallic wire arrays, a material's electric permittivity (a measure of the tendency of the electric charge within the material to distort in the presence of an electric field) can be "tuned" to a desired value (negative, zero, or positive) at a certain wavelength. Metallic SRRs consist of one or two rings or squares with a gap in them that can be used to engineer a material's magnetic permeability (the tendency of a magnetic field to arise in the material in response to an external magnetic field). When an SSR is placed in an external magnetic field that is oscillating at the SSR's resonant frequency, electric current flows around the ring, inducing a tiny magnetic effect known as the magnetic dipole moment. The magnetic dipole moment induced in the SRR can be adjusted to be either in or out of phase with the external oscillating field, leading to either a positive or a negative magnetic permeability. In this way, artificial magnetism can be achieved even if the metal used to construct the SRR is nonmagnetic.

By combining metallic wire arrays and SRRs in such a manner that both ε and μ are negative, materials can be created with a negative refractive index. Refractive index is a measure of the bending of a ray of light when passing from one medium into another (for example, from air into water or from one layer of glass into another). In normal refraction with positive-index materials, light entering the second medium continues past the normal (a line perpendicular to the interface between the two media), but it is bent either toward or away from the normal depending on its angle of incidence (the angle at which it propagates in the first medium with respect to the normal) as well as on the difference in refractive index between the two media. However, when light passes from a positive-index medium to a negative-index medium, the light is refracted on the same side of the normal as the incident light. In other words, light is bent "negatively" at the interface between the two media; that is, negative refraction takes place.

Negative-index materials do not exist in nature, but according to theoretical studies conducted by Russian physicist Victor G. Veselago in 1968, they were anticipated to exhibit many exotic phenomena, including negative refraction. In 2001 negative refraction was first experimentally demonstrated by American physicist Robert Shelby and his colleagues at microwave wavelengths, and the phenomenon was subsequently extended to optical wavelengths. Other fundamental phenomena, such as Cherenkov radiation and the Doppler effect, are also reversed in negative-index materials.

In addition to electric permittivity, magnetic permeability, and refractive index, engineers can manipulate the anisotropy, chirality, and nonlinearity of a metamaterial. Anisotropic metamaterials are organized so that their properties vary with direction. Some composites of metals and dielectrics exhibit extremely large anisotropy, which allows for negative refraction and new imaging systems, such as superlenses. Chiral metamaterials have a handedness; that is, they cannot be superimposed onto their mirror image. Such metamaterials have an effective chirality parameter κ that is nonzero. A sufficiently large κ can lead to a negative refractive index for one direction of circularly polarized light, even when ε and μ are not simultaneously negative. Nonlinear metamaterials have properties that depend on the intensity of the incoming wave. Such metamaterials can lead to novel tunable materials or produce unusual conditions, such as doubling the frequency of the incoming wave.

The unprecedented material properties provided by metamaterials allow for novel control of the propagation of light, which has led to the rapid growth of a new field known as transformation optics. In transformation optics, a metamaterial with varying values of permittivity and permeability is constructed such that light takes a specific desired path. One of the most remarkable designs in transformation optics is the invisibility cloak. Light smoothly wraps around the cloak without introducing any scattered light, thus creating a virtual empty space inside the cloak where an object becomes invisible. Such a cloak was first demonstrated at microwave frequencies by engineer David Schurig and colleagues in 2006.

Owing to negative refraction, a flat slab of negative-index material can function as a lens to bring light radiating from a point source to a perfect focus. This metamaterial is called a superlens, because by amplifying the decaying evanescent waves that carry the fine features of an object, its imaging resolution does not suffer from the diffraction limit of conventional optical microscopes. In 2004, electrical engineers Anthony Grbic and George Eleftheriades built a superlens that functioned at microwave wavelengths, and in 2005, Xiang Zhang and colleagues experimentally demonstrated a superlens at optical wavelengths with a resolution three times better than the traditional diffraction limit.

The concepts of metamaterials and transformation optics have been applied not only to the manipulation of electromagnetic waves but also to acoustic, mechanic, thermal, and even quantum mechanical systems. Such applications have included the creation of a negative effective mass density and negative effective modulus, an acoustic "hyperlens" with resolution greater than the diffraction limit of sound waves, and an invisibility cloak for thermal flows.

Frequency Bands

Terahertz

Terahertz metamaterials interact at terahertz frequencies, usually defined as 0.1 to 10 THz. Terahertz radiation lies at the far end of the infrared band, just after the end of the microwave band. This corresponds to millimeter and submillimeter wavelengths between the 3 mm (EHF band) and 0.03 mm (long-wavelength edge of far-infrared light).

Photonic

Photonic metamaterial interact with optical frequencies (mid-infrared). The sub-wavelength period distinguishes them from photonic band gap structures.

Tunable

Tunable metamaterials allow arbitrary adjustments to frequency changes in the refractive index. A tunable metamaterial expands beyond the bandwidth limitations in left-handed materials by constructing various types of metamaterials.

Plasmonic

Plasmonic metamaterials exploit surface plasmons, which are produced from the interaction of light with metal-dielectrics. Under specific conditions, the incident light couples with the surface plasmons to create self-sustaining, propagating electromagnetic waves known as surface plasmon polaritons.

Applications

Metamaterials are under consideration for many applications. Metamaterial antennas are commercially available.

In 2007, one researcher stated that for metamaterial applications to be realized, energy loss must be reduced, materials must be extended into three-dimensional isotropic materials and production techniques must be industrialized.

Antennas

Metamaterial antennas are a class of antennas that use metamaterials to improve performance. Demonstrations showed that metamaterials could enhance an antenna's radiated power. Materials that can attain negative permeability allow for properties such as small antenna size, high directivity and tunable frequency.

Absorber

A metamaterial absorber manipulates the loss components of metamaterials' permittivity and magnetic permeability, to absorb large amounts of electromagnetic radiation. This is a useful feature for photodetection and solar photovoltaic applications. Loss components are also relevant in applications of negative refractive index (photonic metamaterials, antenna systems) or transformation optics (metamaterial cloaking, celestial mechanics), but often are not utilized in these applications.

Superlens

A *superlens* is a two or three-dimensional device that uses metamaterials, usually with negative refraction properties, to achieve resolution beyond the diffraction limit (ideally, infinite resolution). Such a behaviour is enabled by the capability of double-negative materials to yield negative phase velocity. The diffraction limit is inherent in conventional optical devices or lenses.

Cloaking Devices

Metamaterials are a potential basis for a practical cloaking device. The proof of principle was demonstrated on October 19, 2006. No practical cloaks are publicly known to exist.

RCS (Radar Cross Section) Reducing Metamaterials

Conventionally, the RCS has been reduced either by Radar absorbent material (RAM) or by purpose shaping of the targets such that the scattered energy can be redirected away from the source. While RAMs have narrow frequency band functionality, purpose shaping limits the aerodynamic performance of the target. More recently, metamaterials or metasurfaces are synthesized that can redirect the scattered energy away from the source using either array theory or generalized Snell's law. This has led to aerodynamically favorable shapes for the targets with the reduced RCS.

Seismic Protection

Seismic metamaterials counteract the adverse effects of seismic waves on man-made structures.

Sound Filtering

Metamaterials textured with nanoscale wrinkles could control sound or light signals, such as changing a material's color or improving ultrasound resolution. Uses include nondestructive material testing, medical diagnostics and sound suppression. The materials can be made through a high-precision, multi-layer deposition process. The thickness of each layer can be controlled within a fraction of a wavelength. The material is then compressed, creating precise wrinkles whose spacing can cause scattering of selected frequencies.

Theoretical Models

All materials are made of atoms, which are dipoles. These dipoles modify light velocity by a factor n (the refractive index). In a split ring resonator the ring and wire units act as atomic dipoles: the wire acts as a ferroelectric atom, while the ring acts as an inductor L, while the open section acts as a capacitor C. The ring as a whole acts as an LC circuit. When the electromagnetic field passes through the ring, an induced current is created. The generated field is perpendicular to the light's magnetic field. The magnetic resonance results in a negative permeability; the refraction index is negative as well. (The lens is not truly flat, since the structure's capacitance imposes a slope for the electric induction.)

Several (mathematical) material models frequency response in DNGs. One of these is the Lorentz model, which describes electron motion in terms of a driven-damped, harmonic oscillator. The Debye relaxation model applies when the acceleration component of the Lorentz mathematical model is small compared to the other components of the equation. The Drude model applies when the restoring force component is negligible and the coupling coefficient is generally the plasma frequency. Other component distinctions call for the use of one of these models, depending on its polarity or purpose.

Three-dimensional composites of metal/non-metallic inclusions periodically/randomly embedded in a low permittivity matrix are usually modeled by analytical methods, including mixing formulas and scattering-matrix based methods. The particle is modeled by either an electric dipole parallel to the electric field or a pair of crossed electric and magnetic dipoles parallel to the electric and magnetic fields, respectively, of the applied wave. These dipoles are the leading terms in the multipole series. They are the only existing ones for a homogeneous sphere, whose polarizability can be easily obtained from the Mie scattering coefficients. In general, this procedure is known as the "point-dipole approximation", which is a good approximation for metamaterials consisting of composites of electrically small spheres. Merits of these methods include low calculation cost and mathematical simplicity.

Other first principles techniques for analyzing triply-periodic electromagnetic media may be found in Computing photonic band structure.

Institutional Networks

MURI

The Multidisciplinary University Research Initiative (MURI) encompasses dozens of Universities and a few government organizations. Participating universities include UC Berkeley, UC Los Angeles, UC San Diego, Massachusetts Institute of Technology, and Imperial College in London. The sponsors are Office of Naval Research and the Defense Advanced Research Project Agency.

MURI supports research that intersects more than one traditional science and engineering discipline to accelerate both research and translation to applications. As of 2009, 69 academic institutions were expected to participate in 41 research efforts.

Metamorphose

The Virtual Institute for Artificial Electromagnetic Materials and Metamaterials "Metamorphose VI AISBL" is an international association to promote artificial electromagnetic materials and metamaterials. It organizes scientific conferences, supports specialized journals, creates and manages research programs, provides training programs (including PhD and training programs for industrial partners); and technology transfer to European Industry.

Biomaterials

Biomaterials are used to make devices to replace a part or a function of the body in safe, reliably economically, and physiologically acceptable manner. A variety of devices and materials are used in the treatment of disease or injury. Commonplace examples include suture needles, plates, teeth fillings, etc.

Term Definitions

- Biomaterial: A synthetic material used to make devices to replace part of a living system or to function in intimate contact with living tissue.
- Biological Material: A material that is produced by a biological system.
- Bio-compatibility: Acceptance of an artificial implant by the surrounding tissues and by the body as a whole.

Fields of Knowledge to Develop Biomaterials

1. Science and engineering: (Materials Science) structure-property relationships of synthetic and biological materials including metals, ceramics, polymers, composites, tissues (blood and connective tissues), etc.
2. Biology and Physiology: Cell and molecular biology, anatomy, animal and human physiology, histopathology, experimental surgery, immunology, etc.
3. Clinical Sciences: (All the clinical Specialties) density, maxillofacial, neurosurgery, obstetrics and gynecology, ophthalmology, orthopedics, plastic and reconstructive surgery, thoracic and cardiovascular surgery, veterinary medicine and surgery, etc.

Uses of Biomaterials

Uses of Biomaterials Example	Example
Replacement of diseased and damaged part	Artificial hip joint, kidney dialysis machine
Assist in healing	Sutures, bone plates and screws
Improve function	Cardiac pacemaker, intra-ocular lens
Correct functional abnormalities C	Cardiac pacemaker
Correct cosmetic problem	Mastectomy augmentation, chin augmentation
Aid to diagnosis	Probes and catheters
Aid to treatment	Catheters, drains

Biomaterials in Organs

Organ	Example
Heart	Cardiac pacemaker, artificial heart valve, Totally artificial heart
Lung	Oxy-generator machine
Eye	Contact lens, intraocular lens
Ear	Artificial stapes, cochlea implant
Bone	Bone plate, intra-medullary rod
Kidney	Kidney dialysis machine
Bladder C	Catheter and stent

Materials for use in the Body

Materials	Advantages	Disadvantages	Examples
Polymers (nylon, silicon Rubber, polyester, PTFE, etc)	Resilient Easy to Fabricate	Not strong Deforms with time May degrade	Blood vessels, Sutures, ear, nose, Soft tissues
Metals (Ti and its alloys Co-Cr alloys, stainless Steels)	Strong Tough ductile	May corrode, dense, Difficult to make	Joint replacement, Bone plates and Screws, dental root Implant, pacer, and suture
Ceramics (Aluminum Oxide, calcium phosphates, including hydroxyapatite carbon)	Very biocompatible Inert strong in compression	Difficult to make Brittle Not resilient	Dental coating Orthopedic implants Femoral head of hip
Composites (Carbon-carbon, wire Or fiber reinforced Bone cement)	Compression strong	Difficult to make	Joint implants Heart valves

The science of biomedical materials involves a study of the composition and properties of materials and the way in which they interact with the environment in which they are placed.

Selection of Biomedical Materials

The process of material selection should ideally be for a logical sequence involving:

1. Analysis of the problem;
2. Consideration of requirement;
3. Consideration of available material and their properties leading to; and
4. Choice of material.

The choice of a specific biomedical material is now determined by consideration of the following:

1. A proper specification of the desired function for the material;
2. An accurate characterization of the environment in which it must function, and the effects that environment will have on the properties of the material;
3. A delineation of the length of time the material must function; and
4. A clear understanding of what is meant by safe for human use.

As the number of available materials increases, it becomes more and more important to be protected from unsuitable products or materials, which haven't been thoroughly evaluated.

Most manufacturers of materials operate an extensive quality assurance program and materials are thoroughly tested before being released to the general practitioner.

1. Standard Specifications: Many standard specification tests of both national and international standards organizations (ISO) are now available, which effectively maintain quality levels. Such specifications normally give details for:

(a) the testing of certain products,

(b) the method of calculating the results, and

(c) the minimum permissible result, which is acceptable.

2. Laboratory Evaluation: Laboratory tests, some of which are used in standard specification, can be used to indicate the suitability of certain materials. It is important that methods used to evaluate materials in laboratory give results, which can be correlated with clinical experience.

3. Clinical Trials: Although laboratory tests can provide many important and useful data on materials, the ultimate test is the controlled clinical trial and verdict of practitioners after a period of use in general practice. Many materials produce good results in the laboratory, only to be found lacking when subjected to clinical use. The majority of manufacturers carry out extensive clinical trials of new materials, normally in cooperation with a university or hospital department, prior to releasing a product for use by general practitioners.

Materials Science

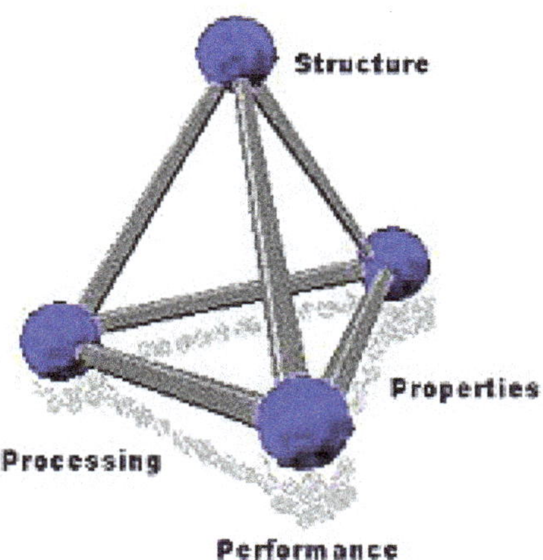

The Materials Science Tetrahedron shows the four main areas in which materials are studied. It often includes Characterization at the center.

Materials science is an interdisciplinary field involving the study of different types of materials and the applications of knowledge about these materials to various areas of science and engineering. It combines elements of applied physics and chemistry, as well as chemical, mechanical, civil and electrical engineering. Materials science and materials engineering are often combined into a larger field of study.

Materials used in early human history included metals, glasses, and clay-based ceramics. The past century has witnessed a surge in the development of new materials, including plastics, advanced ceramics, semiconductors, superconductors, liquid crystals, Bose-Einstein condensates, and nanoscale substances, with a wide range of applications. Furthermore, materials science has grown to include testing these more exotic forms of condensed matter and developing new physics theories to explain their behavior. Consequently, materials science has been propelled to the forefront at many academic institutions and research facilities.

Materials research at the basic level can lead to unprecedented influence on society. For example, semiconductor materials, which are ubiquitous in cars, telephones, computers, clocks, kitchen appliances, children's toys, satellites, telescopes, and more, were a product of materials science research—into the electronic properties of the element germanium. Further research led to the replacement of germanium with the less costly silicon and to diverse approaches to modifying silicon's properties by implanting other elements, such as phosphorous or boron, into the silicon matrix. Since their discovery in 1947, semiconductors have been steadily improved through materials science research driven by ever-increasing performance demands from the computer industry.

Efforts to apply ethical considerations to Materials Science quickly reach what is a common barrier between ethics and the combined fields of science and technology. An individual scientist, for example, who would want to conduct research toward such a noble goal as developing a light-weight and durable structural plastic that is readily recyclable must first either find and join a research group that is already funded to support such research or find an independent funding source for such research.

Fundamentals of Materials Science

In materials science, the researcher conducts a systematic investigation of each material, in terms of its structure, properties, processing, and performance. The research often leads to new applications of known materials and the creation of new materials with desired properties.

On a fundamental level, this field relates the properties and performance of a material to its atomic-scale structure and the different phases it can go through. The major factors that determine the structure and properties of a material are the nature of its constituent chemical elements and the way in which the material was processed into its final form. These factors, related through the laws of thermodynamics, govern the material's microstructure, and thus its properties.

An old adage in materials science says: "materials are like people; it is the defects that make them interesting". Given the limits of today's technology, that is good, because the manufacture of a perfect crystal of a material is physically impossible. Instead, materials scientists manipulate a material's defects to create materials with the desired properties. On an atomic scale, the defects in a crystal could mean that atoms of one element may be missing or replaced by atoms of other elements.

Not all materials have a regular crystalline structure. Glasses and some ceramics—unlike many natural materials— are amorphous, that is, they do not possess any long-range order in their atomic arrangements. Engineering these materials is much more difficult than engineering crystalline materials. Polymers may exhibit varying degrees of crystallinity, and studying them requires a combination of elements of chemical and statistical thermodynamics to give thermodynamic (rather than mechanical) descriptions of physical properties.

Materials in Industry

Radical advances in understanding and manipulating materials drive the creation of new products and even new industries. At the same time, stable industries employ materials scientists to make incremental improvements and troubleshoot issues with currently used materials. Industrial applications of materials science include the design of materials and their cost-benefit tradeoffs in industrial production.

Techniques used for processing materials include:

- casting
- rolling
- welding
- ion implantation
- crystal growth
- thin-film deposition
- sintering
- glassblowing

Techniques used for analyzing (characterizing) materials include:

- electron microscopy
- X-ray diffraction
- calorimetry
- nuclear microscopy (HEFIB)
- Rutherford backscattering
- neutron diffraction

The overlap between physics and materials science has lent itself naturally the development of the interface field of materials physics, which is concerned with the physical properties of materials. The approach is generally more macroscopic and applied than in condensed matter physics.

Classes of Materials

Materials science encompasses various classes of materials, some of which overlap. Examples are:

1. Ionic crystals (crystals in which the atoms are held together by ionic bonds)
2. Covalent crystals (crystals in which the atoms are held together by covalent bonds)
3. Vitreous (glassy) materials
4. Metals
5. Intermetallics
6. Polymers
7. Composite materials

8. Biomaterials (materials derived from or intended for use with biological systems)
9. Electronic and magnetic materials (materials such as semiconductors used to create integrated circuits, storage media, sensors, and other devices)
10. Ceramics and refractories (high-temperature materials, including reinforced carbon-carbon (RCC), polycrystalline silicon carbide, and transformation-toughened ceramics)

Each class of materials may involve a separate field of study.

Subfields of Materials Science

- Nanotechnology: As commonly understood, nanotechnology is the field of applied science and technology concerned with the formation, study, and control of materials having a width ranging from less than 1 nanometer (10–9 meter) to 100 nanometers. These materials are generally engineered on a molecular scale. On a more rigorous level, nanoscience involves the study of materials whose defining properties are present only at the nanoscale.
- Crystallography: This is the study of the arrangement of atoms in a crystalline solid and the relationship between the crystalline structures and their physical properties. It includes the determination of defects associated with crystal structures.
- Materials characterization: Information needed for understanding and defining the properties of materials is acquired through such techniques as diffraction of X-rays, electrons, or neutrons, and various forms of spectroscopy, chromatography, thermal analysis, or electron microscopy.
- Metallurgy: This involves the study of metals and their alloys, including their extraction, microstructure, and processing.
- Tribology: This is the study of the wear of materials due to friction and other factors.
- Surface science: It involves study of the structures and interactions occurring at the interfaces of solids and gases, solids and liquids, and solids and solids.
- Glass science: It involves the study of noncrystalline materials, including inorganic glasses, vitreous metals, and non-oxide glasses.

Some practitioners consider rheology a subfield of materials science, because it can cover any material that flows. Modern rheology, however, typically deals with non-Newtonian fluid dynamics, so it is often considered a subfield of continuum mechanics.

Topics that form the Basis of Materials Science

- Thermodynamics, statistical mechanics, chemical kinetics, and physical chemistry: to understand phase stability and physical and chemical transformations.
- Chemical bonding: to understand the bonds between atoms of the material.
- Mechanics of materials: to understand the mechanical properties of materials and their structural applications.

- Solid-state physics and quantum mechanics: to understand the electronic, thermal, magnetic, chemical, structural, and optical properties of materials.
- Solid-state chemistry and polymer science: to understand the properties of polymers (including plastics), colloids, ceramics, and liquid crystals.
- Biology: for the integration of materials into biological systems.
- Continuum mechanics and statistics: for the study of fluid flows and ensemble systems.
- Diffraction and wave mechanics: for the characterization of materials.

Materials Engineering

Materials that are used as raw material for any sort of construction or manufacturing in an organized way of engineering application are known as Engineering Materials. For example, the computer or the pen we use, are manufactured through controlled engineering processes. These gadgets make use of materials like HDPE, PP, Pb-Silica glass, copper, aluminium, tin, etc. in their fabrication. Civil construction works like bridges, dams, houses, roads, pavements are carried out with raw materials like stone, chips, cement, clay, paint, bars, etc.

Everything we use in our daily life can be tailored to use for specific cases. This can be done efficiently if we know the property of each material beforehand. Hence, materials have been extensively tested for their properties and classified into broad groups. From this grouping one can know about the gross property of any group of material.

Basically Engineering Materials Can be classified into two categories:

1. Metals
2. Non-Metals

Metals

Metals are polycrystalline bodies which are having number of differentially oriented fine crystals. Normally major metals are in solid states at normal temperature. However, some metals such as mercury are also in liquid state at normal temperature. All metals are having high thermal and electrical conductivity. All metals are having positive temperature coefficient of resistance. Means resistance of metals increases with increase in temperature. Examples of metals – Silver, Copper, Gold, Aluminum, Iron, Zinc, Lead, Tin etc.

Metals can be further divided into two groups:

Ferrous Metals

All ferrous metals are having iron as common element. All ferrous materials are having very high permeability which makes these materials suitable for construction of core of electrical machines. Examples: Cast Iron, Wrought Iron, Steel, Silicon Steel, High Speed Steel, Spring Steel etc.

Non-Ferrous Metals

All non-ferrous metals are having very low permeability. Example: Silver, Copper, Gold, Aluminum etc.

Non-Metals

Non-Metal materials are non-crystalline in nature. These exist in amorphic or mesomorphic forms. These are available in both solid and gaseous forms at normal temperature.

Normally all non-metals are bad conductor of heat and electricity.

Examples: Plastics, Rubber, Leathers, Asbestos etc.

As these non-metals are having very high resistivity which makes them suitable for insulation purpose in electrical machines.

Difference between Metals and Non Metals

Sl. No.	Property	Metals	Non-Metals
1.	Structure	All metals are having crystalline structure	All Non-metals are having amorphic & mesomorphic structure
2.	State	Generally metals are solid at normal temperature	State varies material to material. Some are gas state and some are in solid state at normal temperature.
3.	Valance electrons and conductivity	Valance electrons are free to move within metals which makes them good conductor of heat & electricity	Valence electrons are tightly bound with nucleus which are not free to move. This makes them bad conductor of heat & electricity
4.	Density	High density	Low density
5.	Strength	High strength	Low strength
6.	Hardness	Generally hard	Hardness is generally varies
7.	Malleability	Malleable	Non malleable
8.	Ductility	Ductile	Non ductile
9.	Brittleness	Generally non brittle in nature	Brittleness varies material to material
10.	Lustre	Metals possess metallic lustre	Generally do not possess metallic lustre (Except graphite & iodine)

Other Classification of Engineering Materials

Engineering materials can also be classified as below-

1. Metals and Alloys
2. Ceramic Materials
3. Organic Materials

Metals and Alloys

Metals are polycrystalline bodies which are have number of differentially oriented fine crystals. Normally major metals are in solid states at normal temperature. However, some metals such as mercury are also in liquid state at normal temperature.

Pure metals are having very a low mechanical strength, which sometimes does not match with the mechanical strength required for certain applications. To overcome this draw back alloys are used.

Alloys are the composition of two or more metals or metal and non-metals together. Alloys are having good mechanical strength, low temperature coefficient of resistance.

Example: Steels, Brass, Bronze, Gunmetal, Invar. Super Alloys etc.

Ceramic Materials

Ceramic materials are non-metallic solids. These are made of inorganic compounds such as Oxides, Nitrides, Silicates and Carbides. Ceramic materials possess exceptional Structural, Electrical, Magnetic, Chemical and Thermal properties. These ceramic materials are now extensively used in different engineering fields.

Examples: Silica, glass, cement, concrete, garnet, Mgo, Cds, Zno, SiC etc.

Organic Materials

All organic materials are having carbon as a common element. In organic materials carbon is chemically combined with oxygen, hydrogen and other non-metallic substances. Generally organic materials are having complex chemical bonding.

Example: Plastics, PVC, Synthetic Rubbers etc.

Classification of Processing Techniques

The basic aim of processing is to produce the products of the required quality at a reasonable cost. The basic processes can be broadly classified as:

a) Primary Forming Processes

b) Deformative Processes

c) Material Removal Processes

d) Joining Processes

e) Finishing Processes

Most of the engineering materials are processed either individually or in combination by the above mentioned processes. The processes can further be classified as conventional and advanced processes. The specific application area of each will depend on the design requirements and the ability with which a material renders itself to various processing techniques. The selection of a processing technique for any engineering material would broadly depend on the properties (mechanical, physical, chemical) of the material and the required number of parts to be processed.

References

- Engheta, Nader; Richard W. Ziolkowski (June 2006). Metamaterials: Physics and Engineering Explorations. Wiley & Sons. pp. xv, 3–30, 37, 143–50, 215–34, 240–56. ISBN 978-0-471-76102-0
- Properties-of-materials, sorting-materials-into-groups, science: toppr.com, Retrieved 24 July 2018
- Shelby, R. A.; Smith, D. R.; Schultz, S. (2001). "Experimental Verification of a Negative Index of Refraction". Science. 292 (5514): 77–79. Bibcode:2001Sci...292...77S. doi:10.1126/science.1058847. PMID 11292865
- Materials-science: newworldencyclopedia.org, Retrieved 17 March 2018
- Smith, David R. (2006-06-10). "What are Electromagnetic Metamaterials?". Novel Electromagnetic Materials. The research group of D.R. Smith. Archived from the originalon July 20, 2009. Retrieved 2009-08-19
- Engineering-materials, physical-metallurgy: sengerandu.wordpress.com, Retrieved 19 May 2018
- Guenneau, S. B.; Movchan, A.; Pétursson, G.; Anantha Ramakrishna, S. (2007). "Acoustic metamaterials for sound focusing and confinement". New Journal of Physics. 9(11): 399. Bibcode:2007NJPh....9..399G. doi:10.1088/1367-2630/9/11/399
- Classification-of-engineering-materials: electrical4u.com, Retrieved 20 June 2018
- Zouhdi, Saïd; Ari Sihvola; Alexey P. Vinogradov (December 2008). Metamaterials and Plasmonics: Fundamentals, Modelling, Applications. New York: Springer-Verlag. pp. 3–10, Chap. 3, 106. ISBN 978-1-4020-9406-4
- Veselago, V. G. (1968). "The electrodynamics of substances with simultaneously negative values of [permittivity] and [permeability]". Soviet Physics Uspekhi. 10 (4): 509–14. Bibcode:1968SvPhU..10..509V. doi:10.1070/PU1968v010n04ABEH003699

2

Atomic Bonding

An atomic bond is an attractive force between atoms or ions, which enables the formation of chemical compounds. It can be an ionic bond, covalent bond or a metallic bond. The topics elucidated in this chapter cover the essentials of atomic bonding such as Kössel-Lewis approach to chemical bonding, hydrogen bonding, strong chemical bond, etc.

Atomic bonding is chemical bonding. It is the interactions that account for the association of atoms into molecules, ions, crystals, and other stable species that make up the familiar substances of the everyday world. When atoms approach one another, their nuclei and electrons interact and tend to distribute themselves in space in such a way that the total energy is lower than it would be in any alternative arrangement. If the total energy of a group of atoms is lower than the sum of the energies of the component atoms, they then bond together and the energy lowering is the bonding energy.

The ideas that helped to establish the nature of chemical bonding came to fruition during the early 20th century, after the electron had been discovered and quantum mechanics had provided a language for the description of the behavior of electrons in atoms. However, even though chemists need quantum mechanics to attain a detailed quantitative understanding of bond formation, much of their pragmatic understanding of bonds is expressed in simple intuitive models. These models treat bonds as primarily of two kinds—namely, ionic and covalent. The type of bond that is most likely to occur between two atoms can be predicted on the basis of the location of the elements in the periodic table, and to some extent the properties of the substances so formed can be related to the type of bonding.

A key concept in a discussion of chemical bonding is that of the molecule. Molecules are the smallest units of compounds that can exist. One feature of molecules that can be predicted with reasonable success is their shape. Molecular shapes are of considerable importance for understanding the reactions that compounds can undergo, and so the link between chemical bonding and chemical reactivity is discussed briefly in this article.

Although simple models of bonding are useful as rules of thumb for rationalizing the existence of compounds and the physical and chemical properties and structures of molecules, they need to be justified by appealing to more-sophisticated descriptions of bonding. Moreover, there are some aspects of molecular structure that are beyond the scope of the simple theories. To achieve this insight, it is necessary to resort to a fully quantum mechanical description. In practice, these descriptions entail heavy reliance on computers. Such numerical approaches to the chemical bond provide important information about bonding.

After the historical introduction, qualitative models of bonding are discussed, with particular attention given to the formation of ionic and covalent bonds and the correlation of the latter with molecular shapes. The more-sophisticated quantum mechanical approaches to bond formation are then described, followed by a survey of a number of special cases that raise interesting problems or lead to important insights.

Kössel-Lewis Approach to Chemical Bonding

In 1916 Kossel and Lewis succeeded in giving successful explanation based upon the concept of electronic configuration of noble gases about why atoms combine to form molecules. Atoms of noble gases have little or no tendency to combine with each other or with atoms of other elements. This means that these atoms must be having stable electronic configurations. The electronic configurations of noble gases are given in Table.

Name of the element	Symbol	Atomic Number	Electronic Configuration
1. Helium	He	2	$1s^2$
2. Neon	Ne	10	$1s^2 2s^2 2p^6$
3. Argon	Ar	18	$1s^2 2s^2 2p^6 3s^2 3p^6$
4. Krypton	Kr	36	$1s^2 2s^2 2p^6 3s^2 3p^6 3d^{10} 4s^2 4p^6$
5. Xenon	Xe	54	$1s^2 2s^2 2p^6 3s^2 3p^6 3d^{10} 4s^2 4p^6 4d^{10} 5s^2 5p^6$
6. Radon	Rn	86	$1s^2 2s^2 2p^6 3s^2 3p^6 3d^{10} 4s^2 4p^6 4d^{10} 4f^{14} 5s^2 5p^6 5d^{10} 6s^2 6p^6$

Table: Electronic Configuration of Noble Gases

Due to the stable configuration, the noble gas atoms neither have any tendency to gain nor lose electrons and, therefore, their combining capacity or valency is zero. They are so inert that they even do not form diatomic molecules and exist as monoatomic gaseous atoms.

Octet Rule

All atoms other than noble gases have less than eight electrons in their outermost shells. In other words, the outermost shells of these atoms do not have stable configurations. Therefore, they combine with each other or with other atoms to achieve stable noble gas electronic configurations. These elements undergo electronic re-arrangements to attain stable noble gas configurations. Therefore

"The tendency or urge of atoms of various elements to attain stable configuration of eight electrons in their valence shells, is the cause of Chemical combination" and

"The principle of attaining maximum of eight electrons in the valence shell of atoms, is called octet rule."

Lewis introduced simple symbols to denote the electrons present in the outer orbit of atom, these electrons are known as valence electrons. These symbols are known as Electron Dot Symbols and the structure of compound is known as Lewis Dot Structure.

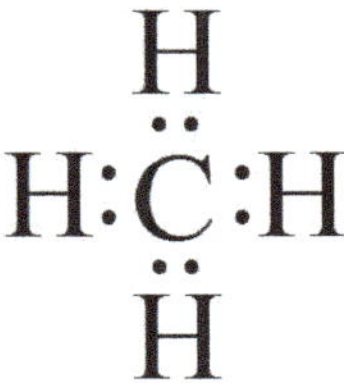

Figure: Dot structure of methane

(Note: Number of dots around the symbol is equal to number of electrons.)

Condition for Writing the Lewis Dot Structures of Molecules

Lewis Structures and Covalent Bond condition for writing the Lewis dot structures of molecules

Conditions for writing the electron dot (or Lewis) structures of covalent molecules are:

- Sharing of an electron pair between the atoms results in the bond formation.
- At the time of bond formation, each bond consists of two electrons which are contributed by each one of the combining atoms.
- When the combining atoms forms bond, because of sharing of electrons they achieve the stable outer shell noble gas configurations. In other words, octets of both the atoms get completed.

Electron dot (or Lewis) structures of covalent molecules are written in accordance with octet rule. According to this, all the atoms in the formula will have a total of eight electrons in their valence shell except the Hydrogen atom. Hydrogen will have only two electrons because only two electrons complete its first shell as in helium. Thus the elements of group 17 (containing seven valence electrons) such as Cl would share one electron to attain stable octet; the elements of group 16 (containing six valence electron) such as O and S would share two electrons; the elements of group 15 (containing five valence electrons) would share three electrons and so on.

For Example, oxygen (with six electrons in the valence shell) completes its octet by sharing its two electrons with two hydrogen atoms to form a water molecule as shown below.

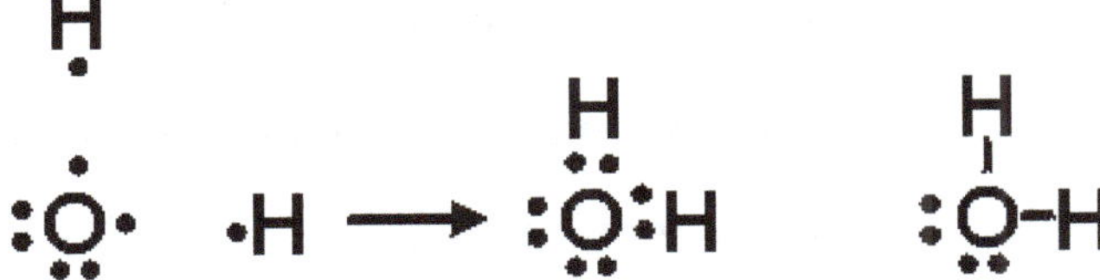
Figure: Lewis Structure of Water Molecule

Similarly, nitrogen has five electrons in its valence shell and shares with three hydrogen atoms to form ammonia.

3H + N → H · N · H → H—N—H

Figure: Lewis Structure of NH3 (Ammonia)

Similarly, carbon has four electrons in its valence shell and shares with four chlorine atoms to form carbon tetrachloride (CCl_4) molecule as shown below.

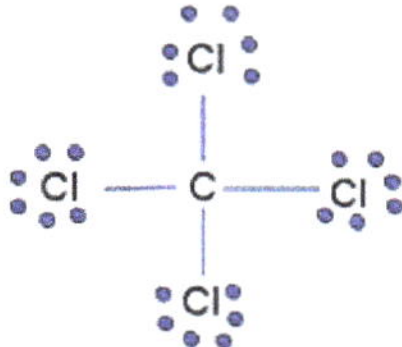

Figure: Lewis Structure of CCl_4 (Carbon Tetrachloride)

Lewis structure for molecules having multiple covalent bonds. If the normal valence of an atom is not satisfied by sharing single electron pair between atoms, the atoms may share more than one electron pair between them.

Writing Lewis Structures

The following steps are adopted for writing the Lewis dot structures or Lewis structures:

Step 1: First, we calculate the required number of electrons for drawing the structure by adding the valence electrons of the combining atoms. For Example, in methane, CH_4 molecule, there are 8 valence electrons (in which 4 belongs to carbon while other 4 to H atoms).

Step 2: Each negative charge i.e. for anions, we add an electron to the valence electrons and for each positive charge that is, for cations we subtract one electron from the valence electrons.

Step 3: Using the chemical symbols of the combining atoms and constructing skeletal structure of the compound, divide the total number of electrons as bonding shared pairs between the atoms in proportion to the total bonds.

Step 4: The central position in the molecule is occupied by the least electronegative atom occupies Hydrogen and fluorine generally occupy the terminal positions.

Step 5: After distributing the shared pairs of electrons for single bonds, the remaining electron pairs are used either for multiple bonds or they constitute lone pairs.

The basic requirement is that each bonded atom gets an octet of electrons.

Example No. 1 Lewis formula for carbon monoxide, CO,

Step 1: Counting the total number of valence electrons of carbon and oxygen atoms:

$C\ (2s^2 2p^2) + O\ (2s^2 2p^4)$

$4 + 6 = 10$ that is, $4(C) + 6(O) = 10$

Step 2: The skeletal structure of carbon monoxide is written as: CO,

Step 3:.Drawing a single bond between C and O and completing octet on O, the remaining two electrons are lone pair on C.

$$:\overset{\oplus}{C}-\overset{\ominus}{O}:::$$

Step 4: This does not complete the octet of carbon, and hence we have triple bond between C and O atom,

$$:\overset{\ominus}{C}\equiv\overset{\oplus}{O}:$$

This satisfies octet rule for both the atoms.

Example No. 2 Lewis Structure of nitrite, No2⁻

Step 1: Counting the total number of valence electrons of one nitrogen atom, two oxygen atoms and the additional one negative charge (equal to one electron).

Total Number of valence electrons are:

N ($2s^2 2p^3$) + 2O ($2s^2 2p^4$) + 1 (negative charge),

5 + 2 6 + 1 = 18

Step 2: The skeletal structure of nitrite ion is written as,

O N O

Step 3: Drawing a single bond between nitrogen and each oxygen atom,

O — N — O

Step 4: Complete the octets of atoms.

$$\left[:\ddot{\underset{\cdot\cdot}{O}}:\ddot{N}::\ddot{\underset{\cdot\cdot}{O}}\right]^-$$

This structure does not complete octet on N if the remaining two electron constitute of a lone pair on it. Therefore, we have double bond between one N and one of the two O atoms.

The Lewis structure is,

$$\left[:\ddot{\underset{\cdot\cdot}{O}}-\ddot{N}=\ddot{O}:\right]^-$$

Ionic Bond

Ionic bond or electrovalent bond is a type of linkage formed from the electrostatic attraction between oppositely charged ions in a chemical compound. Such a bond forms when the valence (outermost) electrons of one atom are transferred permanently to another atom. The atom that loses the electrons becomes a positively charged ion (cation), while the one that gains them becomes a negatively charged ion (anion).

Ionic bonding results in compounds known as ionic, or electrovalent, compounds, which are best exemplified by the compounds formed between nonmetals and the alkali and alkaline-earth metals. In ionic crystalline solids of this kind, the electrostatic forces of attraction between opposite charges and repulsion between similar charges orient the ions in such a manner that every positive ion becomes surrounded by negative ions and vice versa. In short, the ions are so arranged that the positive and negative charges alternate and balance one another, the overall charge of the entire substance being zero. The magnitude of the electrostatic forces in ionic crystals is considerable. Accordingly, these substances tend to be hard and nonvolatile.

An ionic bond is actually the extreme case of a polar covalent bond, the latter resulting from unequal sharing of electrons rather than complete electron transfer. Ionic bonds typically form when the difference in the electro negativities of the two atoms is great, while covalent bonds form when the electro negativities are similar.

Formation of Ionic Bonds

An ionic bond is formed by the complete transference of one or more electrons from the outer energy shell (valency shell) of one atom to the outer energy shell of the other atom. In this way, both the atoms acquire the stable electronic configurations of the nearest noble atom. The atom from which the electrons are transferred i.e., the atom which loses the electrons, acquires a positive charge and becomes cation.

The atom which gains the electrons acquires a negative charge and becomes anion. The electrostatic attraction between the oppositely charged ions results in the formation of an ionic bond or electrovalent bond between the two atoms and the compounds are called ionic compounds or electrovalent compounds.

Factors Influencing the Formation of Ionic Bond

1. Ionization energy

It is defined as the amount of energy required to remove the most loosely bound electron from an isolated gaseous atom of an element. The lesser the ionization energy, the greater is the ease of the formation of a cation.

2. Electron affinity

It is defined as the amount of energy released when an electron is added to an isolated gaseous atom of an element. The higher the energy released during this process, the easier will be the formation of an anion.

Thus, low ionization energy of a metal atom and high electron affinity of a non-metal atom facilitate the formation of an ionic bond between them.

3. Lattice energy

It is defined as the amount of energy released when cations and anions are brought from infinity to their respective equilibrium sites in the crystal lattice to form one mole of the ionic compound. The higher the lattice energy, the greater is the tendency of the formation of an ionic bond. The higher the charges on the ions and smaller the distance between them, the greater is the force of attraction between them.

Structures

Ionic compounds in the solid state form lattice structures. The two principal factors in determining the form of the lattice are the relative charges of the ions and their relative sizes. Some structures are adopted by a number of compounds; for example, the structure of the rock salt sodium chloride is also adopted by many alkali halides, and binary oxides such as magnesium oxide. Pauling's rules provide guidelines for predicting and rationalizing the crystal structures of ionic crystals

Strength of the Bonding

For a solid crystalline ionic compound the enthalpy change in forming the solid from gaseous ions is termed the lattice energy. The experimental value for the lattice energy can be determined using the Born–Haber cycle. It can also be calculated (predicted) using the Born–Landé equation as the sum of the electrostatic potential energy, calculated by summing interactions between cations and anions, and a short-range repulsive potential energy term. The electrostatic potential can be expressed in terms of the interionic separation and a constant (Madelung constant) that takes account of the geometry of the crystal. The further away from the nucleus the weaker the shield. The Born-Landé equation gives a reasonable fit to the lattice energy of, e.g., sodium chloride, where the calculated (predicted) value is −756 kJ/mol, which compares to −787 kJ/mol using the Born–Haber cycle.

Polarization Effects

Ions in crystal lattices of purely ionic compounds are spherical; however, if the positive ion is small and/or highly charged, it will distort the electron cloud of the negative ion, an effect summarised in Fajans' rules. This polarization of the negative ion leads to a build-up of extra charge density between the two nuclei, that is, to partial covalency. Larger negative ions are more easily polarized, but the effect is usually important only when positive ions with charges of 3+ (e.g., Al^{3+}) are involved. However, 2+ ions (Be^{2+}) or even 1+ (Li^{+}) show some polarizing power because their sizes are so small (e.g., LiI is ionic but has some covalent bonding present). Note that this is not the ionic polarization effect that refers to displacement of ions in the lattice due to the application of an electric field.

Comparison with Covalent Bonding

In ionic bonding, the atoms are bound by attraction of oppositely charged ions, whereas, in covalent bonding, atoms are bound by sharing electrons to attain stable electron configurations. In covalent bonding, the molecular geometry around each atom is determined by valence shell electron pair repulsion VSEPR rules, whereas, in ionic materials, the geometry follows maximum packing rules. One could say that covalent bonding is more *directional* in the sense that the energy penalty for not adhering to the optimum bond angles is large, whereas ionic bonding has no such penalty. There are no shared electron pairs to repel each other, the ions should simply be packed as efficiently as possible. This often leads to much higher coordination numbers. In NaCl, each ion has 6 bonds and all bond angles are 90°. In CsCl the coordination number is 8. By comparison carbon typically has a maximum of four bonds.

Purely ionic bonding cannot exist, as the proximity of the entities involved in the bonding allows some degree of sharing electron density between them. Therefore, all ionic bonding has some covalent character. Thus, bonding is considered ionic where the ionic character is greater than the covalent character. The larger the difference in electronegativity between the two types of atoms involved in the bonding, the more ionic (polar) it is. Bonds with partially ionic and partially covalent character are called polar covalent bonds. For example, Na–Cl and Mg–O interactions have a few percent covalency, while Si–O bonds are usually ~50% ionic and ~50% covalent. Pauling estimated that an electronegativity difference of 1.7 (on the Pauling scale) corresponds to 50% ionic character, so that a difference greater than 1.7 corresponds to a bond which is predominantly ionic. Ionic character in covalent bonds can be directly measured for atoms having quadrupolar nuclei (^{2}H, ^{14}N, $^{81,79}Br$, $^{35,37}Cl$ or ^{127}I). These nuclei are generally objects of NQR nuclear quadrupole resonance and NMR nuclear magnetic resonance studies. Interactions between the nuclear quadrupole moments Q and the electric field gradients (EFG) are characterized via the nuclear quadrupole coupling constants,

$$QCC = \frac{e^2 q_{zz} Q}{h}$$

where the eq_{zz} term corresponds to the principal component of the EFG tensor and e is the elementary charge. In turn, the electric field gradient opens the way to description of bonding modes in molecules when the QCC values are accurately determined by NMR or NQR methods.

In general, when ionic bonding occurs in the solid (or liquid) state, it is not possible to talk about a single "ionic bond" between two individual atoms, because the cohesive forces that keep the lattice together are of a more collective nature. This is quite different in the case of covalent bonding, where we can often speak of a distinct bond localized between two particular atoms. However, even if ionic bonding is combined with some covalency, the result is *not* necessarily discrete bonds of a localized character. In such cases, the resulting bonding often requires description in terms of a band structure consisting of gigantic molecular orbitals spanning the entire crystal. Thus, the bonding in the solid often retains its collective rather than localized nature. When the difference in electronegativity is decreased, the bonding may then lead to a semiconductor, a semimetal or eventually a metallic conductor with metallic bonding.

Covalent Bond

Covalent bonding occurs when pairs of electrons are shared by atoms. Atoms will covalently bond with other atoms in order to gain more stability, which is gained by forming a full electron shell. By sharing their outer most (valence) electrons, atoms can fill up their outer electron shell and gain stability. Nonmetals will readily form covalent bonds with other nonmetals in order to obtain stability, and can form anywhere between one to three covalent bonds with other nonmetals depending on how many valence electrons they posses. Although it is said that atoms share electrons when they form covalent bonds, they do not usually share the electrons equally.

Only when two atoms of the same element form a covalent bond are the shared electrons actually shared equally between the atoms. When atoms of different elements share electrons through covalent bonding, the electron will be drawn more toward the atom with the higher electronegativity resulting in a polar covalent bond. When compared to ionic compounds, covalent compounds usually have a lower melting and boiling point, and have less of a tendency to dissolve in water. Covalent compounds can be in a gas, liquid, or solid state and do not conduct electricity or heat well. For each molecule, there are different names for pairs of electrons, depending if it is shared or not. A pair of electrons that is shared between two atoms is called a bond pair. A pair of electrons that is not shared between two atoms is called a lone pair.

Examples of Covalent Bonds

Oxygen Molecule

In the formation of oxygen molecule, each oxygen atom has six electrons in the valence shell and requires two electrons to complete their octet. Therefore the atoms contribute two electrons each for sharing to form oxygen molecule, two electron pair are shared and hence there is a double bond between the two oxygen atoms.

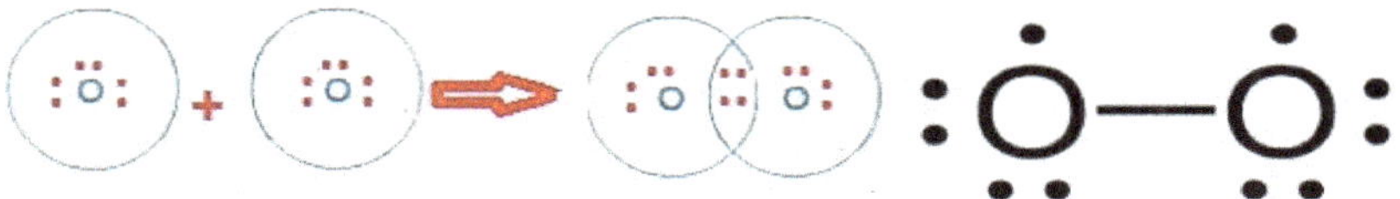

Figure: O2 Molecule with Single bond.

Carbon Dioxide Molecule

Carbon has four valence electrons and oxygen has six. To complete the octets, carbon shares two of its valence electrons with one oxygen atom and two with other oxygen atom.

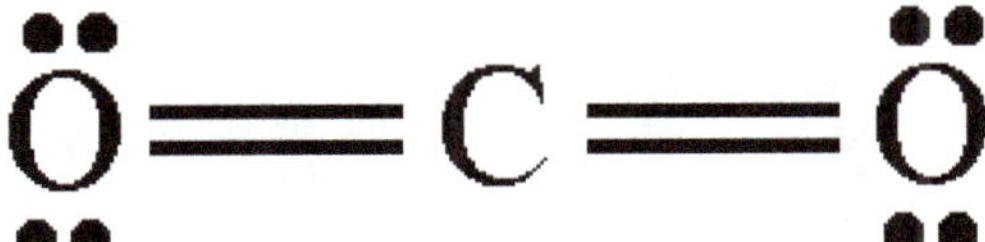

Figure: CO2 Molecule with Double bond.

Thus, there are two double bonds in CO_2 molecule.

Nitrogen Molecule

In the formation of a nitrogen molecule each of the two nitrogen atoms having five valence electrons, provides three electrons to form three electron pairs for sharing. Thus, a triple bond is formed between the two nitrogen atoms.

:N:::N:

:N≡N:

Figure: Nitrogen Molecule with Triple Bond.

Ethylene Molecule

In ethylene, each carbon atom shares two of its valence electron with two hydrogen atom and remaining two electrons with the second carbon atoms. So there is a double bond between the carbon atoms.

```
H       H      H       H
 \     /        \     /
  C::C     or    C=C
 /     \        /     \
H       H      H       H
```

Figure: Ethylene Molecule.

Types of Covalent Bond

Single Bonds

A single bond is when two electrons--one pair of electrons--are shared between two atoms. It is depicted by a single line between the two atoms. Although this form of bond is weaker and has a smaller density than a double bond and a triple bond, it is the most stable because it has a lower level of reactivity meaning less vulnerability in losing electrons to atoms that want to steal electrons.

Example: HCL

Below is a Lewis dot structure of Hydrogen Chloride demonstrating a single bond. As we can see from the picture below, Hydrogen Chloride has 1 Hydrogen atom and 1 Chlorine atom. Hydrogen has only 1 valence electron whereas Chlorine has 7 valence electrons. To satisfy the Octet Rule, each atom gives out 1 electron to share with each other; thus making a single bond.

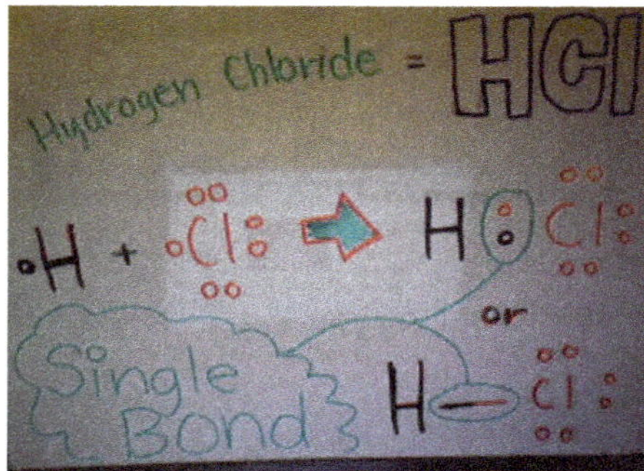

Double Bonds

A Double bond is when two atoms share two pairs of electrons with each other. It is depicted by two horizontal lines between two atoms in a molecule. This type of bond is much stronger than a single bond, but less stable; this is due to its greater amount of reactivity compared to a single bond.

EXAMPLE: CO_2

Below is a Lewis dot structure of Carbon dioxide demonstrating a double bond. As you can see from the picture below, Carbon dioxide has a total of 1 Carbon atom and 2 Oxygen atoms. Each Oxygen atom has 6 valence electrons whereas the Carbon atom only has 4 valence electrons. To satisfy the Octet Rule, Carbon needs 4 more valence electrons. Since each Oxygen atom has 3 lone pairs of electrons, they can each share 1 pair of electrons with Carbon; as a result, filling Carbon's outer valence shell (Satisfying the Octet Rule).

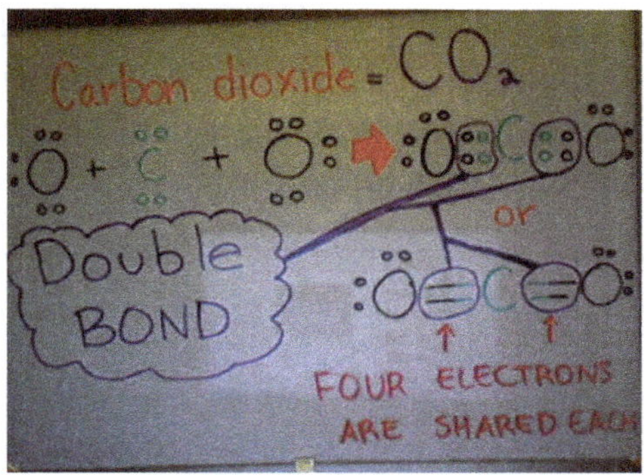

Triple Bond

A Triple bond is when three pairs of electrons are shared between two atoms in a molecule. It is the least stable out of the three general types of covalent bonds. It is very vulnerable to electron thieves.

Example: Acetylene

Below is a Lewis dot structure of Acetylene demonstrating a triple bond. As you can see from the picture below, Acetylene has a total of 2 Carbon atoms and 2 Hydrogen atoms. Each Hydrogen atom has 1 valence electron whereas each Carbon atom has 4 valence electrons. Each Carbon needs 4 more electrons and each Hydrogen needs 1 more electron. Hydrogen shares its only electron with Carbon to get a full valence shell. Now Carbon has 5 electrons. Because each Carbon atom has 5 electrons--1 single bond and 3 unpaired electrons--the two Carbons can share their unpaired electrons, forming a triple bond. Now all the atoms are happy with their full outer valence shell.

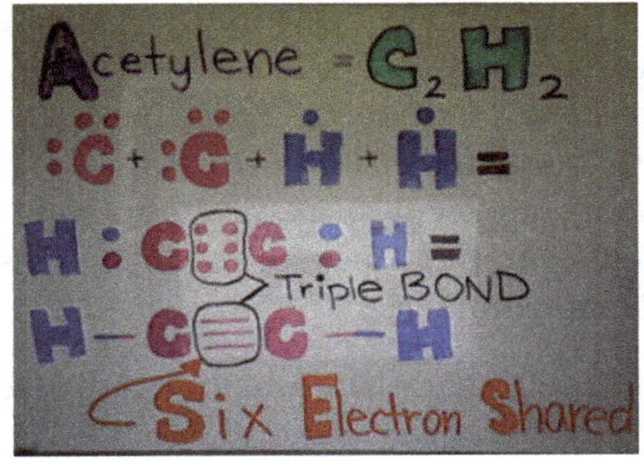

Polar Covalent Bond

A Polar Covalent Bond is created when the shared electrons between atoms are not equally shared. This occurs when one atom has a higher electronegativity than the atom it is sharing with. The atom with the higher electronegativity will have a stronger pull for electrons (Similiar to a Tug-O-War game, whoever is stronger usually wins). As a result, the shared electrons will be closer to the atom with the higher electronegativity, making it unequally shared. A polar covalent bond will result in the molecule having a slightly positive side (the side containing the atom with a lower electronegativity) and a slightly negative side (containing the atom with the higher electronegativity) because the shared electrons will be displaced toward the atom with the higher electronegativity. As a result of polar covalent bonds, the covalent compound that forms will have an electrostatic potential. This potential will make the resulting molecule slightly polar, allowing it to form weak bonds with other polar molecules. One example of molecules forming weak bonds with each other as a result of an unbalanced electrostatic potential is hydrogen bonding, where a hydrogen atom will interact with an electronegative hydrogen, fluorine, or oxygen atom from another molecule or chemical group.

Example: Water, Sulfide, Ozone, etc.

As you can see from the picture above, Oxygen is the big buff creature with the tattoo of "O" on its arm. The little bunny represents a Hydrogen atom. The blue and red bow tied in the middle of the rope, pulled by the two creatures represents--the shared pair of electrons--a single bond. Because the Hydrogen atom is weaker, the shared pair of electrons will be pulled closer to the Oxygen atom.

Nonpolar Covalent Bond

A Nonpolar Covalent Bond is created when atoms share their electrons equally. This usually occurs when two atoms have similar or the same electron affinity. The closer the values of their electron affinity, the stronger the attraction. This occurs in gas molecules; also known as diatomic elements. Nonpolar covalent bonds have a similar concept as polar covalent bonds; the atom with the higher electronegativity will draw away the electron from the weaker one. Since this statement is true--if we apply this to our diatomic molecules--all the atoms will have the same electronegativity since they are the same kind of element; thus, the electro negativities will cancel each other out and will have a charge of 0(A.K.A. Nonpolar covalent bond).

Examples of gas molecules that have a nonpolar covalent bond: Hydrogen gas atom, Nitrogen gas atoms, etc.

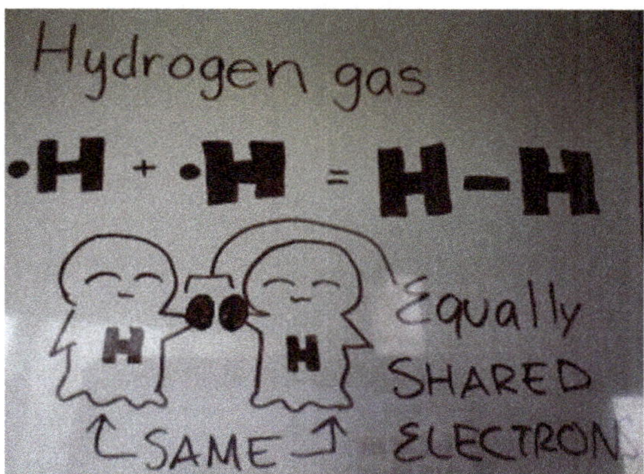

As you can see from the picture above, Hydrogen gas has a total of 2 Hydrogen atoms. Each Hydrogen atom has 1 valence electron. Since Hydrogen can only fit a max of 2 valence electrons in its orbital, each Hydrogen atom only needs 1 electron. Each atom has 1 valence electron, so they can just share, giving each atom two electrons each.

One and three-electron Bonds

Bonds with one or three electrons can be found in radical species, which have an odd number of electrons. The simplest example of a 1-electron bond is found in the dihydrogen cation, H_2^+. One-electron bonds often have about half the bond energy of a 2-electron bond, and are therefore called "half bonds". However, there are exceptions: in the case of dilithium, the bond is actually stronger for the 1-electron Li^+_2 than for the 2-electron Li_2. This exception can be explained in terms of hybridization and inner-shell effects.

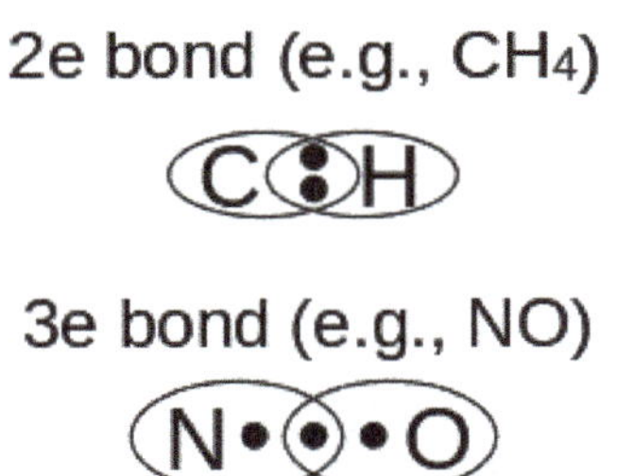

Comparison of the electronic structure of the three-electron bond to the conventional covalent bond.

The simplest example of three-electron bonding can be found in the helium dimer cation, He_2^+ It is considered a "half bond" because it consists of only one shared electron (rather than two); in molecular orbital terms, the third electron is in an anti-bonding orbital which cancels out half of the bond formed by the other two electrons. Another example of a molecule containing a 3-electron bond, in addition to two 2-electron bonds, is nitric oxide, NO. The oxygen molecule, O_2 can also be regarded as having two 3-electron bonds and one 2-electron bond, which accounts for its paramagnetism and its formal bond order of 2. Chlorine dioxide and its heavier analogues bromine dioxide and iodine dioxide also contain three-electron bonds.

Molecules with odd-electron bonds are usually highly reactive. These types of bond are only stable between atoms with similar electronegativities.

Resonance

There are situations whereby a single Lewis structure is insufficient to explain the electron configuration in a molecule, hence a superposition of structures are needed. The same two atoms in such molecules can be bonded differently in different structures (a single bond in one, a double bond in another, or even none at all), resulting in a non-integer bond order. The nitrate ion is one such example with three equivalent structures. The bond between the nitrogen and each oxygen is a double bond in one structure and a single bond in the other two, so that the average bond order for each N–O interaction is $2 + 1 + 1/3 = 4/3$.

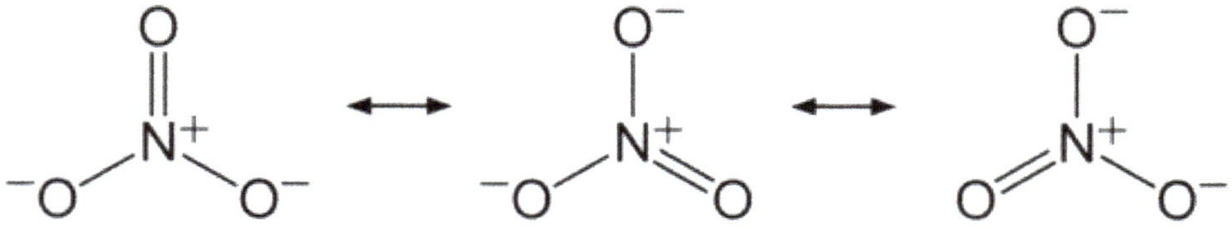

Aromaticity

In organic chemistry, when a molecule with a planar ring obeys Hückel's rule, where the number of π electrons fit the formula $4n + 2$ (where n is an integer), it attains extra stability and symmetry. In benzene, the prototypical aromatic compound, there are 6 π bonding electrons ($n = 1$, $4n + 2 = 6$). These occupy three delocalized π molecular orbitals (molecular orbital theory) or form conjugate π bonds in two resonance structures that linearly combine (valence bond theory), creating a regular hexagon exhibiting a greater stabilization than the hypothetical 1,3,5-cyclohexatriene.

In the case of heterocyclic aromatics and substituted benzenes, the electronegativity differences between different parts of the ring may dominate the chemical behaviour of aromatic ring bonds, which otherwise are equivalent.

Hypervalence

Certain molecules such as xenon difluoride and sulfur hexafluoride have higher co-ordination numbers than would be possible due to strictly covalent bonding according to the octet rule. This is explained by the three-center four-electron bond ("3c–4e") model which interprets the molecular wavefunction in terms of non-bonding highest occupied molecular orbitals in molecular orbital theory and ionic-covalent resonance in valence bond theory.

Electron-deficiency

In three-center two-electron bonds ("3c–2e") three atoms share two electrons in bonding. This type of bonding occurs in electron deficient compounds like diborane. Each such bond (2 per molecule in diborane) contains a pair of electrons which connect the boron atoms to each other in a banana shape, with a proton (nucleus of a hydrogen atom) in the middle of the bond, sharing electrons with both boron atoms. In certain cluster compounds, so-called four-center two-electron bonds also have been postulated.

Quantum Mechanical Description

After the development of quantum mechanics, two basic theories were proposed to provide a quantum description of chemical bonding: valence bond (VB) theory and molecular orbital (MO) theory. A more recent quantum description is given in terms of atomic contributions to the electronic density of states.

Covalency from Atomic Contribution to the Electronic Density of States

In COOP, COHP and BCOOP, evaluation of bond covalency is dependent on the basis set. To overcome this issue, an alternative formulation of the bond covalency can be provided in this way.

The center mass $cm(n,l,m_l,m_s)$ of an atomic orbital $|n,l,m_l,m_s\rangle$, with quantum numbers n, l, m_l, m_s, for atom A is defined as

$$cm^A(n,l,m_l,m_s) = \frac{\int_{E_0}^{E_1} E g^A_{|n,l,m_l,m_s\rangle}(E)dE}{\int_{E_0}^{E_1} g^A_{|n,l,m_l,m_s\rangle}(E)dE}$$

Where $g^A_{|n,l,m_l,m_s\rangle}(E)$ is the contribution of the atomic orbital $|n,l,m_l,m_s\rangle$ of the atom A to the total electronic density of states $g(E)$ of the solid

$$g(E) = \sum_A \sum_{n,l} \sum_{m_l,m_s} g^A_{|n,l,m_l,m_s\rangle}(E)$$

where the outer sum runs over all atoms A of the unit cell. The energy window $[E_0,E_1]$ is chosen in such a way that it encompasses all relevant bands participating in the bond. If the range to select is unclear, it can be identified in practice by examining the molecular orbitals that describe the electron density along the considered bond.

The relative position $C_{nAlA,nBlB}$ of the center mass of $|n_A,l_A\rangle$ levels of atom A with respect to the center mass of $|n_B,l_B\rangle$ levels of atom B is given as,

$$C_{n_Al_A,n_Bl_B} = -\left|cm^A(n_A,l_A) - cm^B(n_B,l_B)\right|$$

where the contributions of the magnetic and spin quantum numbers are summed. According to this definition, the relative position of the A levels with respect to the B levels is,

$$C_{A,B} = -\left|cm^A - cm^B\right|$$

where, for simplicity, we may omit the dependence from the principal quantum number n in the notation referring to $C_{nAlA,nBlB}$.

In this formalism, the greater the value of $C_{A,B}$, the higher the overlap of the selected atomic bands, and thus the electron density described by those orbitals gives a more covalent A–B bond. The quantity $C_{A,B}$ is denoted as the *covalency* of the A–B bond, which is specified in the same units of the energy E.

Coordinate Covalent Bond

Coordinate covalent bonding is also called as the Dative bonding. It is similar to that of the covalent bond. But the only difference is that the shared pair of electron form one atom or one group of atoms.

When shared pair of the electron comes from only atom and not one each from the two atoms involved in the bonding then it is called as co-ordinate or dative bonding. The atoms which donates the electron to form co-ordinate covalent bonding or dative bonding is called as the donor atom, while the atom which accepts the pair of electron for bonding is called as the acceptor atoms.

Thus co-ordinate bond can also be explained as the bond formed between the donor and acceptor atoms. A co-ordinate bond is represented by an arrow starting from the donor atoms and ending in the acceptor atom. In some cases the donor may be a molecule, an atom in the molecule can donate the pair of electron, in that case the Lewis base can act as a donor atom donating a pair of electron and the acceptor would be the Lewis acid.

A co-ordinate covalent bond has all the characteristic of the covalent bond. They have low boiling and melting point. Since the shared pair is between two atoms there are no columbic forces of attraction. They do not conduct electricity in the liquid or in the dissolved state. The compounds are that much soluble in water.

Coordinate Covalent Bond Examples

Lewis acid base reaction is an excellent example of the co-ordinate covalent bond. For example the bond:

$$H_3N: \rightarrow BF_3$$

is a coordinate bond. Here nitrogen acts a donor atom.

The lone pair of electron in the nitrogen is donated to the vacant p orbital of the boron. Here ammonia is Lewis base and BF3 is Lewis acid.

The formation of the hydronium ion when the water reacts with proton is also an example of the co-ordinate bonding:

$$H_2O + H^+ \rightarrow H_3O^+$$

There two lone pair of electron in the oxygen atom which is used to form coordinate bond with the proton. Here water acts the donor atom, more specifically it is the oxygen atom in the water that acts as the donor atom and the proton accepts the electron.

All the coordination complex has the co-ordinate bond. In the complex the donor atom is called as ligand. For example consider the complex $[Ag(NH_3)_2]Cl$. Here the ammonia act as the donor atom or ligand and the metal acts the Lewis acid or acceptor.

Ammonium Ion Coordinate Bond

It is formed by the combination of the ammonia molecule and a hydrogen ion. In ammonia, the nitrogen atom has a lone pair of electrons after completing its octet. It donates this lone pair to the hydrogen ion.

Thus the nitrogen atom becomes the donor. The hydrogen atom becomes the acceptor. The linkage between N and H atoms is called coordinate bond. It is represented by an arrow →.

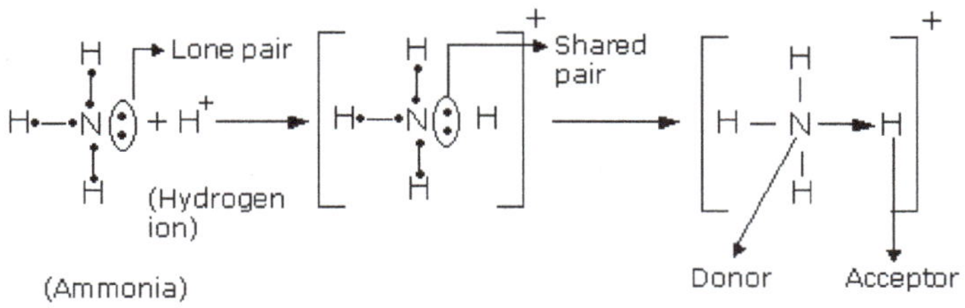

Lone pair
Shared pair
(Hydrogen ion)
(Ammonia)
Donor
Acceptor

Hydronium Ion Coordinate Bond

It is formed by the combination of water molecule and hydrogen ion.

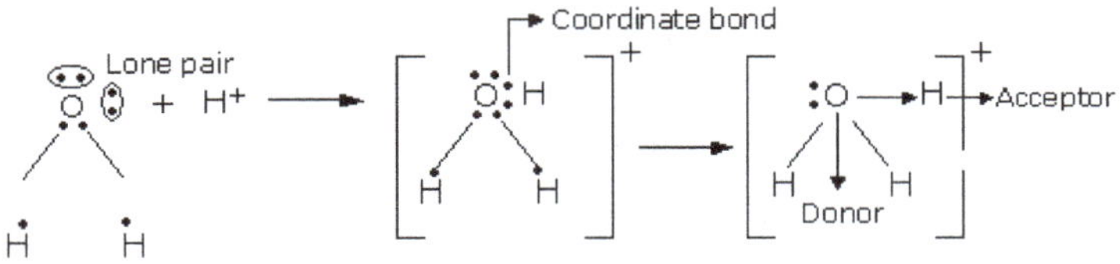

Coordinate bond
Lone pair
Acceptor
Donor

- The oxygen atom in a water molecule has two lone pairs of electrons.
- It donates one pair to the hydrogen ion.
- Oxygen is thus the donor and hydrogen ion, the acceptor.
- The hydrogen ion carries over its charge to the hydronium ion.

Secondary Bonding

Secondary bonds are bonds of a different kind to the primary ones. They are weaker in nature and are broadly classified as Van der Waal's forces and hydrogen bonds. These bonds are due to atomic or molecular dipoles, both permanent and temporary.

Van der Waal's forces are of two types. The first type is as a result of electrostatic attraction between two permanent dipoles. Permanent dipoles are formed in asymmetric molecules where there are permanent positive and negative regions due to difference in electro negativities of the constituent elements. For example, water molecule is made of one oxygen and two hydrogen atoms. Since each hydrogen requires one electron and oxygen requires two electrons to complete their respective noble gas configurations, thus when these atoms approach each other they share a pair of electrons between each hydrogen and the oxygen atom. This way all three achieve stability through the bonds formed. But since oxygen is a highly electronegative atom, therefore the shared electron cloud is attracted more towards it than the hydrogen atoms, giving rise to a permanent dipole. When this water molecule approaches another water molecule, a partial bond is formed between the partially positive hydrogen atom of one molecule and the partially negative oxygen of another. This partial bond is due to an electric dipole and thus is called a Van der Waal's bond.

The second type of Van der Waal's bond is formed due to temporary dipoles. A temporary dipole is formed in a symmetric molecule but which has fluctuations of charges giving rise to partial dipole moments for only a few moments. This can also be seen in atoms of inert gases. For instance, a molecule of methane has one carbon atom and four hydrogen atoms joined together by single covalent bonds between the carbon and the hydrogen atoms. Methane is a symmetric molecule but when it is solidified, the bonds between the molecules are of weak Van der Waal's forces and thus such a solid cannot exist for a long time without tremendously cared for laboratory conditions.

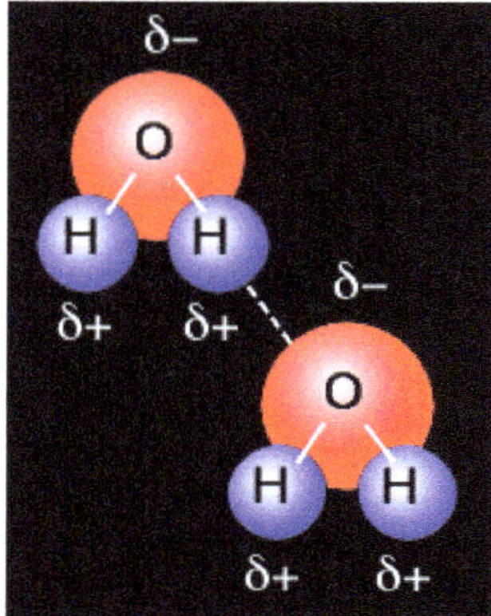

Hydrogen Bonding between two Water Molecules

Hydrogen Bonding

A hydrogen bond is the electromagnetic attraction created between a partially positively charged hydrogen atom attached to a highly electronegative atom and another nearby electronegative atom. A hydrogen bond is a type of dipole-dipole interaction; it is not a true chemical bond. These attractions can occur between molecules (intermolecularly) or within different parts of a single molecule (intramolecularly).

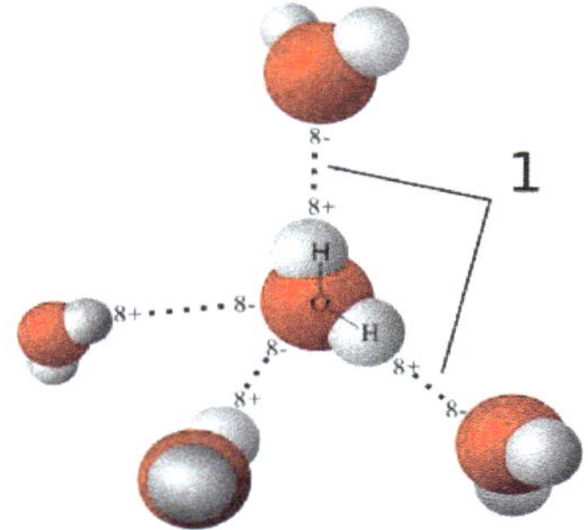

Hydrogen bonding in waterThis is a space-filling ball diagram of the interactions between separate water molecules.

Hydrogen Bond Donor

A hydrogen atom attached to a relatively electronegative atom is a hydrogen bond donor. This electronegative atom is usually fluorine, oxygen, or nitrogen. The electronegative atom attracts the electron cloud from around the hydrogen nucleus and, by decentralizing the cloud, leaves the hydrogen atom with a positive partial charge.

Bond Strength

Hydrogen bonds can vary in strength from weak (1–2 kJ mol^{-1}) to strong (161.5 kJ mol^{-1} in the ion HF^{-}_{2}). Typical enthalpies in vapor include:

- F–H···:F (161.5 kJ/mol or 38.6 kcal/mol), illustrated uniquely by HF_2^-, bifluoride.
- O–H···:N (29 kJ/mol or 6.9 kcal/mol), illustrated water-ammonia.
- O–H···:O (21 kJ/mol or 5.0 kcal/mol), illustrated water-water, alcohol-alcohol.
- N–H···:N (13 kJ/mol or 3.1 kcal/mol), illustrated by ammonia-ammonia.
- N–H···:O (8 kJ/mol or 1.9 kcal/mol), illustrated water-amide.
- HO–H···: OH_3^+ OH^+_3 (18 kJ/mol or 4.3 kcal/mol).

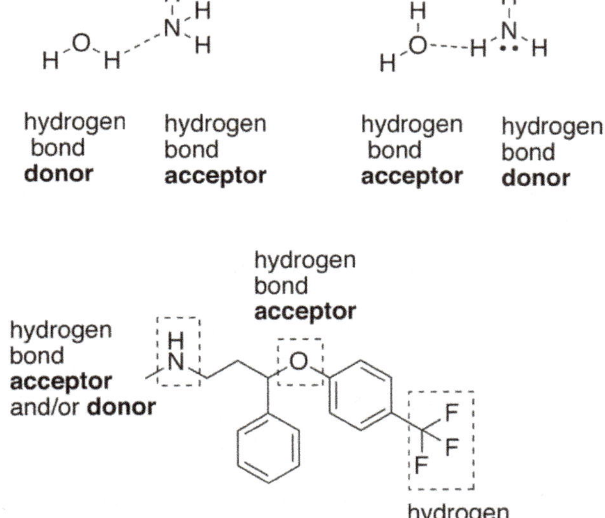

Examples of hydrogen bond donating (donors) and hydrogen bond accepting groups (acceptors)

Structural Details

The X-H distance is typically ≈110 pm, whereas the H···Y distance is ≈160 to 200 pm. The typical length of a hydrogen bond in water is 197 pm. The ideal bond angle depends on the nature of the hydrogen bond donor. The following hydrogen bond angles between a hydrofluoric acid donor and various acceptors have been determined experimentally:

Acceptor···donor	VSEPR geometry	Angle (°)
HCN···HF	linear	180
H_2CO···HF	trigonal planar	120
H_2O···HF	pyramidal	46
H_2S···HF	pyramidal	89
SO_2···HF	trigonal	142

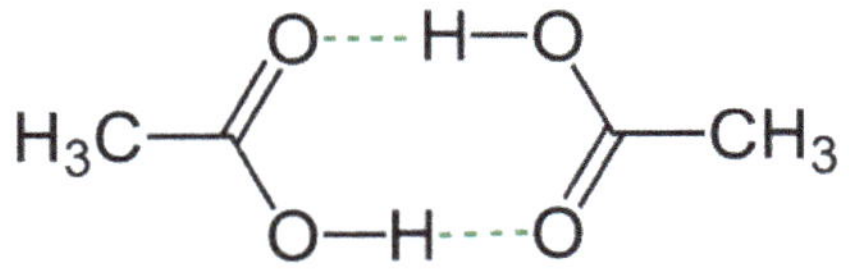

Cyclic dimer of acetic acid; dashed **green** lines represent hydrogen bonds

Spectroscopy

Strong hydrogen bonds are revealed by downfield shifts in the ^{1}H NMR spectrum. For example, the acidic proton in the enol tautomer of acetylacetone appears at δ15.5, which is about 10 ppm downfield of a conventional alcohol.

In the IR spectrum, hydrogen bonding shifts the X-H stretching frequency to lower energy (i.e. the vibration frequency decreases). This shift reflects a weakening of the X-H bond. Certain hydrogen bonds - improper hydrogen bonds - show a blue shift of the X-H stretching frequency and a decrease in the bond length.

Theoretical Considerations

Hydrogen bonding is of continuing theoretical interest. According to a modern description O:H-O integrates both the intermolecular O:H lone pair ":" nonbond and the intramolecular H-O polar-covalent bond associated with O-O repulsive coupling.

Quantum chemical calculations of the relevant interresidue potential constants (compliance constants) revealed large differences between individual H bonds of the same type. For example, the central interresidue N−H···N hydrogen bond between guanine and cytosine is much stronger in comparison to the N−H···N bond between the adenine-thymine pair.

Theoretically, the bond strength of the hydrogen bonds can be assessed using NCI index, non-covalent interactions index, which allows a visualization of these non-covalent interactions, as its name indicases, using the electron density of the system.

From interpretations of the anisotropies in the Compton profile of ordinary ice that the hydrogen bond is partly covalent. However, this interpretation was challenged.

Most generally, the hydrogen bond can be viewed as a metric-dependent electrostatic scalar field between two or more intermolecular bonds. This is slightly different from the intramolecular bound states of, for example, covalent or ionic bonds; however, hydrogen bonding is generally still a bound state phenomenon, since the interaction energy has a net negative sum. The initial theory of hydrogen bonding proposed by Linus Pauling suggested that the hydrogen bonds had a partial covalent nature. This interpretation remained controversial until NMR techniques demonstrated information transfer between hydrogen-bonded nuclei, a feat that would only be possible if the hydrogen bond contained some covalent character.

Hydrogen Bonds in Small Molecules

Water

A ubiquitous example of a hydrogen bond is found between water molecules. In a discrete water molecule, there are two hydrogen atoms and one oxygen atom. Two molecules of water can form a hydrogen bond between them that is to say oxygen-hydrogen bonding; the simplest case, when only two molecules are present, is called the water dimer and is often used as a model system. When more molecules are present, as is the case with liquid water, more bonds are possible because the oxygen of one water molecule has two lone pairs of electrons, each of which can form a hydrogen bond with a hydrogen on another water molecule. This can repeat such that every water molecule is H-bonded with up to four other molecules, as shown in the figure (two through its two lone pairs, and two through its two hydrogen atoms). Hydrogen bonding strongly affects the crystal structure of ice, helping to create an open hexagonal lattice. The density of ice is less than the density of water at the same temperature; thus, the solid phase of water floats on the liquid, unlike most other substances.

Liquid water's high boiling point is due to the high number of hydrogen bonds each molecule can form, relative to its low molecular mass. Owing to the difficulty of breaking these bonds, water has a very high boiling point, melting point, and viscosity compared to otherwise similar liquids not conjoined by hydrogen bonds. Water is unique because its oxygen atom has two lone pairs and two hydrogen atoms, meaning that the total number of bonds of a water molecule is up to four.

The number of hydrogen bonds formed by a molecule of liquid water fluctuates with time and temperature. From TIP4P liquid water simulations at 25 °C, it was estimated that each water molecule participates in an average of 3.59 hydrogen bonds. At 100 °C, this number decreases to 3.24 due to the increased molecular motion and decreased density, while at 0 °C, the average number of hydrogen bonds increases to 3.69. A more recent study found a much smaller number of hydrogen bonds: 2.357 at 25 °C. The differences may be due to the use of a different method for defining and counting the hydrogen bonds.

Where the bond strengths are more equivalent, one might instead find the atoms of two interacting water molecules partitioned into two polyatomic ions of opposite charge, specifically hydroxide (OH^-) and hydronium (H_3O^+). (Hydronium ions are also known as "hydroxonium" ions.)

$$H{-}O^- \; H_3O^+$$

Indeed, in pure water under conditions of standard temperature and pressure, this latter formulation is applicable only rarely; on average about one in every 5.5×10^8 molecules gives up a proton to another water molecule, in accordance with the value of the dissociation constant for water under such conditions. It is a crucial part of the uniqueness of water.

Because water may form hydrogen bonds with solute proton donors and acceptors, it may competitively inhibit the formation of solute intermolecular or intramolecular hydrogen bonds. Consequently, hydrogen bonds between or within solute molecules dissolved in water are almost always unfavorable relative to hydrogen bonds between water and the donors and acceptors for hydrogen bonds on those solutes. Hydrogen bonds between water molecules have an average lifetime of 10^{-12} seconds, or 10 picoseconds.

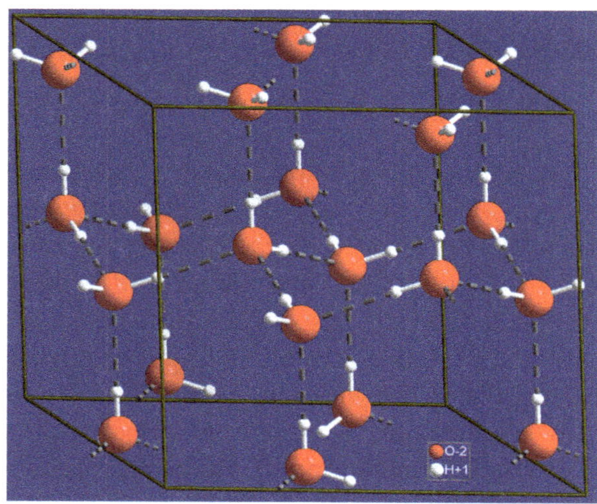

Crystal structure of hexagonal ice. Gray dashed lines indicate hydrogen bonds

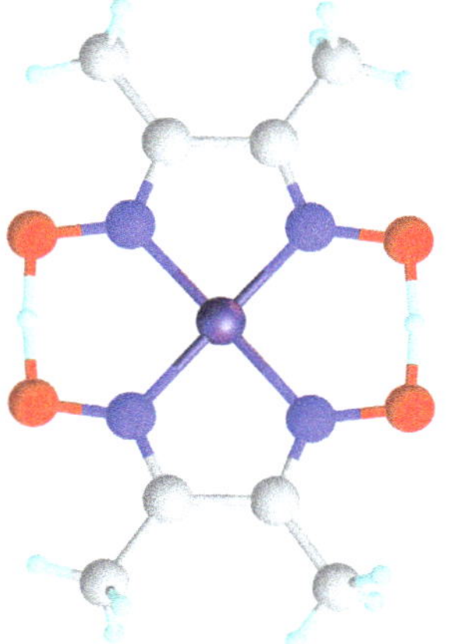

Structure of nickel bis(dimethylglyoximate), which features two linear hydrogen-bonds.

Bifurcated and Over-coordinated Hydrogen Bonds in Water

A single hydrogen atom can participate in two hydrogen bonds, rather than one. This type of bonding is called "bifurcated" (split in two or "two-forked"). It can exist, for instance, in complex natural or synthetic organic molecules. It has been suggested that a bifurcated hydrogen atom is an essential step in water reorientation.

Acceptor-type hydrogen bonds (terminating on an oxygen's lone pairs) are more likely to form bifurcation (it is called overcoordinated oxygen, OCO) than are donor-type hydrogen bonds, beginning on the same oxygen's hydrogens.

Other Liquids

For example, hydrogen fluoride—which has three lone pairs on the F atom but only one H atom—can form only two bonds; (ammonia has the opposite problem: three hydrogen atoms but only one lone pair).

H–F···H–F···H–F

Further Manifestations of Solvent Hydrogen Bonding

- Increase in the melting point, boiling point, solubility, and viscosity of many compounds can be explained by the concept of hydrogen bonding.
- Negative azeotropy of mixtures of HF and water.
- The fact that ice is less dense than liquid water is due to a crystal structure stabilized by hydrogen bonds.
- Dramatically higher boiling points of NH_3, H_2O, and HF compared to the heavier analogues PH_3, H_2S, and HCl, where hydrogen-bonding is absent.
- Viscosity of anhydrous phosphoric acid and of glycerol.
- Dimer formation in carboxylic acids and hexamer formation in hydrogen fluoride, which occur even in the gas phase, resulting in gross deviations from the ideal gas law.
- Pentamer formation of water and alcohols in apolar solvents.

Hydrogen Bonds in Polymers

Hydrogen bonding plays an important role in determining the three-dimensional structures and the properties adopted by many synthetic and natural proteins.

DNA

In these macromolecules, bonding between parts of the same macromolecule cause it to fold into a specific shape, which helps determine the molecule's physiological or biochemical role. For example, the double helical structure of DNA is due largely to hydrogen bonding between its base pairs (as well as pi stacking interactions), which link one complementary strand to the other and enable replication.

Proteins

In the secondary structure of proteins, hydrogen bonds form between the backbone oxygens and amide hydrogens. When the spacing of the amino acid residues participating in a hydrogen bond occurs regularly between positions i and $i + 4$, an alpha helix is formed. When the spacing is less, between positions i and $i + 3$, then a 3_{10} helix is formed. When two strands are joined by hydrogen bonds involving alternating residues on each participating strand, a beta sheet is formed. Hydrogen bonds also play a part in forming the tertiary structure of protein through interaction of R-groups.

The role of hydrogen bonds in protein folding has also been linked to osmolyte-induced protein stabilization. Protective osmolytes, such as trehalose and sorbitol, shift the protein folding equilibrium toward the folded state, in a concentration dependent manner. While the prevalent explanation for osmolyte action relies on excluded volume effects, that are entropic in nature, recent circular dichroism (CD) experiments have shown osmolyte to act through an enthalpic effect. The molecular mechanism for their role in protein stabilization is still not well established, though several mechanism have been proposed. Recently, computer molecular dynamics simulations suggested that osmolytes stabilize proteins by modifying the hydrogen bonds in the protein hydration layer.

Several studies have shown that hydrogen bonds play an important role for the stability between subunits in multimeric proteins. For example, a study of sorbitol dehydrogenase displayed an important hydrogen bonding network which stabilizes the tetrameric quaternary structure within the mammalian sorbitol dehydrogenase protein family.

A protein backbone hydrogen bond incompletely shielded from water attack is a dehydron. Dehydrons promote the removal of water through proteins or ligand binding. The exogenous dehydration enhances the electrostatic interaction between the amide and carbonyl groups by de-shielding their partial charges. Furthermore, the dehydration stabilizes the hydrogen bond by destabilizing the nonbonded state consisting of dehydrated isolated charges.

Wool, being a protein fibre, is held together by hydrogen bonds, causing wool to recoil when stretched. However, washing at high temperatures can permanently break the hydrogen bonds and a garment may permanently lose its shape.

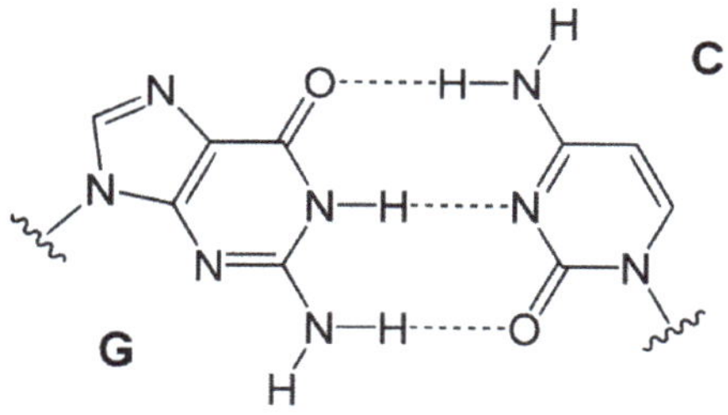

Hydrogen bonding between guanine and cytosine, one of two types of base pairs in DNA

Cellulose

Hydrogen bonds are important in the structure of cellulose and derived polymers in its many different forms in nature, such as such as cotton and flax.

Synthetic Polymers

Many polymers are strengthened by hydrogen bonds within and between the chains. Among the synthetic polymers, a well characterized example is nylon, where hydrogen bonds occur in the repeat unit and play a major role in crystallization of the material. The bonds occur between carbonyl and amine groups in the amide repeat unit. They effectively link adjacent chains, which help reinforce the material. The effect is great in aramid fibre, where hydrogen bonds stabilize the linear chains laterally. The chain axes are aligned along the fibre axis, making the fibres extremely stiff and strong.

The hydrogen-bond networks make both natural and synthetic polymers sensitive to humidity levels in the atmosphere because water molecules can diffuse into the surface and disrupt the network. Some polymers are more sensitive than others. Thus nylons are more sensitive than aramids, and nylon 6 more sensitive than nylon-11.

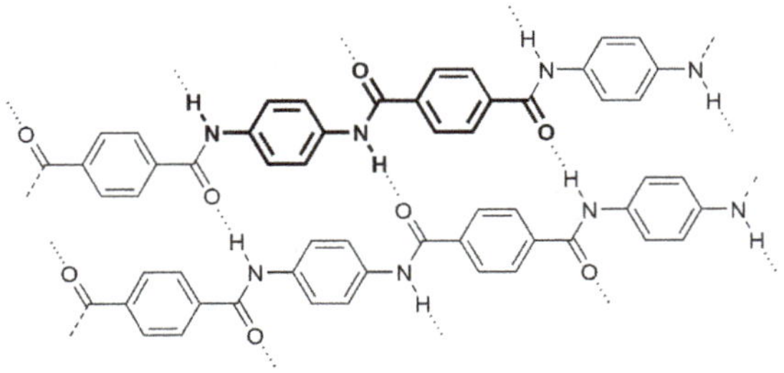

Para-aramid structure

Symmetric hydrogen bond

A symmetric hydrogen bond is a special type of hydrogen bond in which the proton is spaced exactly halfway between two identical atoms. The strength of the bond to each of those atoms is equal. It is an example of a three-center four-electron bond. This type of bond is much stronger than a "normal" hydrogen bond. The effective bond order is 0.5, so its strength is comparable to a covalent bond. It is seen in ice at high pressure, and also in the solid phase of many anhydrous acids such as hydrofluoric acid and formic acid at high pressure. It is also seen in the bifluoride ion $[F-H-F]^-$.

Symmetric hydrogen bonds have been observed recently spectroscopically in formic acid at high pressure (>GPa). Each hydrogen atom forms a partial covalent bond with two atoms rather than one. Symmetric hydrogen bonds have been postulated in ice at high pressure (Ice X). Low-barrier hydrogen bonds form when the distance between two heteroatoms is very small.

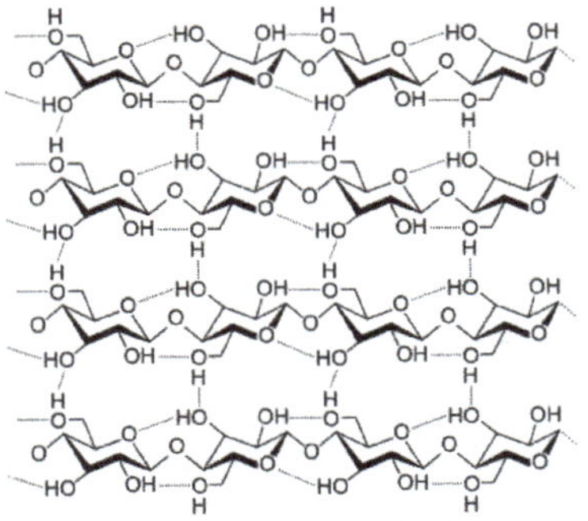

A strand of cellulose (conformation I_α), showing the hydrogen bonds (dashed) within and between cellulose molecules

Dihydrogen Bond

The hydrogen bond can be compared with the closely related dihydrogen bond, which is also an intermolecular bonding interaction involving hydrogen atoms. These structures have been known for some time, and well characterized by crystallography; however, an understanding of their relationship to the conventional hydrogen bond, ionic bond, and covalent bond remains unclear. Generally, the hydrogen bond is characterized by a proton acceptor that is a lone pair of electrons in nonmetallic atoms (most notably in the nitrogen, and chalcogen groups). In some cases, these proton acceptors may be pi-bonds or metal complexes. In the dihydrogen bond, however, a metal hydride serves as a proton acceptor, thus forming a hydrogen-hydrogen interaction. Neutron diffraction has shown that the molecular geometry of these complexes is similar to hydrogen bonds, in that the bond length is very adaptable to the metal complex/hydrogen donor system.

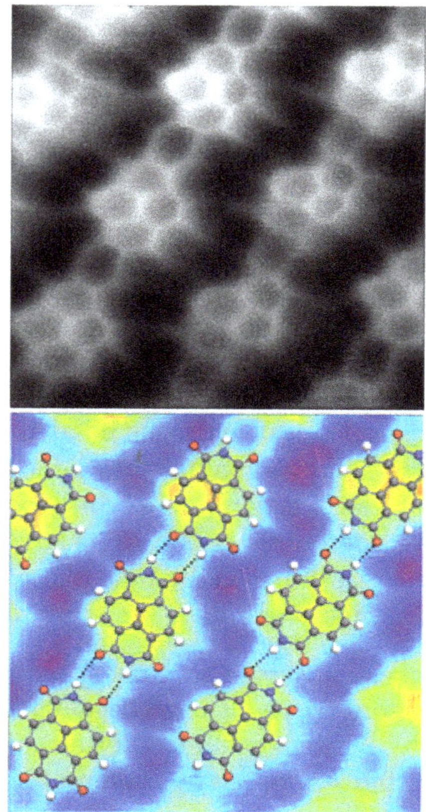

AFM image of napthalenetetracarboxylic diimide molecules on silver-terminated silicon, interacting via hydrogen bonding, taken at 77 K. ("Hydrogen bonds" in the top image are exaggerated by artifacts of the imaging technique.)

Dynamics Probed by Spectroscopic Means

The dynamics of hydrogen bond structures in water can be probed by the IR spectrum of OH stretching vibration. In the hydrogen bonding network in protic organic ionic plastic crystals (POIPCs), which are a type of phase change material exhibiting solid-solid phase transitions prior to melting, variable-temperature infrared spectroscopy can reveal the temperature dependence of hydrogen bonds and the dynamics of both the anions and the cations. The sudden weakening of hydrogen bonds during the solid-solid phase transition seems to be coupled with the onset of orientational or rotational disorder of the ions.

Application to Drugs

Hydrogen bonding is a key to the design of drugs. According to Lipinski's rule of five the majority of orally active drug tend to have between five and ten hydrogen bonds. These interactions exist between nitrogen–hydrogen and oxygen–hydrogen centers. As with many other rules of thumb, many exceptions exist.

Bond Parameter

Bond Length

The distance between two atoms participating in a bond, known as the bond length, can be determined experimentally. X-ray diffraction of molecular crystals allows for the determination of the three-dimensional structure of molecules and the precise measurement of internuclear distances. Various spectroscopic methods also exist for estimating the bond length between two atoms in a molecule.

Bonds are not Static Structures

The bond length is the average distance between the nuclei of two bonded atoms in a molecule. This is because a chemical bond is not a static structure, but the two atoms actually vibrate due to thermal energy available in the surroundings at any non-zero Kelvin temperature. A bond can be modeled as two balls connected by a spring: stretching or compressing the spring initiates a back-and-forth motion with respect to the equilibrium positions of the balls. Measured bond lengths are the distance between those unperturbed, or equilibrium, positions of the balls, or atoms.

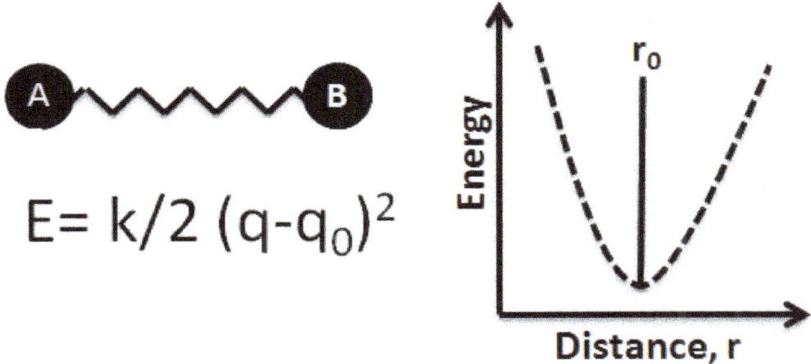

Ball-and-spring model of a chemical bondA bond between two atoms can be thought of as a spring with two balls attached to it. Any stretch or compression of the spring will initiate oscillations of the atoms with respect to their equilibrium (unperturbed) positions. The potential energy function for this system is also indicated. The minimum energy occurs at the equilibrium distance ro, which is where the bond length is measured.

Bond Length Values

Bond lengths have traditionally been expressed in Ångstrom units, but picometers are sometimes preferred (1 Å = 10^{-10} m = 100 pm). Bonds lengths are typically in the range of 1-2 Å, or 100-200 pm. Even though the bond vibrates, equilibrium bond lengths can be determined experimentally to within ±1 pm. The actual distance between two atoms in a molecule depends on factors such as the orbital hybridization and the electronic nature of its components.

Bonds involving hydrogen can be quite short; the shortest bond of all, H–H, is only 74 pm. The covalent radius of an atom is determined by halving the bond distance between two identical atoms. Based on data for the H_2 molecule, the covalent radius of H is 37 pm. Covalent radii can be used to estimate the bond distance between two different atoms; it is the sum of the individual covalent radii.

Periodic Trends

Generally, when we consider a bond between a given atom and a varying atomic bonding partner, the bond length decreases across a period in the periodic table, and increases down a group. This trend is identical to that of the atomic radius.

Atoms with multiple bonds between them have shorter bond lengths than singly bonded ones; this is a major criterion for experimentally determining the multiplicity of a bond. For example, the bond length of [latex]C - C[/latex] is 154 pm; the bond length of [latex]C = C[/latex] is 133 pm.

Bond Enthalphy

Bond enthalpy (which is also known as bond-dissociation enthalpy, average bond energy, or bond strength) describes the amount of energy stored in a bond between atoms in a molecule. Specifically, it's the energy that needs to be added for the homolytic or symmetrical cleavage of a bond in the gas phase. A homolytic or symmetrical bond breaking event means that when the bond is broken, each atom that originally participated in the bond gets one electron and becomes a radical, as opposed to forming an ion.

Chemical bonds form because they're thermodynamically favorable, and breaking them inevitably requires adding energy. For this reason, bond enthalpy values are always positive, and they usually have units of kJ / mol or $kcal / mol$. The higher the bond enthalpy, the more energy is needed to break the bond and the stronger the bond. To determine how much energy will be released when we form a new bond rather than break it, we simply make the bond enthalpy value negative.

Because bond enthalpy values are so useful, average bond enthalpies for common bond types are readily available in reference tables. While in reality the actual energy change when forming and breaking bonds depends on neighboring atoms in a specific molecule, the average values available in the tables can still be used as an approximation.

Using Bond Enthalpies to Estimate Enthalpy of Reaction

Once we understand bond enthalpies, we use them to estimate the enthalpy of reaction. To do this, we can use the following procedure:

Step 1: Identify which bonds in the reactants will break and find their bond enthalpies.

Step 2: Add up the bond enthalpy values for the broken bonds.

Step 3: Identify which new bonds form in the products and list their negative bond enthalpies. Remember we have to switch the sign for the bond enthalpy values to find the energy released when the bond forms.

Step 4: Add up the bond enthalpy values for the formed product bonds.

Step 5: Combine the total values for breaking bonds (from Step 2) and forming bonds (from Step 4) to get the enthalpy of reaction.

Example: Hydrogenation of propene

Let's find the enthalpy of reaction for the hydrogenation of propene, our example from the beginning of the article.

Step 1: Identify bonds broken.

This reaction breaks a C=C bond and an H–H bond.

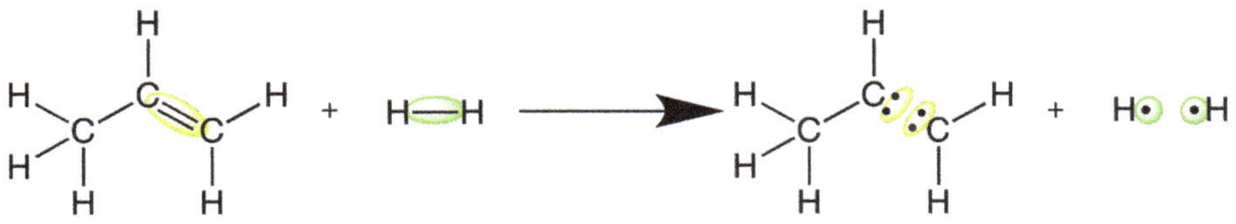

Using a reference table, we find that the C=C bond enthalpy is $610kJ/mol$. and the H- H bond enthalpy is $436kJ/mol$.

Step 2: Find total energy to break bonds.

Combining the values from Step 1 gives us:

Energy added to break bonds= $610kJ/mol+436kJ/mol=1046kJ/mol$

as the total energy required to break the necessary bonds in propene and hydrogen gas.

Step 3: Identify bonds formed.

This reaction forms a C–C, minus, C bond and two new C–H, minus, H bonds.

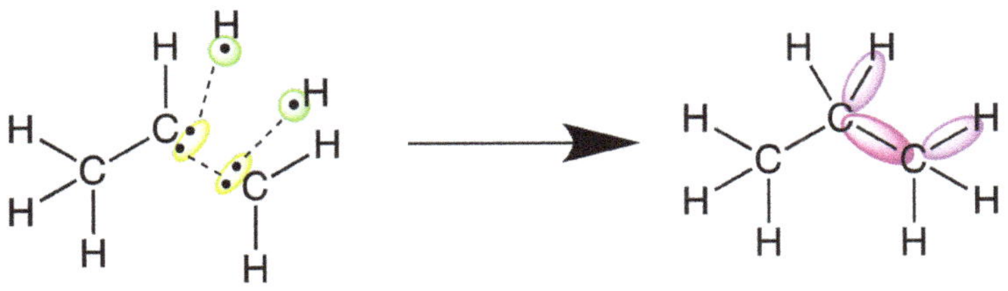

Using a reference table (and keeping in mind that we're forming rather than breaking bonds), we find that the C–C, bond enthalpy is 346kJ/mol, and the bond enthalpy for each of the C–H bonds is 413kJ/mol. To find the energy released when forming these bonds, we will multiply the bond enthalpies by -1. Also, since there are two C-H bonds, we'll need to multiply this value by 2.

Step 4: Find total energy released to form new bonds.

Combining the values from Step 3 gives us:

Energy released to make product bonds= $-346kJ/mol+(2\times-413kJ/mol)=1172kJ/mol$

for the total energy that will be released by forming the new bonds.

Step 5: Add up energy for bonds broken and formed.

From Step 2 and Step 4, we have $1046kJ$ of energy required to break bonds and $-1172kJ$ of energy released from forming bonds. Combining these values, we get for the enthalpy of reaction:

$$\Delta H_{rXn} = energy\ added\ to\ break\ reactant\ bonds + energy\ released\ when\ making\ product\ bonds$$

$$= 1046kJ/mol + (-1172kJ/mol)$$

$$= -126kJ/mol$$

Since the enthalpy of reaction for the hydrogenation of propene is negative, we know that the reaction is exothermic.

Bond Order

Bond order is a measurement of the number of electrons involved in bonds between two atoms in a molecule. It is used as an indicator of the stability of a chemical bond.

Bond order according to Valence Bond theory

Lewis drew small diagrams to illustrate this camaraderie, which are now called Lewis structures. A Lewis structure describes the structure of a molecule by connecting the atoms with lines. The lines represent the number of electrons that have been shared between two or more atoms. Thus, when two atoms share two electrons, we depict it by connecting them with two lines. The number of lines, or more precisely, the number of chemical bonds that comprise a molecule, is called its bond order.

For instance, the bond order of carbon dioxide and methane is 4, which can easily be discerned by examining their Lewis structures. Notice how the magnitude of electrons shared between each pair adequately fills the valence shell of both atoms. Hydrogen only requires a single electron, as its shell is filled when it contains 2 electrons, not 8.

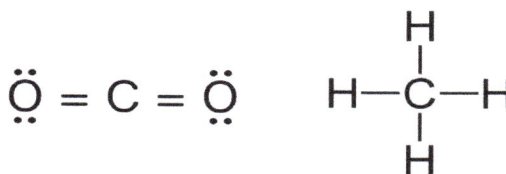

Lewis structures of carbon dioxide and methane.

There also exist molecules that can be described by more than one Lewis structure, such as sulfur dioxide. The bond order of such a molecule is the average of the bond orders of all the possible structures that describe it. The bond order of sulfur dioxide is therefore 1.5, not 3.

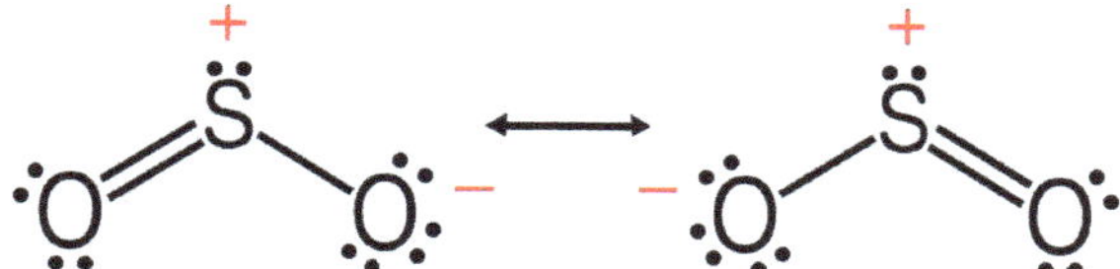

Sulfur dioxide exhibits two Lewis structures.

However, computing bond order by simply referring to the number of lines in the Lewis structure of a molecule is only acceptable under the Valence Bond (VB) theory. When it comes to the Molecular Orbital (MO) theory, the alternative theory that describes molecular bonding, the bond order might be the same, but the implications are drastically different.

Bond Order According to Molecular Orbital Theory

However, before determining the number of electrons in certain orbitals, one must fill these orbitals with electrons first. To fill the orbitals, one must know the rules according to which orbitals are occupied. Without understanding this rule, computing a molecule's bond order would be impossible. I'm sure there are many clever tricks or shortcuts to arrive at the number, but by learning them, you would be deprived of important conceptual knowledge.

People aware of the rules can refer to this expression to compute the bond order of a molecule:

Those who aren't aware have no option but to learn them. If it helps, one can simply learn the rules for filling atomic orbitals. The rules to fill molecular orbitals are the same, except that each "bonding" orbital is followed by an "anti-bonding" orbital. While atomic orbitals are filled as 1s2s2p... molecular orbitals are filled as 1s1s*2s2s*2p.... The asterisked orbitals represent anti-bonding orbitals. Unfortunately, the rules will not be fully explained here, as it would cause us to unnecessarily digress.

People aware of the rules can compute the bond order of, say, oxygen by using the above expression. In total, a single molecule of oxygen consists of 12 valence electrons. Now, according to the rules, the electrons must be arranged in this manner:

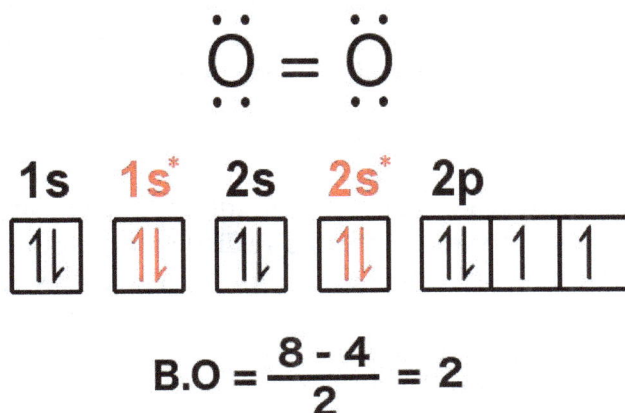

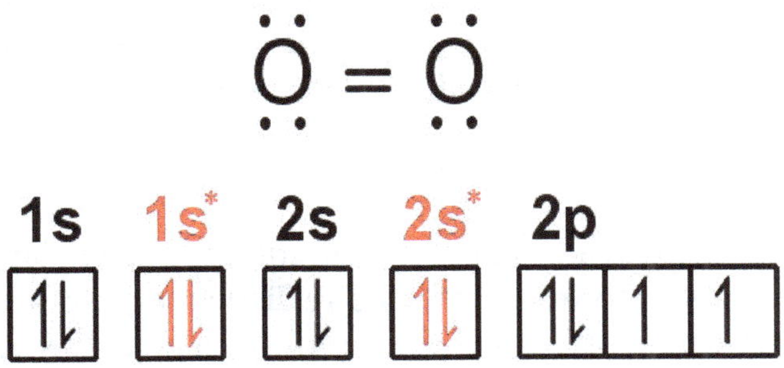

O = O

1s 1s* 2s 2s* 2p

$$B.O = \frac{8 - 4}{2} = 2$$

We see that there are 8 electrons distributed in bonding orbitals, while 4 are in anti-bonding orbitals. Substitute the numbers in the expression and you will find that the bond order of oxygen is 2.

The two orbitals are like Jekyll and Hyde – they inhabit the same geometry, but the anti-bonding orbital exists at a notoriously high energy level, which denies the combination of electrons any stability, whereas the bonding orbital exists at an energy where electrons can comfortably bind, allowing the resulting molecule to attain stability. In fact, the energy of a bonding orbital is lower than the energies of the individual levels that the electrons inhabit in a single atom. This means that the electrons would rather combine and form a molecule than exist in unpaired.

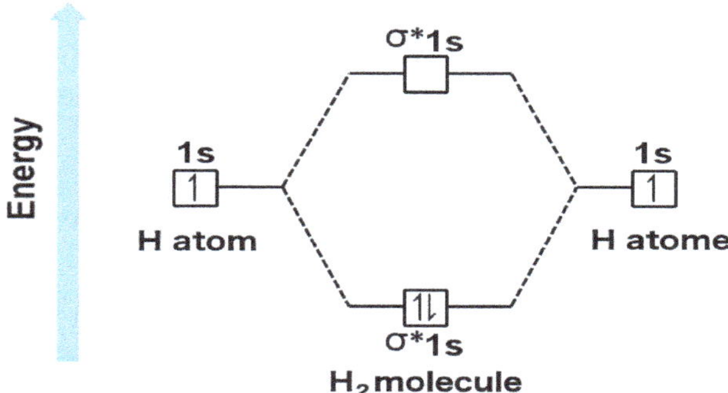

The bonding orbital of a hydrogen molecule exists at a lower energy than the individual levels of single hydrogen atoms. This means that the electrons would rather combine and form a molecule than exist in an unpaired state.

To summarize, if you're not familiar with the MO theory, you can compute the bond order of a molecule by drawing its Lewis structure and then checking the total number of electrons that have been shared (there exist strict rules for drawing Lewis structures as well, but I will assume the reader is aware of them. If not, refer to this link). However, if you're familiar with the MO theory, you can compute the bond order by first filling the bonding and anti-bonding molecular orbitals with valence electrons according to the rules, and then refer to the expression.

The bond order of a molecule gives us a measure or index of the strength of the bonds that bind it. The bonds bind the atoms like a rubber band binding your two hands. A double-bond would mean that the rubber band is now folded in half and tied around your hands. Due to the strength of this new double-bond, your hands have moved closer to each other. The molecule is now more stable. Similarly, three bonds or three folds would move your hands even closer. Furthermore, the stronger the bond, the more energy is required to break it — the hands are much easier to free when bound by a single-fold rubber band than a double- or triple-fold band. The energy required to break the bond is referred to as the molecules' bond energy.

Valence Bond theory vs Molecular Orbital theory

The Valence Bond theory provides a crude representation of molecular structures, whereas the Molecular Orbital theory gives a more accurate and realistic representation of how molecules are formed. In the former, one can observe how the electrons depicted by dots above the atoms are localized, meaning that their location is definite. On the other hand, MO theory is based on the quantum mechanical theory of atoms. It takes into account the probabilistic or non-localized nature of electrons and the distinct energy levels involved.

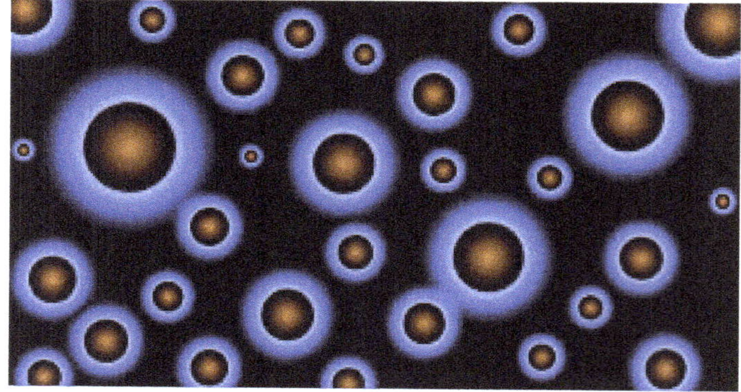

There is no winner among the two: VB theory is easier to understand, but doesn't explain the minute, yet highly crucial details that MO theory explains with the help of highly esoteric and sophisticated concepts. However, this comprehension of geometry and the detail of a molecular structure come at the expense of the ease with which a layman can visualize it.

This is why the implications of MO theory can often be drastically different. For instance, the bond order of an oxygen molecule is found to be 2 in both theories, but VB theory doesn't explain the paramagnetic properties that oxygen exhibits. Experiments have demonstrated how liquid oxygen is weakly affected by a magnetic field. An atom exhibits any type of magnetism when it contains unpaired electrons. If you refer to the Lewis structure of oxygen, you will find that all the electrons are paired, which would render the atom diamagnetic or unaffected by a magnetic field. However, MO theory reveals oxygen's structure in its true detail. The MO theory rightly predicts the presence of unpaired electrons in oxygen's orbitals, and therefore its paramagnetism.

Bond Angle

A bond angle is the angle between two bonds originating from the same atom in a covalent species.

Example:

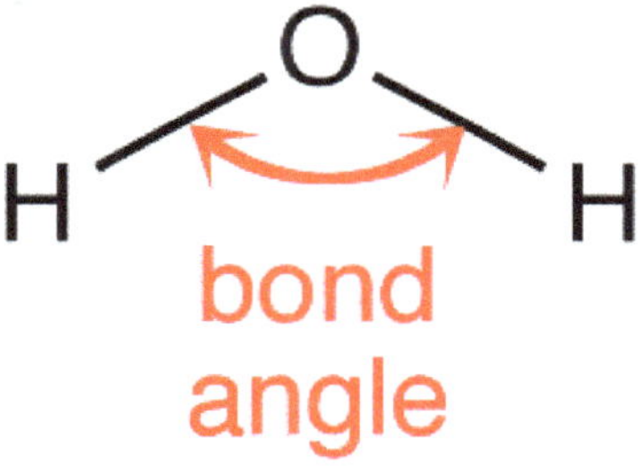

Example:

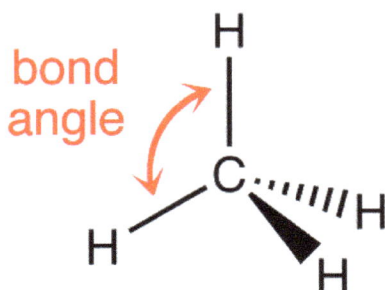

Geometrically, a bond angle is an angle between two converging lines.

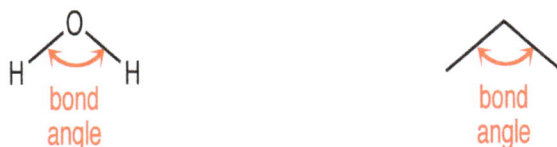

Example:

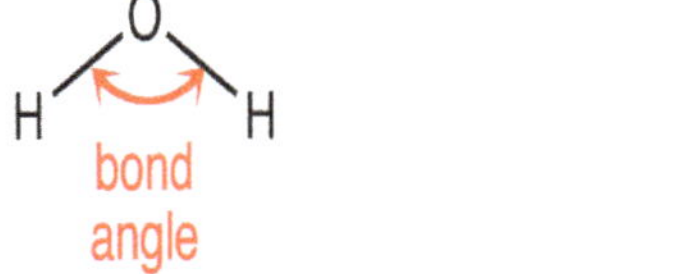

References

- Stranks, D. R.; Heffernan, M. L.; Lee Dow, K. C.; McTigue, P. T.; Withers, G. R. A. (1970). Chemistry: A structural view. Carlton, Vic.: Melbourne University Press. p. 184. ISBN 0-522-83988-6
- Chemical-bonding, science: britannica.com, Retrieved 15 May 2018
- Hughbanks, Timothy; Hoffmann, Roald (2002-05-01). "Chains of trans-edge-sharing molybdenum octahedra: metal-metal bonding in extended systems". Journal of the American Chemical Society. 105 (11): 3528–3537. doi:10.1021/ja00349a027
- 3-major-factors-influencing-the-formation-of-ionic-bond-22845: yourarticlelibrary.com, Retrieved 17 July 2018
- Campbell, Neil A.; Williamson, Brad; Heyden, Robin J. (2006). Biology: Exploring Life. Boston, MA: Pearson Prentice Hall. ISBN 0-13-250882-6. Retrieved 2012-02-05
- Coordinate-covalent-bond, physical-chemistry: chemistry.tutorvista.com, Retrieved 19 March 2018
- Langmuir, Irving (1919-06-01). "The Arrangement of Electrons in Atoms and Molecules". Journal of the American Chemical Society. 41 (6): 868–934. doi:10.1021/ja02227a002
- Primary-and-Secondary-Bonds: owlcation.com, Retrieved 10 April 2018
- Friebolin, H., "Basic One- and Two- Dimensional NMR Spectroscopy, 4th ed.," VCH: Weinheim, 2008. ISBN 978-3-527-31233-7
- What-is-bond-order, pure-sciences: scienceabc.com, Retrieved 11 June 2018
- John R. Sabin (1971). "Hydrogen bonds involving sulfur. I. Hydrogen sulfide dimer". J. Am. Chem. Soc. 93 (15): 3613–3620. doi:10.1021/ja00744a012

3

Properties of Materials

Materials exhibit diverse properties such as mechanical, chemical, thermal, magnetic, optical and electrical properties. These properties determine the usability of materials in engineering applications. An elaborate study of the varied properties of materials has been provided in this chapter.

Physical Properties

A material undergoes transition under the influence of temperature and pressure, and these changes are physical in nature, because their molecules remain intact.

Effects of Temperature on Substances

When temperature rises, a typical substance changes from solid to liquid and then to vapor, at a constant pressure. Some substance has several crystal forms in the solid state. The glassy state is also considered a solid. Transitions from one solid to another solid form, from solid to liquid, from liquid to vapor, from vapor to solid etc. are called phase transitions.

Phase transitions from solid to liquid, and from liquid to vapor absorb heat. The temperature of a system usually does not change as long as two phases are present. The (phase) transition temperature from solid to liquid is called the melting point whereas the temperature at which the vapor pressure of a liquid equals 1 atm (101.3 kPa) is called the boiling point. Thus, the measured boiling point depends on the atmosphere pressure. For compounds that decompose at high temperature, boiling point can be either specified at lower pressure or be replaced by the decomposition temperature. Thus, conditions as well as the value of boiling point listed in literature must be taken into account for application considerations. Boiling points of mixtures change with composition.

However, a glassy material becomes soft in a wide range of temperatures. The temperature at which the material becomes soft (behave molten like) is called glassy temperature, but it may be a range of temperatures. Behavior of a substance as the temperature changes must be carefully considered in its applications. Behavior of a mixture as temperature rises is different from its components. There is no theoretical way to predict the behavior of a mixture from its components, even if its exact composition is known. Addition of one or more materials usually changes the melting or glassy temperature of a substance. Thus, we often employ a blend (mixture) of materials whose behavior is acceptable within the desirable range of temperatures. Antifreeze for automobile radiator and deicing liquid for airplanes are examples of this application.

The behavior of mixtures as temperature and pressure change often requires a phase diagram to represent, and there are several models of two-component systems. A phase diagram of a many-component system requires a lengthy study.

- Phase Transition at Constant Pressure
    - Temperature
- Vapor
    - Boiling point
    - Heat of vaporization
- Liquid
    - Melting point
    - Heat of fussion
- Solid

Combined Effect of Temperature and Pressure on the Behavior of Material

When the temperature remains constant, the pressure also affects the behavior of a material. The volume of a gas changes as the pressure changes even if temperature remains the same. When temperature is close to the melting point of a substance, a liquid may solidify or a solid may melt as the pressure changes. A diagram showing the temperature and pressure combined effect on a system is called a phase diagram.

One-component phase diagrams for water and carbon dioxide are given here.

At pressure below 5.1 atm, solid and gas carbon dioxide coexist, but the vapor pressure depends on the temperature. The variation of vapor pressure is represented by a line, which separates the Dry ice phase from the CO_2 Gas phase. The vapor pressure of dry ice at 194.6 K (-78.5°C) the pressure is 1 atm, and at 216.7 K (-56.4°C) the pressure is 5.11 atm. The line separates the conditions for the formation of solid and vapor. A similar curve borders between liquid and gas CO_2, whereas a line separates dry ice from the liquid phase. At 216.7 K, vapor pressures of solid and liquid CO_2 are the same, 5.11 atm. At, 5.11 atm and 216.7 K, all three phases coexist, and the condition is called the triple point.

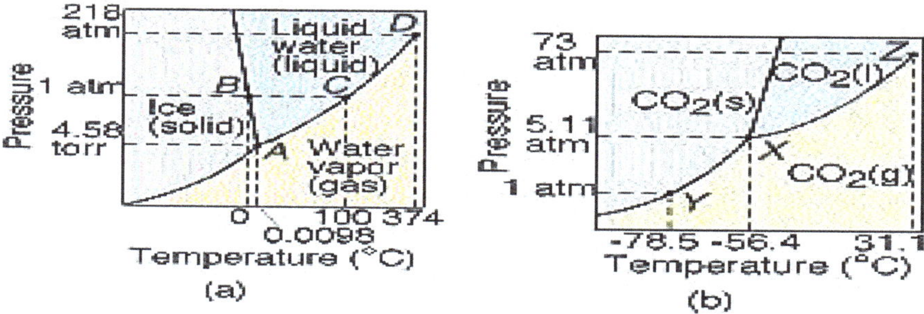

Thermal Expansion Coefficient

A substance expands on heating. For a rod, the lengthening of a unit length per degree Kelvin is the linear thermal expansion coefficient. This factor affects the substance performance in machines or structural assemblies. Thermal expansion causes tight fitted parts to break and moving part to jam, in any machine. The problem is serious if different material is used. When a large body of glass is subject to local heating or cooling, it breaks up due to expansion or shrinkage. Thermal expansion also causes distortion, and some thermometers are made of two strips of different metals. Thermal properties must be considered in any engineering constructions such as railroad, bridges, pipelines, and buildings, especially in areas where temperatures go to extreme values.

Linear thermal expansion coefficient per K at room temperature of some substances	
Aluminum	24
Copper	17
Diamond	1
Glass	11
Pyrex glass	3
Rubber, hard	80

Electric resistance of some Elements	
Metal	Resistance /ohm m
Al	$2.4x10^{-8}$
Au	$2.05x10^{-8}$
W	$5.65x10^{-6}$
Ge	0.46
Si	0.03
S (yellow)	$2x10^{15}$
P (white)	109 Note the range of 10^{15}

Heat and Electric Conductance

Transmissions of energy and electric charge across a body of material give rise to heat and electric conductance respectively. The rate of flow across a unit-area section when the temperature or electric potential difference applied to the wire of unit length is called the thermal or electric conductance coefficient. Metals are usually good conductors of both, and their conductance coefficients are high. Insulation material for heat and electricity should have low conductance, whearas metals have high conductance.

The reciprocal of electric conductance is called electric resistance; thus, the higher the conductance, the lower the resistance. Electric resistance for some familiar materials are given in the table here. Note the large range of 10^{15} among these substances. Aluminium and copper are very good conductors, and their resistances are very low, in the order of 10^{-8}, almost 100 times smaller than that of tungsten, W. Germanium, Ge, and silicon, Si, are typical semiconductors, whereas sulfur and phosphorous are insulation material.

Magnetic Properties

A magnetic field strength is measured in Tesla (T) and gauss (G, 1 T = 10,000 G). The Earth magnetic field is 0.5 G. When a material is placed into a magnetic field H, a magnetic field of different intensity B is produced inside the material. The ratio B/H is called the magnetic susceptibility. The higher the magnetic susceptibility, the easier the material is magnetized. Most substances are diamagnetic. The magnetic fields (B) within the bodies of these substances when they are placed in a magnetic field (H) are less than that of an empty space (vacuum).; thus their magnetic susceptibilities (B/H ratio) are less than 1. When a body of paramagnetic substance is placed in a magnetic field, the intensity of the field within the body is slightly larger than that of the applied field. The magnetic susceptibilities of paramagnetic substances are slightly greater than 1.

Iron, cobalt and nickel are some ferromagnetic substances, there are some other alloys and oxides that behave this way. They possess a spontaneous magnetic moment. A magnetic field is present in these materials even in the absence of an external magnetic field. However, ferromagnetism is temperature dependent, and above the so called Curie temperatures of the substances, magnetism vanishes. The Curie temperature or Curie point of a substance is unique. The Curie points for Fe, Co, and Ni are 1043, 1400, and 630 K respectively.

Ferromagnetism are due to the presence of magnetic domains in the substance, and when these domains line up parallel to each other, they give a net magnetic field. If the domains line up antiparallel to each other at the Curie point, the substance is said to be antiferromagnetic. The magnetic susceptibility reaches a maximum at Curie temperature for antferromagnetic material. For example, F_eO, MnO, C_oO, N_iO, F_eF_2, F_eCl_2, a-Mn, $C_{r2}O_3$ etc. are some of the antiferromagnetic substances.

Ferromagnetic substances play important roles in recording tapes and disks for audio, video, and computer signals. Furthermore, ferromagnetic materials are used in permanent magnets, which are used in motors, antenna, and speakers. Recent development in strong magnets enables communication equipment and computers to be miniaturized.

Density

The mass per unit volume (cm^3 = mL, m^3 etc.) of is called density, an intensive property. Often, specific gravity is given. Specific gravity is the ratio of density of a substance compared to that of water. As a ratio, it has no units. Since density of water is 1.00 g/mL, specific gravity is the density in g/mL. Other units to use are kg/L or 10^3 kg m^{-3}. Specific gravity for a few common substances are given here: Au, 19.3; mercury, 13.6; alcohol, 0.7893; benzene, 0.8786. Do you know which element has the highest density?

Dielectric Constant

The dielectric constant e of a medium is its ability to reduce the force F of attraction of charged (q_1 and q_2) particles separated at distance r, compared to vacuum. It is usually defined by the equation, $F = q_1 q_2 / (e r)$. A substance with large dielectric constant placed between two plates to which an electric voltage has been applied will result in a weak electric field within it. Water, due to its polar nature, has a rather large dielectric constant, 80.4. At the atomic scale, water molecules weaken the attraction between Na+ and Cl- ions, resulting in dissolving it. Dielectric constants for some familiar substances are: H_2O, 80.4; methanol, 33.6; benzene, 2.3; H_2 at 20 K, 1.23.

Heat Capacity

The amount of energy required to raise the temperature of a substance by 1 K is the heat capacity. If the substance has a unit mass, the amount is referred to as specific heat capacity, or specific heat. For example, it takes 1 cal (4.184 J) to raise the temperature of 1 g water by 1 K. Thus, the specific heat for water is 1 cal g^{-1} K^{-1} (75 J mol^{-1} $K^{-1)}$. Specific heat of water is large compared to most other substances, for example: Cu, 24.4 J mol^{-1} K^{-1}. This large heat capacity of water affect the weather, making temperatures in areas close to large bodies of water more steadier than large dry land.

Refractive Index

Refractive index of some common substances	
water	1.3
benzene	1.5011
ethanol	1.359
quartz	1.5*
NaCl solid	1.5
*wavelength dependent	

The ratio of light speed in vacuum to its speed in the medium is refractive index. Light travel slower in any medium than in vacuum. Thus, refractive index is always greater than unity (1), and light beam usually bents when entering from air to another medium. This value depends on the wavelength of the light used, and the property is important for material used in optics instrument such as eye glasses. The higher the refractive index, the thinner the glasses to achieve the same power. The difference in refractive indexes between two liquid gives rise to the visible boundary between layers.

Difference in refractive indexes of lights of different wavelengths can be separated using a prism. Refractive indexes for some familiar substances are given in a box. It should also be kept in mind that index of refraction changes with dissolved substance and concentration.

Solubility

The amount of substance dissolved in 100 mL of solvent is called solubility. However, units for solubility might be specified in some other fashion. Solubility depends on temperature, and the variation can be used to separate components in a solid mixture. Sodium acetate trihydrate, $CH_3COONa\cdot 3H_2O$, when heated will melt in the sense that it dissolves in its water of crystallization. This liquid remains liquid till about -15 C (258 K), and when crystallization does take place after triggered by cold hand, heat is released providing a source of heat. This property provides a winter hand warmer pack for skiers or winter travelers.

Washing and cleaning also involve solubility, and the formulation of an effective cleaning agent requires the knowledge of many substances. Substances can be classified according to polarity. Water, ammonia (NH_3), and methanol (CH_3OH) are polar, because their molecules have negative and positive sites, whereas methane (CH_4), gasoline, and motor oil are non-polar. Regarding solubility, a rule of thumb reads like dissolves like, which means that polar solvents dissolve polar substances and non-polar solvents dissolve non-polar substances. An organic compound with a polar group attached to non-polar chain will bring water molecules to a non-polar surface, and hence it can be used as a detergent or wetting agent. This rule of thumb has potential for both domestic and industrial applications.

Optical Activity

The ability of certain substances to rotate the plane of polarized light as it passes through a crystal, liquid, or solution is generally referred to as the optical activity. Substances possessing this activity usually lack a center of symmetry, and they have two isomers as mirror images of each other. The two isomers, called dextrorotatory (d, right hand) or laevorotory (l, left hand) isomers, rotate the polarized light in opposite directions. Thus, equi-molar or racemic mixture of the two appears optically inactive. For example, sugar, tartaric acid, and aminoacids are optically active compounds.

Viscosity and Surface Tension

Viscosity and surface tension are properties of liquid state. The former is a measure of its resistance to flow. Molasses, glycerin, oil, softened glass have high viscosity, and water, gasoline, ethanol have low viscosity. The SI units for viscosity is N.s /m2, but the unit poise (P, a cgs unit) have been used for a long time, and is more common, and 1 P = 0.1 N¸s m^{-2}. Viscosity usually decreases with increase in temperature, and softened glass has a viscosity more than 1014 N¸s m^{-2}.

Surface tension results from intermolecular attraction, the higher of which, the higher surface tension. Energy required to stretch the surface by some defined unit is called surface tension, and whose unit is $N.m/m^2$ (= N/m). Like viscosity, surface tension decrease with increase temperature. Surface tension causes the dew and raindrops to be round, and the idea to manufacture perfect spheres in zero gravity zone is making use of surface tension. Soap reduces surface tension of water and soapy water forms bubbles when air is blown into it.

Electrical Properties

It is well known that one of the subatomic particles of an atom is the electron. The electrons carry a negative electrostatic charge and under certain conditions can move from atom to atom. The direction of movement between atoms is random unless a force causes the electrons to move in one direction. This directional movement of electrons due to an electromotive force is what is known as electricity.

Electrical Conductivity

Electrical conductivity is a measure of how well a material accommodates the movement of an electric charge. It is the ratio of the current density to the electric field strength. Its SI derived unit is the Siemens per meter, but conductivity values are often reported as percent IACS. IACS is an acronym for International Annealed Copper Standard, which was established by the 1913 International Electrochemical Commission. The conductivity of the annealed copper (5.8001 x 107S/m) is defined to be 100% IACS at 20°C . All other conductivity values are related back to this conductivity of annealed copper. Therefore, iron with a conductivity value of 1.04 x 107 S/m, has a conductivity of approximately 18% of that of annealed copper and this is reported as 18% IACS. An interesting side note is that commercially pure copper products now often have IACS conductivity values greater than 100% IACS because processing techniques have improved since the adoption of the standard in 1913 and more impurities can now be removed from the metal.

Conductivity values in Siemens/meter can be converted to % IACS by multiplying the conductivity value by 1.7241 $x10^{-6}$. When conductivity values are reported in microSiemens/centimeter, the conductivity value is multiplied by 172.41 to convert to the % IACS value.

Electrical conductivity is a very useful property since values are affected by such things as a substances chemical composition and the stress state of crystalline structures. Therefore, electrical conductivity information can be used for measuring the purity of water, sorting materials, checking for proper heat treatment of metals, and inspecting for heat damage in some materials.

Electrical Resistivity

Electrical resistivity is the reciprocal of conductivity. It is the is the opposition of a body or substance to the flow of electrical current through it, resulting in a change of electrical energy into heat, light, or other forms of energy. The amount of resistance depends on the type of material. Materials with low resistivity are good conductors of electricity and materials with high resistivity are good insulators.

The SI unit for electrical resistivity is the ohm meter. Resistivity values are more commonly reported in micro ohm centimeters units. As mentioned above resistivity values are simply the reciprocal of conductivity so conversion between the two is straightforward. For example, a material with two micro ohm centimeter of resistivity will have ½ microSiemens/centimeter of conductivity. Resistivity values in microhm centimeters units can be converted to % IACS conductivity values with the following formula:

172.41 / resistivity = % IACS

Temperature Coefficient of Resistivity

Electrical conductivity values (and resistivity values) are typically reported at 20° C. This is done because the conductivity and resistivity of material is temperature dependant. The conductivity of most materials decreases as temperature increases. Alternately, the resistivity of most material increases with increasing temperature. The amount of change is material dependant but has been established for many elements and engineering materials.

The reason that resistivity increases with increasing temperature is that the number of imperfection in the atomic lattice structure increases with temperature and this hampers electron movement. These imperfections include dislocations, vacancies, interstitial defects and impurity atoms. Additionally, above absolute zero, even the lattice atoms participate in the interference of directional electron movement as they are not always found at their ideal lattice sites. Thermal energy causes the atoms to vibrate about their equilibrium positions. At any moment in time many individual lattice atoms will be away from their perfect lattice sites and this interferes with electron movement.

When the temperature coefficient is known, an adjusted resistivity value can be computed using the following formula:

$$R_1 = R_2 * [1 + a * (T_1 - T_2)]$$

Where: R_1 = resistivity value adjusted to T_1

R_2 = resistivity value known or measured at temperature T2

a = Temperature Coefficient

T_1 = Temperature at which resistivity value needs to be known

T_2 = Temperature at which known or measured value was obtained

For example, suppose that resistivity measurements were being made on a hot piece of aluminum. Normally when measuring resistivity or conductivity, the instrument is calibrated using standards that are at the same temperature as the material being measured, and then no correction for temperature will be required. However, if the calibration standard and the test material are at different temperatures, a correction to the measured value must be made. Presume that the instrument was calibrated at 20°C (68°F) but the measurement was made at 25°C (77°F) and the resistivity value obtained was 2.706×10^{-8} ohm meters. Using the above equation and the following temperature coefficient value, the resistivity value corrected for temperature can be calculated.

$$R_1 = R_2 * [1 + a * (T_1 - T_2)]$$

Where: R_1 = ?

$R_2 = 2.706 \times 10^{-8}$ ohm meters (measured resistivity at 25°C)

a = 0.0043/ °C

T1 = 20°C

T2 = 25°C

$R1 = 2.706 \times 10^{-8}$ohm meters * [1 + 0.0043/ oC * (20 °C – 25 °C)]

$R1 = 2.648 \times 10^{-8}$ohm meters

Note that the resistivity value was adjusted downward since this example involved calculating the resistivity for a lower temperature.

Since conductivity is simply the inverse of resistivity, the temperature coefficient is the same for conductivity and the equation requires only slight modification. The equation becomes:

$$s_1 = s_2 / [1 + a * (T_1 - T_2)]$$

Where: s_1 = conductivity value adjusted to T_1

s_2 = conductivity value known or measured at temperature T_2

a = Temperature Coefficient

T_1 = Temperature at which conductivity value needs to be known

T_2 = Temperature at which known or measured value was obtained

In this example let's consider the same aluminum alloy with a temperature coefficient of 0.0043 per degree centigrade and a conductivity of 63.6% IACS at 25°C. What will the conductivity be when adjusted to 20°C?

s_1= 63.6% IACS / [1 + 0.0043 * (20°C – 25°C)]

s_1= 65.0% IASC

The temperature coefficient for a few metallic elements is shown below.

Material	Temperature Coefficient (/°C)
Nickel	0.0059
Iron	0.0060
Molybdenum	0.0046
Tungsten	0.0044
Aluminum	0.0043
Copper	0.0040
Silver	0.0038
Platinum	0.0038
Gold	0.0037
Zinc	0.0038

Energy Band Structures in Solids

When atoms come together to form a solid, their valence electrons interact due to Coulomb forces, and they also feel the electric field produced by their own nucleus and that of the other atoms. In addition, two specific quantum mechanical effects happen. First, by Heisenberg's uncertainty principle, constraining the electrons to a small volume raises their energy, this is called promotion. The second effect, due to the Pauli exclusion principle, limits the number of electrons that can have the same property (which include the energy). As a result of all these effects, the valence electrons of atoms form wide valence bands when they form a solid. The bands are separated by gaps, where electrons cannot exist. The precise location of the bands and band gaps depends on the type of atom (e.g., Si vs. Al), the distance between atoms in the solid, and the atomic arrangement (e.g., carbon vs. diamond).

In semiconductors and insulators, the valence band is filled, and no more electrons can be added, following Pauli's principle. Electrical conduction requires that electrons be able to gain energy in an electric field; this is not possible in these materials because that would imply that the electrons are promoted into the forbidden band gap.

In metals, the electrons occupy states up to the Fermi level. Conduction occurs by promoting electrons into the conduction band, that starts at the Fermi level, separated by the valence band by an infinitesimal amount.

Conduction in Terms of Band and Atomic Bonding Models

Conduction in metals is by electrons in the conduction band. Conduction in insulators is by electrons in the conduction band and by holes in the valence band. Holes are vacant states in the valence band that are created when an electron is removed.

In metals there are empty states just above the Fermi levels, where electrons can be promoted. The promotion energy is negligibly small so that at any temperature electrons can be found in the conduction band. The number of electrons participating in electrical conduction is extremely small.

In insulators, there is an energy gap between the valence and conduction bands, so energy is needed to promote an electron to the conduction band. This energy may come from heat, or from energetic radiation, like light of sufficiently small wavelength.

A working definition for the difference between semiconductors and insulators is that in semiconductors, electrons can reach the conduction band at ordinary temperatures, where in insulators they cannot. The probability that an electron reaches the conduction band is about $\exp(-E_g/2kT)$ where E_g is the band gap and kT has the usual meaning. If this probability is, say, $< 10^{-24}$ one would not find a single electron in the conduction band in a solid of 1 cubic centimeter. This requires $E_g/2kT > 55$. At room temperature, $2kT = 0.05$ eV; thus $E_g > 2.8$ eV can be used as the condition for an insulator.

Besides having relatively small E_g, semiconductors have covalent bond, whereas insulators usually are partially ionic bonded.

Electron Mobility

Electrons are accelerated in an electric field E, in the opposite direction to the field because of their negative charge. The force acting on the electron is -eE, where e is the electric charge. This force produces a constant acceleration so that, in the absence of obstacles (in vacuum, like inside a TV tube) the electron speeds up continuously in an electric field. In a solid, the situation is different. The electrons scatter by collisions with atoms and vacancies that change drastically their direction of motion. Thus electrons move randomly but with a net drift in the direction opposite to the electric field. The drift velocity is constant, equal to the electric field times a constant called the mobility μ,

$$v_d = -\mu_e E$$

which means that there is a friction force proportional to velocity. This friction translates into energy that goes into the lattice as heat. This is the way that electric heaters work.

The electrical conductivity is:

$$\sigma = n|e|\mu_e$$

where n is the concentration of electrons (n is used to indicate that the carriers of electricity are negative particles).

Electrical Resistivity of Metals

The resistivity then depends on collisions. Quantum mechanics tells us that electrons behave like waves. One of the effects of this is that electrons do not scatter from a perfect lattice. They scatter by defects, which can be:

- Atoms displaced by lattice vibrations.
- Vacancies and interstitials.
- Dislocations, grain boundaries.
- Impurities.

One can express the total resistivity ρ_{tot} by the Matthiessen rule, as a sum of resistivities due to thermal vibrations, impurities and dislocations. Illustrates how the resistivity increases with temperature, with deformation, and with alloying.

Electrical Characteristics of Commercial Alloys

The best material for electrical conduction (lower resistivity) is silver. Since it is very expensive, copper is preferred, at an only modest increase in ρ. To achieve low r it is necessary to remove gases occluded in the metal during fabrication. Copper is soft so, for applications where mechanical strength is important, the alloy CuBe is used, which has a nearly as good ρ. When weight is important one uses Al, which is half as good as Cu. Al is also more resistant to corrosion.

When high resistivity materials are needed, like in electrical heaters, especially those that operate at high temperature, nichrome (N_iC_r) or graphite are used.

Intrinsic Semiconduction

Semiconductors can be intrinsic or extrinsic. Intrinsic means that electrical conductivity does not depend on impurities, thus intrinsic means pure. In extrinsic semiconductors the conductivity depends on the concentration of impurities.

Conduction is by electrons and holes. In an electric field, electrons and holes move in opposite direction because they have opposite charges. The conductivity of an intrinsic semiconductor is:

$$\sigma = n|e|\mu_e + \rho|e|\mu_h$$

where ρ is the hole concentration and μ_h the hole mobility. One finds that electrons move much faster than holes:

$$\mu_e > \mu_h$$

In an intrinsic semiconductor, a hole is produced by the promotion of each electron to the conduction band. Thus:

$$n = p$$

Thus, $\sigma = 2n|e|\ (\mu_e + \mu_h)$ (only for intrinsic semiconductors).

Extrinsic Semiconduction

Unlike intrinsic semiconductors, an extrinsic semiconductor may have different concentrations of holes and electrons. It is called p-type if p>n and n-type if n>p. They are made by doping, the addition of a very small concentration of impurity atoms. Two common methods of doping are diffusion and ion implantation.

Excess electron carriers are produced by substitutional impurities that have more valence electron per atom than the semiconductor matrix. For instance phosphorous, with 5 valence electrons, is an electron donor in Si since only 4 electrons are used to bond to the Si lattice when it substitutes for a Si atom. Thus, elements in columns V and VI of the periodic table are donors for semiconductors in the IV column, Si and Ge. The energy level of the donor state is close to the conduction band, so that the electron is promoted (ionized) easily at room temperature, leaving a hole (the ionized donor) behind. Since this hole is unlike a hole in the matrix, it does not move easily by capturing electrons from adjacent atoms. This means that the conduction occurs mainly by the donated electrons (thus n-type).

Excess holes are produced by substitutional impurities that have fewer valence electrons per atom than the matrix. This is the case of elements of group II and III in column IV semiconductors, like B in Si. The bond with the neighbors is incomplete and so they can capture or accept electrons from adjacent silicon atoms. They are called acceptors. The energy level of the acceptor is close to the valence band, so that an electron may easily hop from the valence band to complete the bond leaving a hole behind. This means that conduction occurs mainly by the holes (thus p-type).

Dielectric Behavior

A dielectric is an electrical insulator that can be made to exhibit an electric dipole structure (displace the negative and positive charge so that their center of gravity is different).

Capacitance

When two parallel plates of area A, separated by a small distance l, are charged by +Q, –Q, an electric field develops between the plates,

$$E = D/\varepsilon\varepsilon_o$$

where $D = Q/A$. ε_o is called the vacuum permittivity and e the relative permittivity, or dielectric constant (e = 1 for vacuum). In terms of the voltage between the plates, $V = E\,l$,

$$V = Dl/\varepsilon\varepsilon_o = Q\,l/A\varepsilon\varepsilon_o = Q / C$$

The constant $C= A\varepsilon\varepsilon_o/l$ is called the capacitance of the plates.

The dipole moment of a pair of positive and negative charges (+q and –q) separated at a distance d is $p = qd$. If an electric field is applied, the dipole tends to align so that the positive charge points in the field direction. Dipoles between the plates of a capacitor will produce an electric field that opposes the applied field. For a given applied voltage V, there will be an increase in the charge in the plates by an amount Q' so that the total charge becomes $Q = Q' + Q_o$, where Q_o is the charge of a vacuum capacitor with the same V. With Q' = PA, the charge density becomes $D = D_o E + P$, where the polarization $P = \varepsilon_o (\varepsilon-1) E$.

Types of Polarization

Three types of polarization can be caused by an electric field:

- Electronic polarization: the electrons in atoms are displaced relative to the nucleus.
- Ionic polarization: cations and anions in an ionic crystal are displaced with respect to each other.
- Orientation polarization: permanent dipoles (like H2O) are aligned.

Frequency Dependence of the Dielectric Constant

Electrons have much smaller mass than ions, so they respond more rapidly to a changing electric field. For electric field that oscillates at very high frequencies (such as light) only electronic polarization can occur. At smaller frequencies, the relative displacement of positive and negative ions can occur. Orientation of permanent dipoles, which require the rotation of a molecule can occur only if the oscillation is relatively slow (MHz range or slower). The time needed by the specific polarization to occur is called the relaxation time.

Dielectric Strength

Very high electric fields (>108 V/m) can free electrons from atoms, and accelerate them to such high energies that they can, in turn, free other electrons, in an avalanche process (or electrical discharge). This is called dielectric breakdown, and the field necessary to start the is called the dielectric strength or breakdown strength.

Dielectric Materials

Capacitors require dielectrics of high e that can function at high frequencies (small relaxation times). Many of the ceramics have these properties, like mica, glass, and porcelain). Polymers usually have lower e.

Ferroelectricity

Ferroelectric materials are ceramics that exhibit permanent polarization in the absence of an electric field. This is due to the asymmetric location of positive and negative charges within the unit cell. Two possible arrangements of this asymmetry results in two distinct polarizations, which can be used to code "0" and "1" in ferroelectric memories. A typical ferroelectric is barium titanate, $BaTiO_3$, where the Ti^{4+} is in the center of the unit cell and four O^{2-} in the central plane can be displaced to one side or the other of this central ion.

Piezoelectricity

In a piezolectric material, like quartz, an applied mechanical stress causes electric polarization by the relative displacement of anions and cations.

Biological Properties

A living organism has a material structure to provide an environment for complicated chemistry of living. Chemical and physical reactions provide energy to maintain living functions and to renew structural material. Thus, consideration of biological properties is a natural extension of physical and chemical properties.

To a large extend, biological functions of any materials are related to their chemical and physical properties. However, reactions in biological systems are catalyzed by enzymes. Furthermore, products of one reaction may be reactants for another in a complicate scheme of reactions to maintain live. Malfunction of a reaction causes trouble, leading to disease or death. Thus, biological properties deserve special consideration.

Absorption and Transport

Generally speaking, food, medicine, and toxin can be given to a person by means of feeding, absorption, inhalation and injection. Food must be absorbed in the digestive track and transported from to the targeted organs, tissues and cells to provide energy for life. Injected substances may already be in organs or tissues, but they may not be in the targeted cells. In any cases, materials in question must pass through membranes, either by enzyme aided (active) transport or by diffusion (passive transport). Transportation of a substance in a biological system is a very complicated process.

Absorption and transport of a substance often involve its solubility in the medium. Substances prefer to dissolve in water type fluid are said to be hydrophilic, whereas those prefer to dissolve in oily fluids are lipophilic. Absorption and transport of substances depend on their solubility in water and lipid media. The ultimate effect on cells, tissues, and organs must take place at the molecular level. However, effects at the molecular level are often not observable, and the symptoms of these effects may appear to be unrelated to the material in question.

Solid substances such as fiber, gold and charcoal, not absorbed and used in any biological function pass out as feces. Absorbed soluble substances, but not utilized by animals are excreted with urine.

Material for Biological Structure

Bullets and shell fragments anchored in bones of war veterans were not removed, because doctors considered them biologically inert. Stainless steel parts replace joints and bones today, because they are inert.

Both inorganic and organic materials are involved in structures of living organism. Lignin, cellulose, muscles, skins, and cell walls are mostly organic, whereas bones, teeth, and shells involve mostly inorganic substances. These substances may serve only as structural materials in biological systems, and if so they can be replaced by biological inert substances.

However some subsystems of structural organs are responsible for vital biological functions. For example, bone marrow is responsible for blood regeneration. Thus, replacement of biological structure material involves many disciplines.

Biological Materials and Biomaterial

Plasma, membrane, tissue, protein, lipid, enzyme, the digestive system, and the central nervous systems are some examples of biological materials, for which properties for consideration include growth and decay, turn over time, biological half-life, retention time, composition and its change, and active ingredient. These are manifestation of physical and chemical properties of biological materials. However, biological properties allow us to identify and solve the biological problems. Biological materials had been studied by biologists, chemists, and engineers from the macroscopic, molecular, and functional viewpoints.

In contrast, replacement or implant materials that imitate living tissues or organs are called biomaterial, which can be divided into two categories: soft tissue and hard tissue replacement biomaterial. The former includes sutures, surgical tapes, adhesives, skin implants etc. The latter include metals (steel, aluminum, titanium, cobalt-based alloys, and titanium-based alloys); ceramics (made up of Al_2O_3, TiO_2, SiO_2, Fe_2O_3 etc.); carbon (graphite and glassy carbon); and polymers.

The chemistry of living is complex, and properties of biological materials towards biomaterials are of great interest. The general reaction of biological materials towards foreign biomaterials is expel (or rejection). Living tissues form a thin layer around the inert biomaterial, but materials that irritate the tissues causes inflammation. Most pure metals evoke severe tissue reaction due to their redox reactions. However, aluminum and titanium are metals of choice, because the formation of a thin oxide layer on their surface made them inert. Similarly, ceramics are compatible to body fluid because they are made of the metal oxides. The nature of the surface also affects the biological properties, rough ones enable tight attachment of tissues.

Biological Activity

Biological activities of materials can be divided according to biological functions. Substances that provide nutrition, energy, and structural need are called food, whereas those that disrupt the normal functions are called toxins. Substances used to correct the abnormal biological functions are called medicines.

In recent years, a lot of research had gone into finding quantitative structure-activity relationships (QSAR) of various substances aimed at improving drug design. As they provide an indication of some biological activity, we list some categories here:

- Anti-infective agents,
- Antibiotics,
- Anti-tumor agents,
- Cardiovascular agents,
- Anti-allergic,
- Anti-ulcer,
- Anti-inflammatory, and
- Anti-arthritics.

Agents affecting the central nervous system includes analgesics, anesthetics, antidepressants, convulsants, anti-convulsants, neuroleptics, and psychotomimetics. There are also steroids and hormones which interacts with genes in the nuclei of cells causing complicated developments.

Mechanical Properties

The mechanical properties of a material are those which effect the mechanical strength and ability of a material to be molded in suitable shape. Some of the typical mechanical properties of a material are listed below-

- Strength
- Toughness
- Hardness
- Hardenability
- Brittleness
- Malleability
- Ductility
- Creep and Slip
- Resilience
- Fatigue

Strength

It is the property of a material which opposes the deformation or breakdown of material in presence of external forces or load. Materials which we finalize for our engineering products, must have suitable mechanical strength to be capable to work under different mechanical forces or loads.

Toughness

It is the ability of a material to absorb the energy and gets plastically deformed without fracturing. Its numerical value is determined by the amount of energy per unit volume. Its unit is Joule/ m3. Value of toughness of a material can be determined by stress-strain characteristics of a material. For good toughness, materials should have good strength as well as ductility.

For example: brittle materials, having good strength but limited ductility are not tough enough. Conversely, materials having good ductility but low strength are also not tough enough. Therefore, to be tough, a material should be capable to withstand both high stress and strain.

Hardness

It is the ability of a material to resist to permanent shape change due to external stress. There are various measure of hardness – Scratch Hardness, Indentation Hardness and Rebound Hardness.

- Scratch Hardness

  Scratch Hardness is the ability of materials to the oppose the scratches to outer surface layer due to external force.

- Indentation Hardness

  It is the ability of materials to oppose the dent due to punch of external hard and sharp objects.

- Rebound Hardness

  Rebound hardness is also called as dynamic hardness. It is determined by the height of "bounce" of a diamond tipped hammer dropped from a fixed height on the material.

Hardenability

It is the ability of a material to attain the hardness by heat treatment processing. It is determined by the depth up to which the material becomes hard. The SI unit of hardenability is meter (similar to length). Hardenability of material is inversely proportional to the weld-ability of material.

Brittleness

Brittleness of a material indicates that how easily it gets fractured when it is subjected to a force or load. When a brittle material is subjected to a stress it observes very less energy and gets fractures without significant strain. Brittleness is converse to ductility of material. Brittleness of material is temperature dependent. Some metals which are ductile at normal temperature become brittle at low temperature.

Malleability

Malleability is a property of solid materials which indicates that how easily a material gets deformed under compressive stress. Malleability is often categorized by the ability of material to be formed in the form of a thin sheet by hammering or rolling. This mechanical property is an aspect of plasticity of material. Malleability of material is temperature dependent. With rise in temperature, the malleability of material increases.

Ductility

Ductility is a property of a solid material which indicates that how easily a material gets deformed under tensile stress. Ductility is often categorized by the ability of material to get stretched into a wire by pulling or drawing. This mechanical property is also an aspect of plasticity of material and is temperature dependent. With rise in temperature, the ductility of material increases.

Creep and Slip

Creep is the property of a material which indicates the tendency of material to move slowly and deform permanently under the influence of external mechanical stress. It results due to long time exposure to large external mechanical stress with in limit of yielding. Creep is more severe in material that are subjected to heat for long time. Slip in material is a plane with high density of atoms.

Resilience

Resilience is the ability of material to absorb the energy when it is deformed elastically by applying stress and release the energy when stress is removed. Proof resilience is defined as the maximum energy that can be absorbed without permanent deformation. The modulus of resilience is defined as the maximum energy that can be absorbed per unit volume without permanent deformation. It can be determined by integrating the stress-strain cure from zero to elastic limit. Its unit is joule/m^3.

Fatigue

Fatigue is the weakening of material caused by the repeated loading of the material. When a material is subjected to cyclic loading, and loading greater than certain threshold value but much below the strength of material (ultimate tensile strength limit or yield stress limit), microscopic cracks begin to form at grain boundaries and interfaces. Eventually the crack reaches to a critical size. This crack propagates suddenly and the structure gets fractured. The shape of structure affects the fatigue very much. Square holes and sharp corners lead to elevated stresses where the fatigue crack initiates.

Thermal Property

A thermal property iis any characteristic of a material defining the substance and related to temperature; e.g. thermal conductivity is said to be a thermal property, but electrical conductivity is not. However, all properties, thermal and non-thermal, are temperature dependent, and in this sense included under thermal properties.

The effect of temperature on thermal properties may be large (what may be used to build good thermometers). Standard values are usually given at 20 °C (comfort lab conditions), but other reference conditions are also traditionally used: 0 °C because its ease of reproducing, 15 °C because it is the average temperature in the Earth surface, 20 °C (human comfort), or 25 °C because it is easier to maintain a bath temperature a little over the oscillating ambient temperature, than below. Fortunately, the influence of all those temperature-standards is minor on property values, but care should be paid to make it explicit.

The effect of pressure on thermal properties is very low on condense substances. The standard value for pressure is 100 kPa, although 101.325 kPa, the average pressure in the Earth surface, is sometimes used. The effect of uncertainty in composition of the substance is usually small (e.g. properties of tap water, and even of sea water, may be taken as those of pure water, in many instances), except on some sensitive properties, like for the thermo-optical properties of substances, that are heavily dependent on contamination, or the thermal conductivity of metals, that may vary a lot with small alloys, etc.

Specific Heat Capacity

Specific heat capacity of inorganic building materials (concrete, break, natural stone) is varied within 0.75-0.92 kJ/(kg °C), for wood it is 0.7 kJ/(kg °C). Since water has very high specific heat capacity of 4 kJ/(kg °C), the specific heat capacity of materials increases with the increase of their humidity. Specific heat capacity of composite materials can vary significantly if temperature variations are accompanied by phase changes. Typical example is related to sea ice consisting fresh ice and brine cells. The heating of sea ice is accompanied not only by the increasing of sea ice temperature but also by the melting of fresh ice around brine cells. This process becomes more important with increasing sea ice temperature.

	Solids	Liquids
		Ammonia (4.19); Water (4)
	Plastic (1.76; Cork, Rubber (1.68); Wool (1.63); Cellulose (1.55); Coal, Naphthalene (1.3); Concrete (1.13)	Aquafortis (2.77); Hexane (2.51); Phenol (2.35); Kerosene (2.1); Azote liquid (2.01)
Specific heat capacity, kJ/(kg°C)		Benzene (1.84); Turpentine (1.7); Luboil, Liquid oxygen (1.68); Nitrobenzene (1.38)
	Aluminum, Clay, Brick (0.92); Coke (0.84); Sand (0.8); Glass (0.84-0.42); Slag (0.75); Wood (0.7); Iron, Steel (0.5); Copper (0.385); Zinc (0.38); Lead (0.13	

Table: Mean specific heat capacities of solid and liquid materials in temperature range 0-100° C.

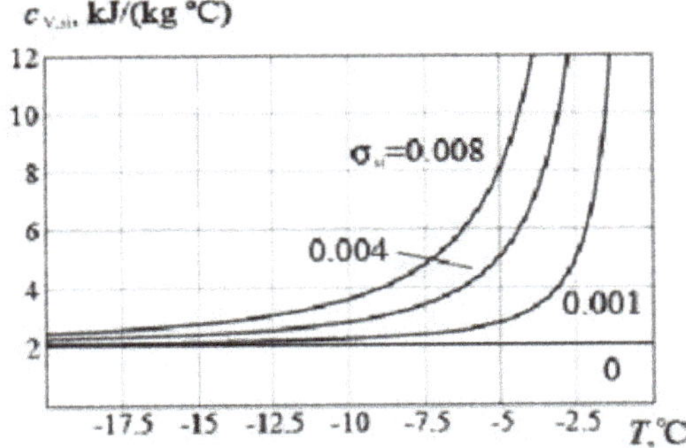

Figure: Specific heat capacity of sea ice versus its temperature for different values of sea ice salinity.

Thermal Conductivity

Thermal conductivity of materials with simple chemical composition is greater than thermal conductivity of materials of complicated chemical composition. Thermal conductivity of materials with crystal structure is higher than thermal conductivity of materials with mixed or amorphous structure. For example, mean thermal conductivities of single crystal of quartz is 7-8 W/(m °C), for sand-rock with impurities it is 2.1-2.9 W/(m °C), and for normal glass with amorphous structure it is 0.76 W/(m °C). Porous materials conduct Porous material has smaller thermal conductivity than continuous materials when the pores are filled by air. Thermal conductivity of materials with small closed pores is smaller than thermal conductivity of the same material with bigger pores under the same overall porosity. It is because heat transfer due to convection is reduced in the material with smaller pores.

Material	Density, kg/m³	Thermal conductivity, W/(m°C)
Foam Plastic	30	0.047
Cork fines	110	0.047
Glass-woo	200	0.35-0.047
Cinder-woo	250	0.076
Felt-wool	300	0.47
Wood cross fibers	600	0.14-0.174
Wood along fibers	600	0.384
Asbestos	600	0.151
Insulation brick	600	0.116-0.209
Textolite	1380	0.244
Dry sand	1500	0.349-0.814
Brick lining	1700	0.698-0.814
Fire brick	1840	1.05
Concrete	2300	1.28

Aluminum	2700	203.5
Cast Iron	7500	46.5-93
Steel	7850	36.5
Stainless steel	7900	17.5
Bronze	8000	64
*Latten	8500	93
Copper	8800	384
Lead	11400	34.9

*Bronze-like yellow alloy used to make church utensils in the middle ages by beating it into thin sheets by virtue of its malleability and ductility.

Melting Point

Crystal materials have certain melting points above which their crystal structure is destroyed. Below their melting point crystal materials are solid and above it they become liquids. The softening of amorphous materials occurs gradually with increasing temperature, evolving into viscous fluids with decreasing viscosity under increasing temperature.

Table: Melting points if some materials.

Materi	Melting point, o C	Material	Melting Point .C
Water	0	Zinc	419
Wolfram	3370	Lead	327
Gold	1063	Tin	232
Iron	1535	Mercury	39
Copper	1083		

Physical properties of many materials depend on the proximity of their temperature to the melting point. This property is characterized by homologous temperature calculated as a ratio of the actual temperature to the melting point. Homologous temperatures of some materials are shown in figure b) in natural range of actual temperature from -100 °C to 100 °C.

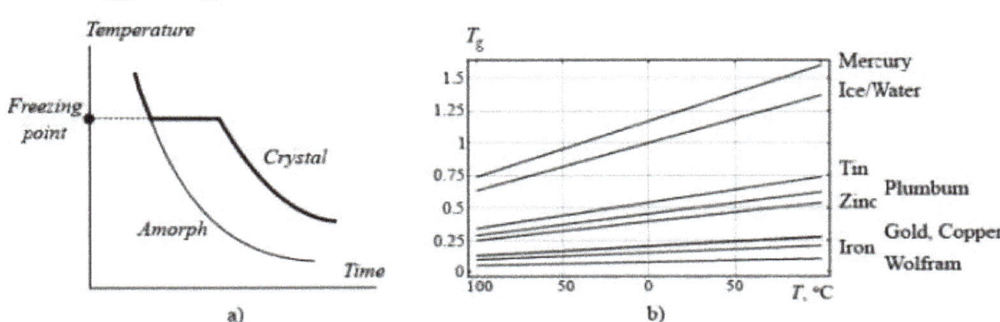

Figure: (a).Temperature-time curves for the cooling of amorphous and crystal materials, (b). Homologous temperature of some materials

Latent Heat

Latent heat is equal to the amount of energy necessary for the melting of unit mass of crystal material at the freezing point. Materials with high latent heat are more stable.

Table: Latent heat of some materials.

Material	Latentc heat, J/kg
Ice	334000
Lead	23100
Copper	214000
Iron	270000
Mercury	11800

Thermal Expansion

When heat is passed through a material, its shape changes. Generally, a material expands when heated. This property of a material is called Thermal Expansion. There can be change in area, volume and shape of the material.

For example, railway tracks often expand and as a result, get misshapen due to extreme heat.

Thermal Stress

The stress experienced by a body due to either thermal expansion or contraction is called thermal stress. It can be potentially destructive in nature as it can make the material explode.

For example, cracks can be seen on roads where the heat is extreme. The crack is a result of thermal stress.

Engineer Properties

Engineering or more often called mechanical properties deals with material's response to tension, compression, shear, and repeated combined actions of these three. Some engineering concepts are required for understanding these properties, and it is helpful for a general reader to acquire some of these concepts in order to understand the material world. On the other hand, engineers have to consider physical and chemical properties in their designs and applications.

Hardness

The Mohs Scale of Hardness

1. Talc
2. Gypsum
3. Calcite
4. Fluorite
5. Aptite
6. Orthoclase
7. Quartz
8. Topaz

9. Corundum
10. Diamond

Ability to abrade or indent one another is referred to as hardness. There are several scales of hardness, and the oldest one is a Mohs hardness scale, which assigned a number consecutively from 1 to 10 for talc, gypsum, calcite, fluorite, aptite, orthoclase, quartz, topaz, corundum and diamond, respectively. Diamond is the hardest of all substances. This scale is not linear, but relative. The hardness is assigned by a scratching test. A modified Mohs scale gives 15 grade of hardness between talc and diamond, whereas the Knoop values of hardness range from 32 for gypsum to 7000 for diamond. Hardness should be prime considerations in all tools and machine parts. Harder substances are used for cutting tools. Due to their hardness, the most important. applications of small synthetic diamonds are in cutting tools. In these applications, their colors and sizes are not important.

Tensile Characteristics

Hooke's Law states that the strain produced in an elastic body is proportional to the stress which causes it. The stress may be in the form of tensile or shear.

A tensile stress pulls the two ends of a rod or cable apart. When a small tensile stress is applied, the rod or cable lengthens elastically and returns to its original length after the tensile stress is removed. The ratio of tensile stress to lengthening (strain) divided by the area of the cross section is called modulus of elasticity. The (maximum) tensile stress or load causing permanent or inelastic deformation is called yield load. At the yield load, the lengthening will continue without adding more tensile stress. If the yield load is not removed, the rod lengthening continues until it breaks. The tensile strength is the yield load divided by area of the cross section of a rod or bar. Thus, tensile strengths have the units of force per unit area. The yield load of a cable or rod is equal to the tensile strength multiplied by the cross section area.

Shear Characteristics

Shear stress is a pair of force applied in opposite direction but in a sliding fashion. In response to the shear stress, a material deforms. Usually, shear stress causes adjacent laminar elements of a solid body to slide over each other of. In the case of homogeneous isotopic elastic medium, the ratio of shear strain to the shear stress is called shear modulus. Shear strength is the stress, usually expressed in force unit per unit area, required to produce fracture when impressed perpendicular upon the cross-section of a material.

Shear strengths are important for bolts, rivets, drive key, cutting, and polishing applications.

Fatigue Strength

We all experience that repeated bending of a steel wire on the same kink-point results in breaking it. Repeated loading and unloading of a specific stress on a piece of material till it fails is called fatigue strength or endurance limit. The stress can be tensile, compressed, shear, bending, or a combination of these. Fatigue strengths are hard to measure because of the varying and many stress types. Usually some percentage of tensile and compression strength are taken as a possible fatigue strength. Engineering designs allow wide margins for safety to avoid failure due to fatigue.

Appearance and Dimensional Properties

Color and surface finish give an object an appearance, which is a very important factor in engineering and commercial products. Combining color and surface finish gives artistic designs appealing to customers.

In terms of color, we are interested in the property of a material to filter and reflect lights. Our reaction to and perception of lights depends on the wavelengths. For engineering, comparison, and communication purposes, we must have standard measurements and specifications of color. Four parameters, three for the relative amounts of red, green, and blue components and one for the brightness, are required to specify a color.

Specific concepts and terms must be developed for the description of surface features. There are more than 20 mathematical parameters applied to surface description and some of the terms are: roughness, irregular features of wave, height, width, lay, and direction on the surface; camber, deviation from straightness; out of flat, measure of macroscopic deviations from flatness of a surface.

Furthermore, the surface properties not only affect visual perception, they affect engineering applications as well. Rough surfaces in moving contacts leads to wearing, whereas over smoothness is avoided because of high cost for their manufacture. The importance of surface is indicated by the many methods developed for its analysis. Optical or laser technology can be used to measure large surface areas, whereas visual examination detects features of the order of millimeters. Finer surface features are examined by microscopy, which enlarge the surface by a factor of 10 to 3000.

Still finer surface features are examined using scanning electron microscopy (SEM), which enlarges a surface up to 10,000,000 times. Some SEM machines are now equipped with an energy dispersive X-ray analyzer (EDX), which identifies the chemical elements in the bombarded area. The composition of the surface layers is often determined by electron spectroscopy for chemical analysis (ESCA). The surface is bombarded by low-energy X-rays to emit photoelectrons, the analysis of which gives the atomic composition of the surface layers. Another method yielding composition information is called secondary ion mass spectroscopy (SIMS), which bombards the surface with ions to remove some of the atoms. These atoms are ionized and they pass through a mass spectrometer for further analysis. Auger electron spectroscopy (AES) bombards the surface with an electron beam, which causes the ejection of Auger electrons. The Auger electron is chemical element specific, and they reveal also composition at the surface.

Dimensional properties deal with size and shape, which is also related to quantities. In a modern society, material requirement must be very carefully specified regarding the details such as the surface texture, flatness, allowable defects, shape, dimensions, camber, tolerance, etc. A slight misinterpretation or miscommunication causes not only lots of money, but also a lot of frustration.

Chemical Property

It is important to have the knowledge of chemical properties of engineering materials. Because most the engineering materials come into contact with other materials and react chemically to each other. Due to this chemical reaction they may suffer from chemical deterioration. Some of the chemical properties of engineering materials are listed below:

1. Chemical composition
2. Atomic bonding
3. Corrosion resistance
4. Acidity or Alkalinity

Chemical Composition

The chemical composition of engineering material indicates the elements which are combined together to form that material. Chemical composition of a material effects the properties of engineering materials very much. The strength, hardness, ductility, brittleness, corrosion resistance, weld ability etc. depends on chemical composition of materials.

Hence, we should also have the knowledge of chemical composition of engineering materials. For Example the Chemical compositions of some materials are listed below:

Sl. No.	Material	Chemical Composition
1.	Steel	Fe, Cr, Ni
2.	Brass	Cu = 90%, Ni = 10%
3.	Bronze	90% Cu, 10% Ni
4.	Invar	Fe = 64%, Ni = 36%
5.	Gun Metal	Cu = 88%, Tin = 10%, Zn = 2%
6.	German Silver or Nickel Silver or Electrum	Cu = 50%, Zn = 30%, Ni = 20%
7.	Nichrome	Ni = 60%, Cr = 15%, Fe = 25%
8.	Phosphor Bronge	Cu = 89 – 95.50% , Tin = 3.50 -10%, P = 1%
9.	Manganin	Cu = 84%, Mn = 12%, Ni = 4%
10.	Constantan	Cu = 60%, Ni = 40%

Atomic Bonding

Atomic bonding represents how atoms are bounded to each other to form the material. Many properties, such as melting point, boiling point, thermal conductivity and electrical conductivity of materials are governed by atomic bonding of materials. Hence, to understand the properties of materials, it is very important to study the atomic bonding of materials. Atomic bonds in materials are of following types:

1. Ionic bond: froms by exchanging of valence electrons between atoms.
2. Covalent bonds: froms by sharing of electrons between atoms.
3. Metallic bonds: found in metals.

Corrosion Resistance

Corrosion is a gradual chemical or electromechemical attack on a metal by its surrounding medium. Due to the corrosion, metal starts getting converted into an oxide, salt or some other compound. Corrosion of a metal is effected by many factors such as air, industrial atmosphere, acid, bases, slat solutions and soils etc. Corrosion has a very adverse effect on materials. Due to corrosion, the strength and life of a material is reduced.

Corrosion resistance of a material is the ability of material to resist the oxidation in atmospheric condition. Generally pure metals such as iron, copper, aluminum etc. gets corroded in slowly in atmosphere. To avoid the corrosion of these metal in pure form, we use these metals in the form of alloys such as stainless steel, brass, bronze, German silver, Gunmetal etc.

Acidity or Alkalinity

Acidity or Alkalinity is an important chemical property of engineering materials. A material is acetic or Alkane, it is decided by the ph value of the material. Ph value of a material varies from 0 to 14. Ph value of 7 is considered to be neutral. Ordinary water is having ph value of 7. The materials which are having ph value below 7 are called Acetic and Materials which are having ph value greater than 7 are called alkane. Acidity of Alkalinity of material indicates that how they react with other materials.

References

- Mechanical-properties-of-engineering-materials: electrical4u.com, Retrieved 18 July 2018
- Propertyp, applychem: science.uwaterloo.ca, Retrieved 08 April 2018
- Thermal-properties-of-materials, physics: byjus.com, Retrieved 28 June 2018
- Electrical, Physical-Chemical, Education-Resources, Community-College, Materials: ndeed.org, Retrieved 31 March July 2018
- Chemical-properties-of-materials: electrical4u.com, Retrieved 11 May 2018

4

Crystalline Defects and Diffusion

Crystalline materials have a number of crystalline defects, which have a significant influence on its properties. The diffusion process by which the random thermally activated movement of atoms results in a net transfer of atoms in a solid is known as atomic diffusion. This chapter discusses the crystalline defects and diffusion through an elucidation of topics such as point defects, line defects, planar defects and amorphous solids.

Crystalline Defect

Crystalline defect is the disruptions in the regularity of the crystal structure in real single crystals. In idealized crystal structures, atoms occupy strictly defined positions, forming regular three-dimensional lattices (crystal lattices). In natural and artificially grown real crystals, various deviations from the regular location of atoms or ions or of groups of atoms or ions are usually observed. Such disturbances may be either on an atomic scale or of macroscopic dimensions, noticeable even to the naked eye. In addition to static defects, there exists another type of deviations from the ideal lattice that are related to thermal oscillations of the particles that make up the lattice (dynamic defects).

Crystal defects are formed during growth of crystals; under the influence of thermal, mechanical, and electrical factors; and upon irradiation by neutrons, electrons, X rays, and ultraviolet radiation (radiation defects).

A distinction is made among point (zero-dimensional) defects, line (one-dimensional) defects, defects that form surfaces in the crystal (two-dimensional), and volumetric (three-dimensional) defects. In a one-dimensional defect the size is much greater in one direction than the distance between neighboring like atoms (the lattice parameter), but in the other two directions it is of the same order. In a two-dimensional defect the dimensions are greater in two directions than is the distance between the closest atoms, and so on.

Departures of a crystalline solid from a regular array of atoms or ions. A "perfect" crystal of NaCl, for example, would consist of alternating Na^+ and Cl^- ions on an infinite three-dimensional simple cubic lattice, and a simple defect (a vacancy) would be a missing Na^+ or Cl^- ion. There are many other kinds of possible defects, ranging from simple and microscopic, such as the vacancy and other structures, to complex and macroscopic, such as the inclusion of another material, or a surface.

Natural crystals always contain defects, due to the uncontrolled conditions under which they were formed. The presence of defects which affect the color can make these crystals valuable as gems, as in ruby (Cr replacing a small fraction of the Al in Al_2O_3). Crystals prepared in the laboratory will also always contain defects, although considerable control may be exercised over their type, concentration, and distribution.

The importance of defects depends upon the material, type of defect, and properties which are being considered. Some properties, such as density and elastic constants, are proportional to the concentration of defects, and so a small defect concentration will have a very small effect on these. Other properties, such as the conductivity of a semiconductor crystal, may be much more sensitive to the presence of small numbers of defects. Indeed, while the term defect carries with it the connotation of undesirable qualities, defects are responsible for many of the important properties of materials, and much of solid-state physics and materials science involves the study and engineering of defects so that solids will have desired properties. A defect-free silicon crystal would be of little use in modern electronics; the use of silicon in devices is dependent upon small concentrations of chemical impurities such as phosphorus and arsenic which give it desired electronic properties.

An important class of crystal defect is the chemical impurity. The simplest case is the substitutional impurity, for example, a zinc atom in place of a copper atom in metallic copper. Impurities may also be interstitial; that is, they may be located where atoms or ions normally do not exist. In metals, impurities usually lead to an increase in the electrical resistivity. Impurities in semiconductors are responsible for the important electrical properties which lead to their widespread use. The energy levels associated with impurities and other defects in nonmetals may also lead to optical absorption in interesting regions of the spectrum.

Even in a chemically pure crystal, structural defects will occur. These may be simple or extended. One type of simple defect is the vacancy, but other types exist. The atom which left a normal site to create a vacancy may end up in an interstitial position, a location not normally occupied. Or it may form a bond with a normal atom in such a way that neither atom is on the normal site, but the two are symmetrically displaced from it. This is called a split interstitial. The name Frenkel defect is given to a vacancy-interstitial pair, whereas an isolated vacancy is a Schottky defect.

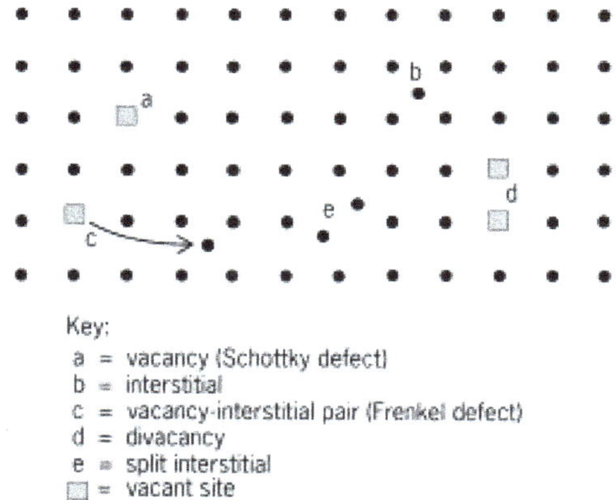

The simplest extended structural defect is the dislocation. An edge dislocation is a line defect which may be thought of as the result of adding or subtracting a half-plane of atoms. A screw dislocation is a line defect which can be thought of as the result of cutting partway through the crystal and displacing it parallel to the edge of the cut. Dislocations are of great importance in determining the mechanical properties of crystals. A dislocation-free crystal is resistant to shear, because atoms must be displaced over high-potential-energy barriers from one equilibrium position to another. It takes relatively little energy to move a dislocation (and thereby shear the crystal), because the atoms at the dislocation are barely in stable equilibrium. Such plastic deformation is known as slip.

For both scientific and practical reasons, much of the research on crystal defects is directed toward the dynamic properties of defects under particular conditions, or defect chemistry. Much of the motivation for this arises from the often undesirable effects of external influences on material properties, and a desire to minimize these effects. Examples of defect chemistry abound, including one as familiar as the photographic process, in which incident photons cause defect modifications in silver halides or other materials. A property of materials in nuclear reactors is another important case.

Point Defect

Point defect means missing of the atoms in the crystal, from the lattice site. In a crystal atoms all the ions, atoms, and molecules are perfectly arranged in a three dimensional space. If we will study these crystals minutely then we will observe some defects in these crystals. Each crystal has some defects present in it. The main reason behind these defects is the impurity of the crystal from which the semiconductor is made. It is almost impossible to make a crystal using a material which is 100% pure.

Classification of Point Defects

Two point defects are intrinsic to the material, meaning that they form spontaneously in the lattice without any external intervention. These two are the vacancy and the self-interstitial, shown schematically in a 2-D representation in the upper panels of figure. The vacancy is simply an atom missing from a lattice site, which would be occupied in a perfect lattice. The self-interstitial is an atom lodged in a position between normal lattice atoms; that is in an interstice. The qualification "self" indicates that the interstitial atom is the same type as the normal lattice atoms.

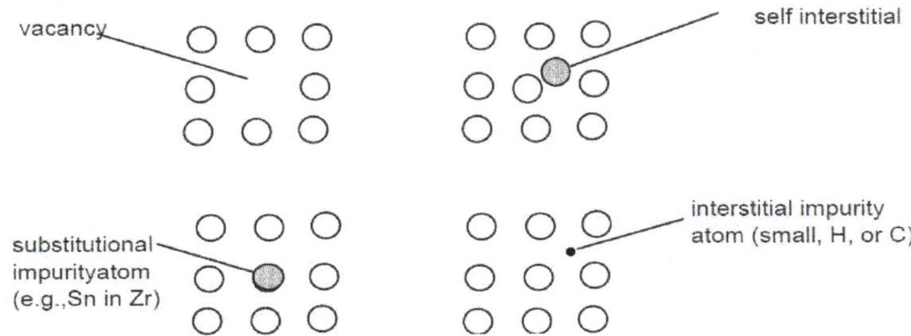

Figure: Point Defects in an Elemental Crystal

The lower two panels in figure show the two basic mechanisms by which a foreign or impurity atom exists in the crystal lattice of a host element. Large impurity atoms, usually of the same category as the host atoms (e.g., both metals, as nickel in iron), replace the host atoms on regular lattice positions. These are called substitutional impurities. The structure of the lattice is not disturbed, only the identity of the atoms occupying the lattice sites are different. Small atoms that are also chemically dissimilar from the host atoms occupy interstitial positions and do not appreciably distort the surrounding host crystal. They are termed interstitial impurities. Typical examples are carbon in iron and hydrogen in zirconium.

Aside from their identities relative to the host atoms, the self-interstitial and the interstitial impurity differ in the way that they reside in the lattice. As shown in figure using the bcc lattice as an example, the self-interstitial, because of its size, displaces a host atom off of its normal lattice position, creating a dumbbell-shaped pair.

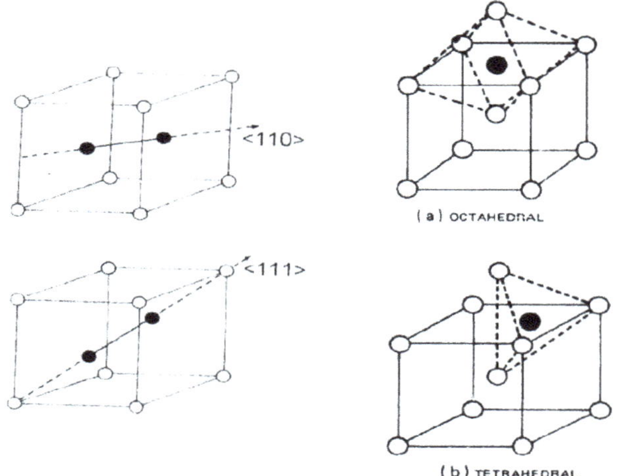

Figure: Interstitials in the bcc lattice. Left: self-interstitials; Right: interstitial impurities

This configuration is also called a split interstitial. The orientation of the dumbbell and the distance between the two atoms are determined by the condition that potential energy of the lattice be a minimum.

The small interstitial impurity atoms, on the other hand, occupy definite sites without significant distortion of the host lattice. These sites are named after the shape of the polyhedron formed by joining the host atoms surrounding the interstitial. The examples shown in figure for the bcc lattice are octahedral and tetrahedral sites. These two interstices offer the most space for the impurity atom to reside in, but which site is occupied is a sensitive function of the interaction energy between the impurity atom and the host atom.

Equilibrium Concentrations of Point Defects in Elemental Crystals

The criterion of chemical equilibrium is the minimization of the Gibbs free energy of the system at constant temperature and pressure. The system in this case consists of atoms on regular lattice sites and the intrinsic point defects randomly distributed on some lattice sites. Even though there is only one element involved, thermodynamically the system consists of two components, the regular atoms and the empty sites in case of a vacancy or atoms on irregular sites in case of the self-interstitial. In thermodynamic analysis, the two types of point defects are treated as independent entities.

The reason that point defects spontaneously form lies in the components of the Gibbs free energy, which, contains enthalpy and entropy contributions, $G = H - TS$. Each of the two properties on the right also contain two terms. The enthalpy change accompanying creation of the point defects is written as $H = E_{PD} + pV$, where EPD, the dominant contribution, is the energy required to create the point defect from the perfect lattice. It is always positive. The pV term represents the work involved in the change in system volume as the atom is moved between the interior site and the surface. Because of E_{PD}, the enthalpy component acts to make G more positive than that of the perfect lattice, and thus tends to suppress the formation of the point defect.

The two parts of the entropy involved in formation of point defects are expressed by the equation $S = S_{PD} + S_{mix}$. The minor component SPD results from the change in the vibrational motion of the atoms around the point defect from that of atoms in the perfect lattice. This term is positive for vacancies and negative for interstitials. The major component of the entropy change accompanying point defect formation is S_{mix}, the entropy of mixing. The reason for this behavior is that S_{mix} is a measure of randomness, and the introduction of point defects into a perfect crystal reduces the system's state of order. The entropy of mixing is responsible for the spontaneous existence of point defects at equilibrium; the magnitude of the positive energy of point defect formation, E_{PD}, governs the concentration of these species at thermal equilibrium.

Quantitative application of thermodynamics to point defect formation is illustrated using the vacancy. The system consists of NV vacancies and N atoms. The vacancy, even though it is literally nothing, is a bona fide component in a thermodynamic sense. Together, these two components occupy $N_S = N + N_V$ lattice sites.

The process to which the thermodynamic properties apply is the creation of a vacancy by moving an interior atom to the surface of the solid, as shown in figure. The process can be described the equilibrium reaction:

$$\text{Null} = V + A_{surf}$$

Where "null" denotes the perfect crystal, V is a single vacancy, and A_{surf} is an atom on the surface of the solid. The equilibrium is maintained because of the equality of the formation and removal rates depicted in figure. The components of the free energy equation, G = H – TS, are considered as property differences between the right and left hand sides of equation $s_{mix} = -k_B \left[N \, \text{In}\left(\frac{N}{N+N_V} \right) + N_V \, \text{In}\left(\frac{N_V}{N+N_V} \right) \right]$. The enthalpy to create NV vacancies is $H_V = N_V h_V$, where h_V is the enthalpy of formation of a single vacancy.

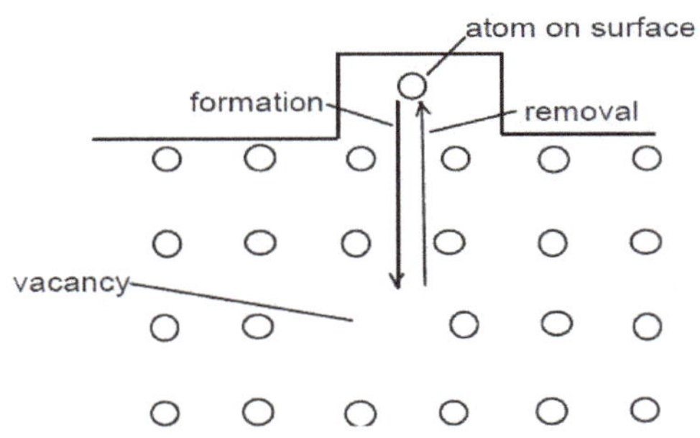

Figure: Vacancy formation

The creation of a vacancy from a region of perfect lattice involves breaking bonds between the atom to be moved and its neighbors and recouping about half of these in its surface position. The common process of vaporization entails breaking of bonds of surface atoms to form free gas atoms. Based on this simplistic picture, it is understandable that the energy of vacancy formation is found to be roughly equal to the heat of vaporization (per atom).

Similarly, the minor pV and S_V terms can be written as $p\Omega N_V$ and $s_V N_V$, respectively, where Ω is the volume per atom and s_V is the change in entropy associated with the formation of a single vacancy.

The principal contributor to the entropy change is that due to randomly mixing N_V vacancies with N atoms.

$$S_{mix} = -k_B \left[N \text{ In}\left(\frac{N}{N+N_V} \right) + N_V \text{ In}\left(\frac{N_V}{N+N_V} \right) \right]$$

Where k is Boltzmann's constant. The Gibbs free energy of the system becomes:

$$G(T) = G(0) + N_V (h_V + p\Omega - \Delta T s_V) - T\Delta S_{miX}$$

where G(0) is the free energy of a reference perfect lattice with N atoms and no vacancies. Even though the term in parentheses is always positive, the ΔTS_{mix} term provides a negative component to G, effectively assuring that the minimum value of G will be one that has some vacancies present. The equilibrium vacancy concentration is obtained from equation $G(T) = G(0) + N_V (h_V + p\Omega - \Delta T s_V) - T\Delta S_{miX}$ using the general criterion for equilibrium in a system at constant temperature and pressure, namely, that G be a minimum:

$$\left. \frac{dG}{dN_V} \right|_N = 0 = h_V + p\Omega - Ts_V + k_B T \, In\left(\frac{N_V}{N+N_V} \right)$$

The last term on the right in this equation is the derivative of equation with respect to N_V, holding N constant. The sum of N and N_V is equal to N_S, so the parenthetical term is the site fraction of vacancies, denoted by x_V. This concentration unit is the solid-state analog of the mole fraction unit that appears in the equilibrium equations in gases and liquids. The above equation is rewritten as:

$$x_V = N_V / N_S = e^{S_V/K_B} e^{-h_V/K_B T} e^{-p\Omega/K_B T}$$

s_V is usually approximated as zero, principally because it is unknown for most elements, and is small for those elements for which it has been measured. The term involving the pressure can be estimated using the following values: p = 100 MPa, Ω = 4x10^{-29} m^3 /atom, kB = 1.38x10^{-23} J/K-atom (8.62x10^{-5} eV/K) and T = 1200 K. This combination yields pΩ/kT = 0.25, so that this factor in equation $x_V = N_V / N_S = e^{S_V/K_B} e^{-h_V/K_B T} e^{-p\Omega/K_B T}$ is ~ 0.8. For most applications, this factor is sufficiently close to unity to be ignored. There are, however, phenomena in which this term in equation $x_V = N_V / N_S = e^{S_V/K_B} e^{-h_V/K_B T} e^{-p\Omega/K_B T}$ is essential to a correct description of the process. Barring such situations, a common approximation to equation $x_V = N_V / N_S = e^{S_V/K_B} e^{-h_V/K_B T} e^{-p\Omega/K_B T}$ is:

$$x_V = e^{-hV/k_B T}$$

Example: The formation energy of vacancies in copper is 100 kJ/mole (about 1 eV/atom). At 1300 K (which is just below the melting point) what fraction of the lattice sites are empty? Using these values in equation $x_V = e^{-hV/k_B T}$, the site fraction of vacancies is x_V = 10^{-4}. This value is too small to influence properties such as the density, but even smaller values of x_v are critical in determining the transport property of self-diffusion.

The process analogous to that shown in figure for forming self-interstitials in elemental crystals is shown in figure

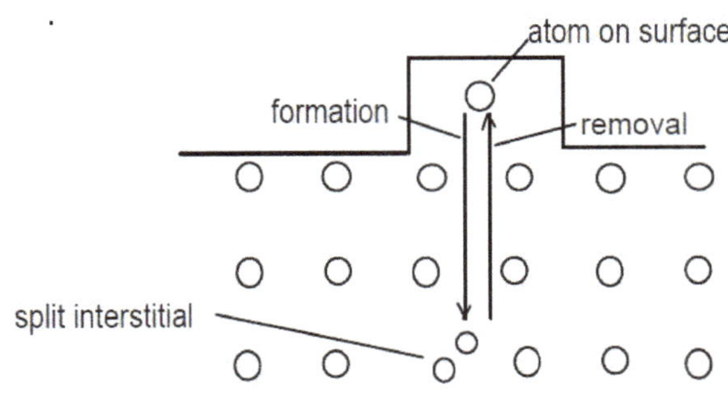

Light Water Reactor Mat ials, aft 2 6 © nald land and Arthur Motta
8/31/2009 6

Figure: Formation of self-interstitials

The formation of self-interstitials is totally independent of vacancy creation. However, the thermodynamic analysis is formally identical to that described above for vacancies. Aside from changes in the sign of the entropy s_i compared to s_v and of the $p\Omega$ term, the equilibrium concentration (site fraction) of self-interstitials is given in approximate form by the following:

$$x_i = e^{-h_i/k_B T}$$

For copper, the interstitial formation enthalpy is hi ~ 300 kJ./mole (~3eV/atom), which is about three times larger than the energy required to create a vacancy in this metal. As a result, the equilibrium concentration of interstitials is very much smaller than that of vacancies (by 8 orders of magnitude at 1300 K). This is true of all elements. Thermally generated interstitials can usually be neglected in most applications. However, in the presence of high-energy radiation, the two types of point defects are created at equal rates, and interstitials cannot be ignored.

An approximate but extremely simple method for treating point defect equilibria is afforded by regarding the process as a pseudo chemical reaction and directly utilizing the well-known theory of chemical reaction equilibria. The "reaction" that produces vacancies is given by equation Null = $V + A_{surf}$. The law of mass action expresses the equilibrium constant for this reaction by:

$$K_V = \frac{\text{activity of vacancies} \times \text{activity of surfaceatoms}}{\text{activity of atoms in prefect lattice}}$$

Because atoms in the undisturbed lattice and on the surface are at concentrations much greater than that of the point defect, they are undisturbed by the formation of the latter. Consequently, their activities can be taken to be unity. The activity of the vacancies, however, is equal to its site fraction x_V. This choice is based on the results of the previous exact analysis that produced equation $x_V = N_V / N_S = e^{S_V/K_B} e^{-h_V/K_B T} e^{-p\Omega/K_B T}$.

The other feature of chemical reaction theory that is adapted for the vacancy formation process is the relation between the equilibrium constant and the free energy change of the reaction. The Gibbs free energy change for forming a single vacancy, g_V, is equal to the sum of the three terms in parentheses in equation $G(T) = G(o) + N_V(h_V + p\Omega - \Delta Ts_V) - T\Delta S_{miX}$. Applying these adaptations of chemical reaction equilibrium theory to the process of vacancy formation yields:

$$K_V = x_V = e^{-gv/kT} = e^{Sv/K_B} e^{-hy/K_BT} e^{-p\Omega/K_BT}$$

which, is identical to equation $x_V = N_V / N_S = e^{Sv/K_B} e^{-hv/K_BT} e^{-p\Omega/K_BT}$ obtained by the exact method. A similar application of chemical equilibrium theory applies to self-interstitials as well.

Point Defects in Ionic Crystals

Self-interstitials and vacancies occur naturally in ionic crystals as well as in elemental solids. However, because the cations and anions carry electrical charges, vacancy and interstitial formation are not independent processes. To create a vacancy on the anion sublattice by moving the anion to the surface, for example, would leave the surface negatively charged and the interior around the vacancy with a net positive charge. This violation of local electrical neutrality precludes such a process. Similar arguments apply to cation vacancies or self-interstitials of either ionic type.

Point defects that preserve local electrical neutrality are shown in figure in a simplified two-dimensional representation.

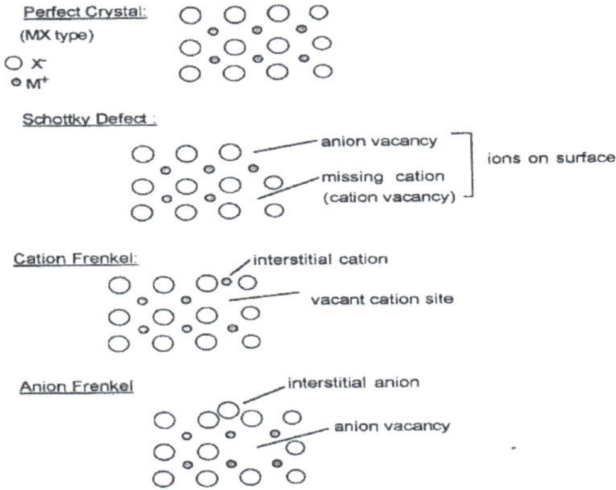

Figure: Point defects in an MX-type ionic crystal

For simplicity, an MX type ionic solid is depicted. The Schottky defect involves simultaneous movement of a cation and an anion to the surface. In an MX_2 type crystal, two anion vacancies would have to be created for each cation vacancy.

The other important point defect in ionic crystals is called a Frenkel defect. It can occur either on the cation sublattice or on the anion sublattice. As shown in figure, an ion is moved from a regular lattice site to a nearby interstitial site, thereby maintaining local electrical neutrality

Schottky and Frenkel defects are created independently. In any particular crystal, one will dominate, while the other will either be absent or a minor contributor. In UO_2, for example, the dominant defect is anion Frenkel, but Schottky defecting occurs to a lesser extent.

The chemical equilibrium treatment of the thermodynamics of defect formation in ionic crystals is a straightforward extension of the method applied to elemental solids. The major advantage of this simplified approach is the avoidance of the complication of calculating the entropy of mixing for a two-component system with its associated point defects.

The reaction producing a Schottky pair is:

$$\text{Null} = V_M + V_X + \text{ions on surface}$$

Where null denotes the perfect crystal and V_M and V_X are vacancies on the cation and anion lattice sites, respectively. Since the activities of the ions in the perfect crystal and on the surface are both unity, the law of mass action for equation Null = $V_M + V_X$ + ions on surface is:

$$K_s = x_{VX}x_{VM} = e^{S_S/k}e^{-\varepsilon_S/KT}$$

K_S is the Schottky equilibrium constant expressed in terms of the entropy s_S and energy ε_S of formation of a Schottky pair. The pressure-volume term that appears in equation $x_V = N_V / N_S = e^{S_V/K_B}e^{-h_V/K_BT}e^{-p\Omega/K_BT}$ for elemental crystals has been neglected for simplicity in equation $K_s = x_{VX}x_{VM} = e^{S_S/k}e^{-\varepsilon_S/KT}$.

The vacancy site fractions are defined in terms of the numbers of defects and numbers of regular sites:

N_{VX} = number of vacant anion sites

N_{SX} = number of anion sites

N_{VM} = number of vacant cation sites

N_{SM} = number of cation sites

So that:

$$x_{VX} = N_{VX}/N_{SX} \quad \text{and} \quad x_{VM} = N_{VM}/N_{SM}$$

For an MX-type crystal, $N_{SX} = N_{SM}$, so that charge neutrality is expressed by:

$$N_{VX} = N_{VM} \text{ or } x_{VX} = x_{VM}$$

Combining equations $K_s = x_{VX}x_{VM} = e^{S_S/k}e^{-\varepsilon_S/KT}$ and $N_{VX} = N_{VM}$ or $x_{VX} = x_{VM}$ yields the individual vacancy fractions:

$$X_{VX} = X_{VM} = \sqrt{K_S}$$

Anion Frenkel defects are produced by the reaction:

$$\text{Null} = V_X + I_X$$

where I_X is an interstitial anion. The law of mass action for this reaction is:

$$K_{FX} = x_{VX}x_{IX} = e^{S_{FX}/K}e^{-\varepsilon_{FX}/kT}$$

where the subscript FX denotes Frenkel defects on the anion sublattice. The condition of charge neutrality is:

$$N_{VX} = N_{IX}$$

This condition must first be converted to site fractions before combining with equation $K_{FX} = x_{VX}x_{IX} = e^{S_{FX}/K}e^{-\varepsilon_{FX}/kT}$. The site fraction of vacancies, x_{VX}, is the same as in equation $x_{VX} = N_{VX}/N_{SX}$ and $x_{VM} = N_{VM}/N_{SM}$. The site fraction of interstitials, however, depends on the number of anion interstitial sites:

$$x_{IX} = N_{IX}/N_{SIX}$$

where

N_{IX} = number of anion interstitials

N_{SIX} = number of anion interstitial sites

When counted on the same basis (say per unit volume, or per unit cell), the numbers of anion interstitial sites need not be equal to the number of regular anion sites. However, assuming for simplicity that $N_{SIX} = N_{SX}$, equation $N_{VX} = N_{IX}$ becomes:

$$x_{VX} = x_{IX}$$

Substituting equation $x_{VX} = x_{IX}$ into equation $K_{FX} = x_{VX}x_{IX} = e^{S_{FX}/K}e^{-\varepsilon_{FX}/kT}$ gives the final result:

$$X_{VX} = X_{IX} = \sqrt{K_{FX}}$$

An entirely analogous derivation applies to cation Frenkel defects.

Point Defects in Stoichiometric UO_2

Uranium dioxide is a complicated material insofar as point defect formation is concerned. First, the uranium cation can assume oxidation states of 3+, 4+, 5+ and 6+. The mixture of oxidation states depends upon the prevailing oxygen pressure. Thus, the nature and concentrations of the point defect in UO_2 depends on a gas-solid equilibrium as well as on the point defect equilibria treated in the previous two sections. In addition, UO_2 as a reactor fuel is never a pure material; as a result of the fission process, sites on the uranium sublattice contain fission products of a multiplicity of valences, ranging from Ba^{2+}, La^{3+}, Zr^{4+} to Nb^{5+}.

In this section, we analyze point defect production in uranium dioxide with three important restrictions: first, all uranium atoms are in their IV oxidation state; second, no impurities are present on the cation (or anion) sublattice; third, the oxide is exactly stoichiometric, meaning that the O/U atom ratio is 2.

Extensive research has demonstrated that the principal defect in UO_2 is the Frenkel anion defect of the type shown generically in figure. Understanding of the defect type goes further than simply this identification. The unit cell of the perfect UO_2 lattice is shown in figure. While there is no ambiguity concerning the anion vacancy (it is simply a missing oxygen ion), the nature of the oxygen interstitial ions is more complicated. Neutron diffraction studies have shown that oxygen interstitial ions occupy the unit cell in pairs, as shown in figure.

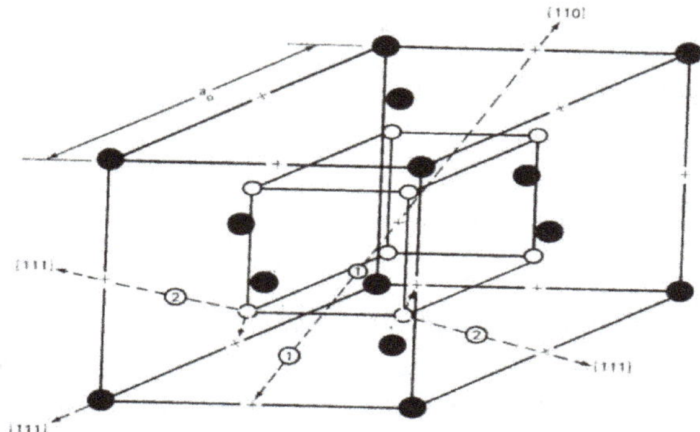

Figure: Location of anion (oxygen) interstitials in UO_2

The two oxygen interstitial ions added to the unit cell are labeled "1" in the diagram. They are disposed along a direction. To reduce the repulsive Coulomb interaction between the interstitial pair and the nearby corner O^{2-} on normal lattice sites, the latter (labeled "2" in the diagram) are pushed off in directions. The separation distances between the "1" interstitials and the "2" original anions from their normal lattice sites are determined by minimizing the energy of the configuration.

Even though UO_2 is an MX_2-type crystal, the anion Frenkel defect formation reaction is still given by equation Null = $V_X + I_X$ with the equilibrium constant given by equation $K_{FX} = x_{VX}x_{IX} = e^{S_{FX}/K}e^{-e_{FX}/kT}$. For the restrictions placed on this process (pure UO_2 with an O/U ratio of exactly 2), the electrical neutrality condition is given by equation $N_{VX} = N_{IX}$. Where the UO_2 analysis differs from the simple anion Frenkel analysis of the preceding section is in the relation between the numbers of point defects and sites to the site fractions that are required in equation $K_{FX} = x_{VX}x_{IX} = e^{S_{FX}/K}e^{-e_{FX}/kT}$. Replacing the anion identification X by that for oxygen, O, the mass action law for the anion Frenkel equilibrium can be written as:

$$K_{FO} = X_{10} \bullet X_{VO} = \frac{N_{10}}{N_{SIO}} \bullet \frac{N_{VO}}{N_{SO}}$$

where the defect and site numbers on the right have been defined in the text above equation $x_{VX} = N_{VX}/N_{SX}$ and $x_{VM} = N_{VM}/N_{SM}$ and above equation $x_{IX} = N_{IX}/N_{SIX}$. The numbers in the denominators of the above equation are to be related to the number of cation (uranium ion) sites, NSU. From the stoichiometry of UO_2, $N_{SO} = 2\ N_{SU}$. Examination of figure shows that the oxygen interstitial pairs can be located on any of the 12 edges of the small cube. Since there are 4 cations in the unit cell, $N_{SIO} = 3\ N_{SU}$. Using these ratios to convert the denominators of the above equation to NSU and taking account of the electrical neutrality condition of equation $x_{IX} = N_{IX}/N_{SIX}$ yields the following result:

$$\frac{N_{10}}{N_{SU}} = \frac{N_{VO}}{N_{SU}} = \frac{\text{defect pairs}}{\text{cation site}} = \frac{\text{moles of defect pairs}}{\text{mole of } UO_2} = \sqrt{6K_{FO}}$$

Example: The properties of the anion Frenkel defect in UO_2 are: s_{FO} = 63 J/mole-K and ε_{FO} = 297 kJ/mole. What is the concentration of anion Frenkel pairs at 2000K? K_{FO} is computed from equation $K_{FX} = x_{VX}x_{IX} = e^{S_{FX}/K}e^{-\varepsilon_{FX}/kT}$ using the gas constant R = 8.314 J/mole-K instead of Boltzmann's constant. The result is K_{FO} = 4.4x10⁻⁵. From equation $\frac{N_{10}}{N_{SU}} = \frac{N_{VO}}{N_{SU}} = \frac{\text{defect pairs}}{\text{cation site}} = \frac{\text{moles of defect pairs}}{\text{mole of }UO_2} = \sqrt{6K_{FO}}$, the concentration of defect pairs is:

$$\frac{\text{moles of defectpairs}}{\text{mole of }UO_2} = \sqrt{6\times 3.4\times 10_{-5}} = 1.4\times 10^{-2}$$

the anion Frenkel defects greatly influence the diffusivity of oxygen ions in UO_2. The secondary defecting process in UO_2 is the Schottky process. This defect produces vacancies on the cation sub-lattice, and is responsible for the diffusion coefficient of uranium ions in UO_2.

Another property affected by the creation of point defects in ionic crystals is the heat capacity. At relatively low temperatures (but not approaching absolute zero), the heat capacity is nearly temperature-independent with a value of 3R per gram atom (R is the gas constant). Per mole of UO_2, this classical value of the heat capacity is $(CV)_{lattice}$ = 9R. This contribution to the heat capacity arises from the vibrations of the atoms in the lattice. The creation of point defects by the anion Frenkel process provides an additional component to the heat capacity, which can be expressed by:

$$CV = (CV)_{lattice} + (CV)_{defects}$$

The extra energy in the solid provided by the presence of the anion Frenkel defects is the concentration given by equation $\frac{N_{10}}{N_{SU}} = \frac{N_{VO}}{N_{SU}} = \frac{\text{defect pairs}}{\text{cation site}} = \frac{\text{moles of defect pairs}}{\text{mole of }UO_2} = \sqrt{6K_{FO}}$ times the energy to produce a mole of defects, which is ε_{FO} (expressed in molar units). The excess energy due to the point defects is:

$$e_{ex} = \varepsilon_{FO}\sqrt{6K_{FO}}$$

The excess heat capacity arising from the point defects is the derivative of e_{ex} with respect to temperature, or, using equation $K_{FX} = x_{VX}x_{IX} = e^{S_{FX}/K}e^{-\varepsilon_{FX}/kT}$ for K_{FO} with k replaced by R:

$$(C_V)_{defect} = \frac{de_{ex}}{dT} = \frac{\sqrt{6}}{2}R\left(\frac{\varepsilon_{FO}}{RT}\right)^2 e^{s_{FO}/2R}\, e^{-\varepsilon_{FO}/2RT}$$

Example: What is the contribution of anion Frenkel defects to the heat capacity of UO2 at 2000 K?

Using the values of s_{FO} and ε_{FO} given in the previous example, the above equation gives: $(C_V)_{defect}$ =2.2R. This is nearly 25% of the classical value of 9R, and is an important contribution to C_V.

Line Defect

Dislocations are linear defects; they are lines through the crystal along which crystallographic registry is lost. Their principle role in the microstructure is to control the yield strength and subsequent plastic deformation of crystalline solids at ordinary temperatures. Dislocations also participate in the growth of crystals and in the structures of interfaces between crystals. They act as electrical defects in optical materials and semiconductors, in which they are almost always undesirable.

The Edge Dislocation

The simplest way to grasp the idea of a dislocation is to imagine how you might go about creating one. We begin by making an edge dislocation, which is the easiest type of dislocation to visualize in a crystal, and follow the recipe laid down by Volterra.

Consider the solid body that is drawn in figure. For the present purpose it does not matter whether the body is crystalline; it may be easier to imagine that it is rubber.

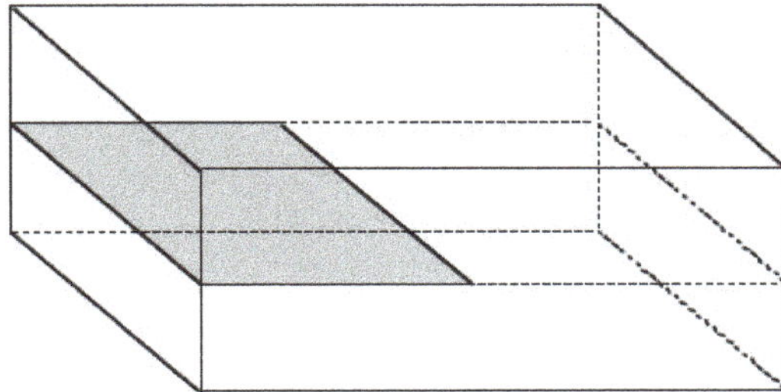

Figure: A block of material with a planar cut indicated by the Shading

To create an edge dislocation in this body we first make a planar cut part way through it, as illustrated by the shaded region in the figure. We then fix the part of the body below the cut, and apply a force to the body above the cut that tends to displace it in the direction of the cut, as illustrated in figure. The upper part slides, or slips over the lower by the vector distance b, which is the relative displacement of the two lips of the cut. The plane of the cut, where slip occurs, is called the slip plane. The cut is finite and constrained at its end, so material accumulates there. The end of the cut, or equivalently, the boundary of the planar region of slip, is a linear discontinuity in the material. The situation after the slip is shown in figure.

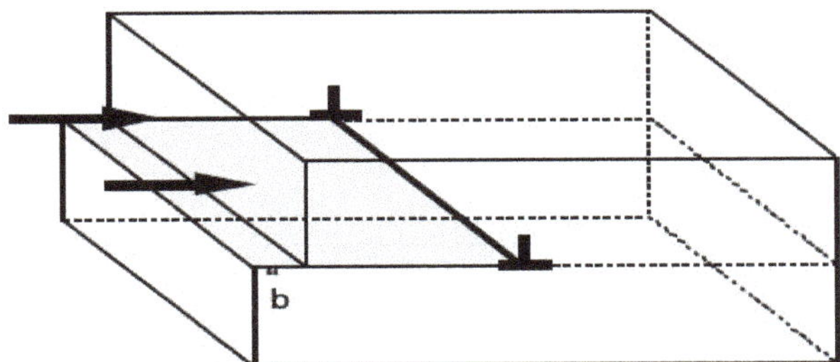

Figure: The upper part of the body has been slipped by the vector b over the shaded area. The terminal line is a discontinuity, marked by the heavy line.

Now suppose that we have a mechanism for welding the material back together which is so efficient that it is impossible to tell after the fact that the weld was ever made. If we match the material above the shaded plane to that below so that there is no physical discontinuity across the plane and weld the lips together, the shading disappears since it is impossible to tell that the material was ever separated. However, matching the material across the plane of slip requires that excess material be gathered at the line at which slip terminates. This line is, therefore, a linear defect in the material. It is called an edge dislocation. It is an isolated defect, as shown in figure since after the material has been re-welded there is no unique way to determine how the dislocation was created. For example, the dislocation would be exactly the same if the material to the right of it on the slip plane were slipped by the vector -b on the same plane. Only the exterior step indicates the origin of the dislocation, and this may be removed, or may have pre-existed the formation of the dislocation.

However it was created, the edge dislocation in figure has the property that it defines an element of slip, b, where the vector b is called the Burgers vector of the dislocation. We can always identify the slip plane of a dislocation like that shown in figure. It is the plane that contains the Burgers vector, b, and the line of the dislocation. However the dislocation actually came to its present position, its net effect is that the material above the slip plane to the left of the dislocation (in the direction of -b) has been displaced by b relative to that below the slip plane. The dislocation is a linear defect whose location is defined by its line and whose nature is characterized by its Burgers vector, b. In the case shown in figure the Burgers vector is perpendicular to the dislocation line. This perpendicularity is characteristic of an edge dislocation.

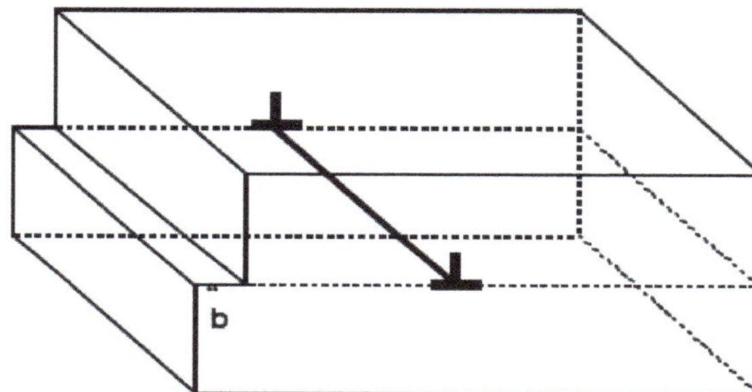

Figure: Isolated edge dislocation after the cut surface has been rejoined.

If a dislocation moves the area that has been slipped grows or shrinks accordingly. Imagine that the dislocation is initially created at the left edge of the slip plane in figure, and is then gradually moved to the right edge. Applying the construction in figure to the initial and final positions of the dislocation, it follows that the motion of the dislocation through the body causes the whole volume of material above the slip plane to be displaced by the vector b with respect to that below it, as shown in figure.

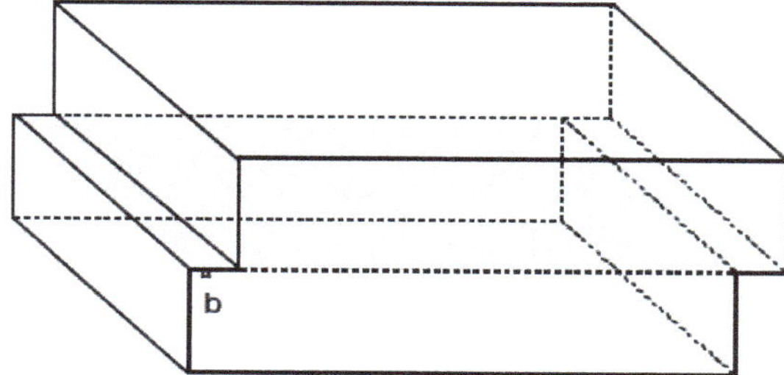

Figure: Final state of the body after an edge dislocation with Burgers' vector b has crossed the whole of the slip plane shown.

Figure illustrates the connection between the motion of a dislocation (in this case, an edge dislocation) and the plastic, or permanent deformation of a material. A change in shape that occurs at constant volume can always be represented geometrically as the sum of elementary deformations of a type known as simple shear. A simple shear is the kind of deformation that deforms a cube into a parallelogram; it changes the angles between initially perpendicular directions in the cube. The shear due to the passage of an edge dislocation is illustrated in figure. While the dislocation translates the top of the crystal rigidly over the bottom to create a discrete step, the Burgers' vector has atomic dimensions, so the step is invisible. Macroscopic deformation is the sum of the slip caused by many dislocations.

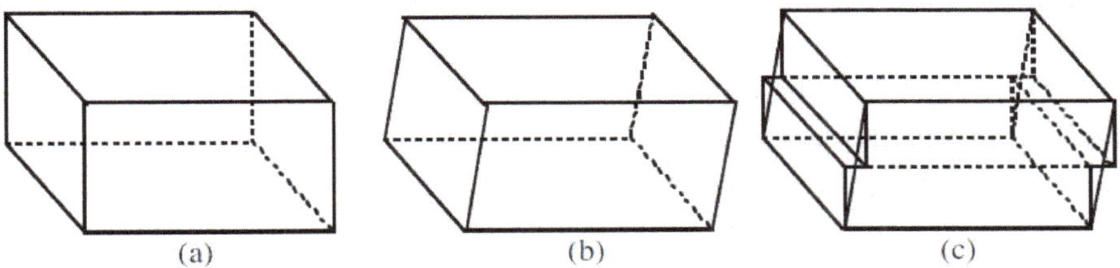

Figure: Figure (b) is obtained from (a) by a simple shear of the top over the bottom. Figure (c) shows how the same shear can be caused by an edge dislocation.

An Edge Dislocation in a Simple Cubic Crystal

The procedure that was used to create the edge dislocation that appears in Figure made no reference to the structure of the solid, and can be used to form an edge dislocation in any material. However, when the material is crystalline the ordered pattern of atoms restricts the values that the Burgers vector, b, can have. The restriction is introduced by the welding step that is used to change the configuration shown in figures. The welding must be so perfect that it is impossible to tell that the two surfaces were ever separated. If the solid is crystalline this can only be true if the crystal structure is continuous across the slip plane after the weld is made. It follows that the relative displacement across the slip plane must equal a lattice vector so that atoms can re-bond without changing their local atomic configurations. Since the relative displacement is equal to the Burgers vector, b, of the dislocation, b must be a lattice vector. If the dislocation is an edge dislocation, b must also be perpendicular to the dislocation line. The geometry of an edge dislocation is relatively easy to visualize when the crystal has a simple cubic crystal structure. The atomic configuration around the dislocation line is more complicated in real crystal structures, but it is not necessary to deal with that complexity to understand the behavior of dislocations at the level. Whenever we need to consider the crystallography of the dislocation we shall assume that the crystal structure is simple cubic.

An edge dislocation in a simple cubic structure is drawn in figure, which shows both a two-dimensional view and a three-dimensional section along the dislocation line. The dislocation can be created by making a cut in the crystal on the dashed plane that terminates at the dislocation line, displacing the material above the cut plane to the left of the dislocation by one lattice spacing, and allowing the atoms to re-bond across the slip plane. This recipe recreates the simple cubic unit cell everywhere except on the dislocation line itself (ignoring the small elastic distortion of the cells that border the dislocation line). Hence the Burgers vector, b, of the dislocation that is drawn in the figure is b = a, where a is a vector along the edge of the cubic unit cell.

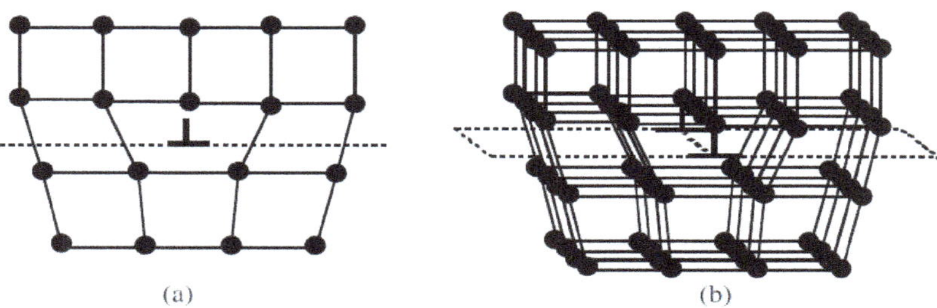

Figure: An edge dislocation in a simple cubic structure. The dotted plane is the slip plane

The process that creates the edge dislocation shown in figure leaves one extra vertical half-plane of atoms above the slip plane. This extra half-plane terminates at the dislocation line, and is compressed there, as shown in the figure. The distortion at the dislocation line is local. The simple cubic arrangement of atoms is essentially restored a few atom spacing away from the dislocation line. The influence of the dislocation on the atomic configuration rapidly decays into a small displacement that decreases in magnitude with the inverse cube of the distance from the dislocation line. The local distortion near the dislocation line (or dislocation core) is indicated in the figure.

In principle, the Burgers vector of a crystal dislocation can be any lattice vector; for example, it is geometrically possible for an edge dislocation to be the termination of any integral number of lattice planes. In reality, however, the Burgers vector is almost invariably equal to the shortest lattice vector in the crystal. The reason is that the energy per unit length of dislocation line, which is called the line energy, or, in a slightly different context, the line tension of the dislocation, increases with the square of the magnitude of b, $|b|^2$. (While we shall not prove this, it is obvious from figure that the local distortion of the crystal would increase dramatically if two or more extra half-planes terminated at the dislocation line.) Let the Burgers vector, b, be the vector sum of smaller lattice vectors, $b = b_1 + b_2$. Unless b_1 and b_2 are perpendicular, $|(b_1+b_2)|^2 > |b_1|^2 + |b_2|^2$, and the dislocation can decrease its energy by splitting into two or more dislocations that have smaller Burgers vectors.

The Burgers Circuit

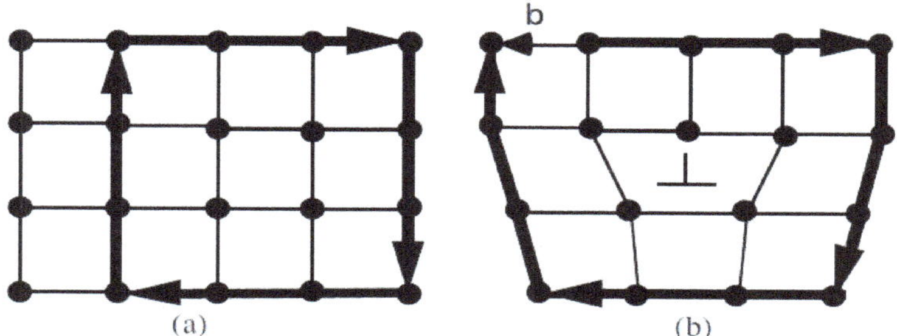

Figure: A Burgers circuit closes in a {100} plane of a cubic crystal, but fails to close by the Burgers vector, b, when the same circuit encloses an edge dislocation.

While it is always possible to find the Burgers vector, *b*, of a dislocation by determining the slip that would be required to make it, this is often inconvenient. A simpler method uses a geometric construction known as the Burgers circuit. To construct the Burgers circuit, choose a direction for the dilocation line and draw a clockwise closed circuit in the perfect crystal by taking unit steps along the lattice vectors. An example is drawn for a {100} plane in a simple cubic crystal in figure. If the same circuit is drawn so that it encloses a dislocation, it fails to close. The vector (from the starting position) that is required to complete the circuit is the Burgers vector, *b*, of the dislocation, and measures the net displacement experienced by an imaginary observer who completes a circuit around the dislocation that would be closed in a perfect crystal.

Motion of an Edge Dislocation: Glide and Climb

The reason that dislocations control the plastic deformation of crystalline solids is that it is relatively easy to move dislocations to produce shear deformation of the sort that is pictured in figure. It would be enormously difficult to shear a crystal by forcing the glide of rigid planes of atoms over one another; one would have to force the simultaneous reconfiguration of every crystal bond that crossed the slip plane. The same result is more easily achieved by moving dislocations stepwise through the crystal. Stepwise dislocation motion requires a much smaller force since each elementary step can be accomplished by reconfiguring only the bonds that neighbor the dislocation line.

The stepwise motion of an edge dislocation in a simple cubic crystal is illustrated in figure. In order for the dislocation to move one lattice spacing to the right it is only necessary to break the bond indicated by the long dash in figure a and establish the bond indicated by the short dash. The new configuration is shown in figure b. Of course one bond must be broken for each plane through which the dislocation threads, so a significant force is still required. But the force is small compared to that required to slip the upper part of the crystal as a rigid body. If the dislocation moves through the crystal in a sequence of individual steps like that shown in figure it causes a net slip of the material above its plane of motion by the Burgers vector, b, and hence causes a rigid displacement of the whole upper part of the crystal.

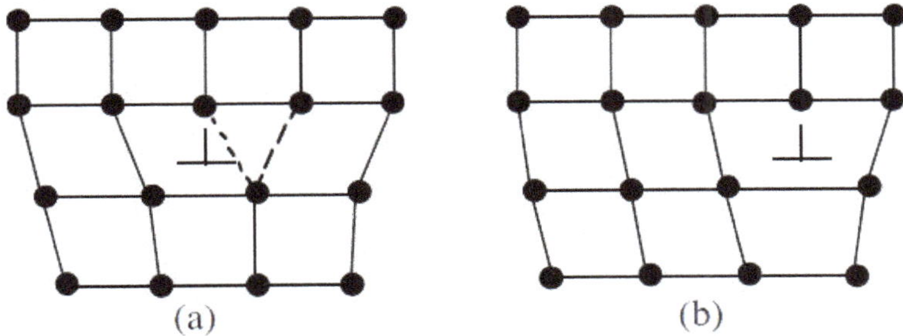

Figure: Glide of an edge dislocation. Only a single bond must be broken per plane for each increment of glide.

The type of motion that is illustrated in figure is called dislocation glide, and is relatively easy to accomplish. However, an edge dislocation cannot glide in an arbitrary direction. It can only glide in a particular plane, the slip plane or glide plane, which contains both the Burgers vector and the dislocation line.

When an edge dislocation moves out of its glide plane its motion is called climb. The climb of a dislocation is difficult at ordinary temperatures since it requires that atoms be absorbed on or liberated from the extra half-plane of atoms that defines the dislocation line. The climb of an edge dislocation is illustrated in figure. The mechanism is slightly different depending on whether the dislocation moves up, which contracts the extra half-plane, or down, which extends it.

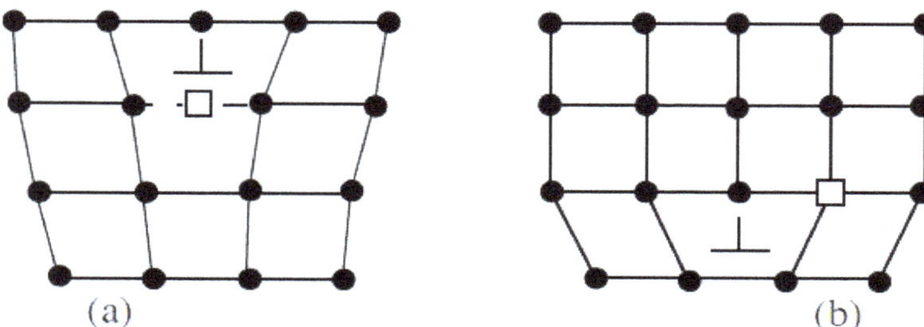

Figure: Climb of an edge dislocation. Movement up out of the plane requires the elimination of atoms by vacancies. Movement down requires the addition of atoms.

If the dislocation climbs up atoms must be liberated from the edge of the extra half-plane. Since the number of atoms is conserved, this requires the absorption of vacancies from the lattice. One vacancy is needed per plane the dislocation threads. If the dislocation climbs down it must add atoms to the extra half-plane, and can only do this by liberating one vacancy per plane into the matrix, as shown in figure b. Both processes are difficult except at high temperature when, as we shall see, the equilibrium concentration of vacancies is high and the exchange of vacancies and atoms is relatively easy.

Because of the difficulty of climb at ordinary temperatures the plastic deformation of real crystals tends to occur through the motion of dislocations on well-defined planes that are the glide planes of the active dislocations. Under a microscope one can often see discrete slip steps on the surface of a crystal that has been deformed. These result from the glide of many dislocations on closely spaced, parallel planes. At high temperature climb becomes possible and the slip planes are less well-defined. When this happens the strength of the crystal (its resistance to plastic deformation) decreases dramatically. For this reason most solids are relatively soft at high temperature.

Screw Dislocations

Dislocations in real crystals rarely have a pure edge character. Their Burgers vectors lie at various angles to their line directions. In the extreme case the Burgers vector is parallel to the dislocation line, which is the characteristic of a screw dislocation. A screw dislocation is difficult to visualize in a crystal, but can be created by a method suggested by Volterra that closely resembles the way the edge dislocation was formed. A screw dislocation of the general Volterra type is shown in figure.

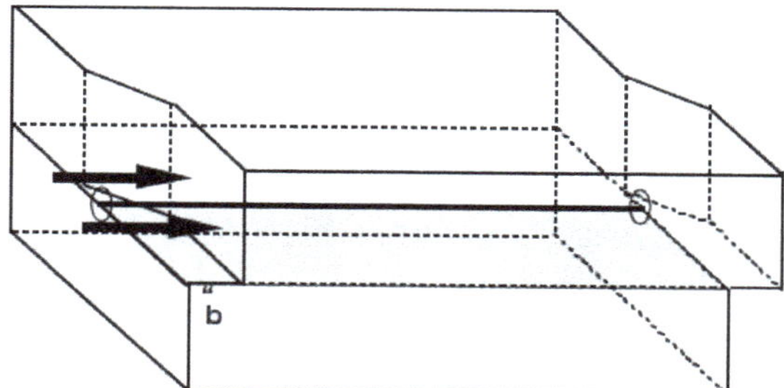

Figure: A method for forming a screw dislocation in a solid

To introduce a screw dislocation we slice the solid part-way through in the direction of its width, as shown in figure. But instead of displacing the material above the cut toward the dislocation line we displace it by a vector, *b* that lies parallel to the dislocation line, as shown in the figure. The direction of the force required to do this is also indicated. The material is then re-welded so that it is continuous across the plane of slip. The residual distortion is concentrated at the dislocation line, which then constitutes an isolated linear defect.

If the body shown in figure is crystalline then the cut surface shown in the figure is a plane of atoms. In order for the crystal to be continuous across the slip plane after it is rejoined, the displacement, b, must be such that this plane of atoms joins continuously onto a crystallographically identical plane. It follows that a closed circuit (Burgers circuit) that encloses a screw dislocation not only fails to close, but produces a net translation by b along the dislocation line, where b is a lattice vector. A circuit that starts on one plane of atoms finishes on another a distance b below. Continuing the circuit causes a displacement by b at each revolution, without the circuit ever leaving the atom plane. The effect of a screw dislocation is to join a set of parallel atom planes so that they become a single plane like one that would be created by extending a plane outward from the thread of a screw. Hence the name: screw dislocation. As in the case of an edge dislocation the line energy of a screw dislocation is proportional to the square of its Burgers vector. Hence the Burgers vector of the screw dislocation is ordinarily the smallest lattice vector that is compatible with the direction of its line.

Screw Dislocations and Plastic Deformation

A screw dislocation differs from an edge not only in its geometry but in the way it accomplishes plastic deformation. The most important qualitative differences concern its direction of motion under an applied force and its relative freedom of movement. Figure suggests the connection between slip and dislocation motion for a screw dislocation. As the screw is location is displaced through the width of the body the material above its plane is slipped in the direction of the Burgers vector, hence along the length of the body. It follows that the longitudinal force shown in the figure acts to drive the screw dislocation sideways. If a screw dislocation is passed through the full width of the body it causes the shear shown in figure, which is the same as that caused by the passage of an equivalent edge dislocation through the length.

In contrast to an edge dislocation, a screw dislocation can glide in any plane. Since the Burgers vector lies parallel to the dislocation line both are in any plane that contains the dislocation line and the screw dislocation can move in any direction perpendicular to its line.

Dislocations in Real Materials: Mixed Dislocations

Dislocations in real materials are most commonly neither pure edge nor pure screw in their character, but are mixed dislocations whose Burgers vectors lie at an intermediate angle to the local direction of the dislocation line. Because of the way in which they interact with other elements of the microstructure dislocation lines are ordinarily curved. Since the dislocation bounds a region that has been slipped by the Burgers vector, b, the Burgers vector is the same at every point on the dislocation line. Hence the character of a curved dislocation changes continuously along its length. For this reason it is often most useful to think of a dislocation as the boundary of a surface over which the crystal has slipped rather than as a defect with a particular local atomic configuration.

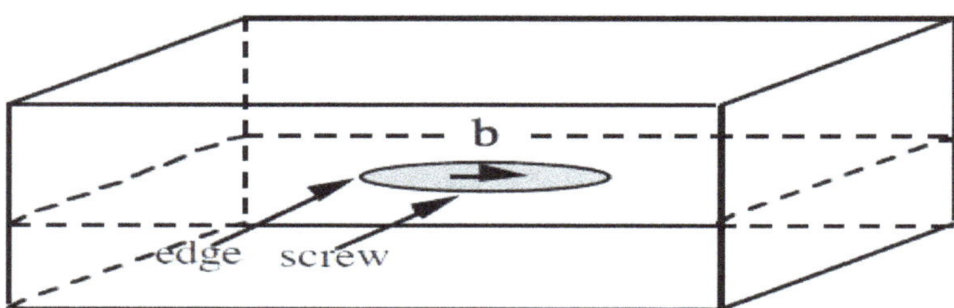

Figure: A dislocation loop in a crystal.

This point is illustrated by the dislocation loop shown in figure. Dislocation loops are relatively common features of the microstructures of structural alloys. The loop shown in the figure is created by a process that is equivalent to a slip of the cylinder of material above the loop by the vector b. The dislocation loop is the boundary of the surface on which slip has occurred. The character of the dislocation changes continuously around the loop. The dislocation has pure edge character at the extremities of the loop along the length of the body, and has pure screw character at the extremities along the width.

The qualitative nature of the force on a curved dislocation, such as the loop in figure, can also be understood by regarding it as the boundary of the slipped region. If the force on the body is oriented to shear the body by the vector b then the dislocation will tend to move so that it increases the area of slip; the loop will expand. If the force acts to shear the body in the direction -b, the dislocation will tend to move so that it shrinks the slipped area; the loop will contract.

Whether edge, screw or mixed in character, all dislocations have three important properties. First, they bound surfaces that divide the crystal so that the part above the surface is slipped by the vector, b, with respect to that below. Second, the slip on the surface bounded by the dislocation is constant and equal to a lattice vector. Hence the Burgers vector, b, is a lattice vector and has the same value at every point on the dislocation line. Third, since the slip is constant, a dislocation cannot simply end within a material; there is always a boundary between a slipped and an unslipped area. The dislocation can terminate at a free surface, as in figures, it can close on itself, as in figure, or it can end at a junction with other dislocations that bound surfaces over which the magnitude or direction of slip is different.

Partial Dislocations

In the original concept of Volterra, the Burgers' vector of a crystal dislocation is a translation vector of the crystal, that is, a vector that connects atom positions so that the crystal can be rejoined perfectly across its glide plane. Such a dislocation is called a total dislocation. The element of slip it induces carries an atom in the glide plane from one atom position to another. Only total dislocations can be true two-dimensional defects. If the Burgers' vector were not a translation vector of the crystal the slipped surfaces would not weld perfectly together across the glide plane and a planar, surface defect would be left behind whenever the dislocation moved. While most crystal dislocations are total dislocations when viewed from sufficiently far away, it is not uncommon to find them dissociated locally into a configuration that can be described as two parallel partial dislocations connected by a planar defect that is called a stacking fault in the crystal. The prototypic example is found in FCC crystals.

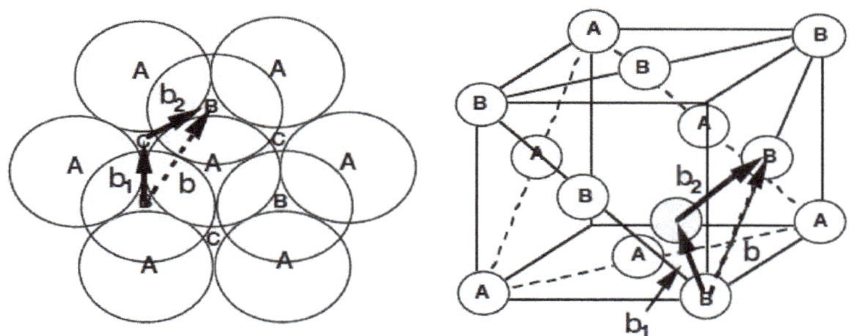

Figure: Illustration of the slip of close-packed planes. The dashed arrow shows slip by a total dislocation, $b = \frac{1}{2}[\bar{1}01]$, which carries $B \rightarrow B$. The solid arrows show the competing process of successive slip by two partial dislocations. The shaded atom in the cell is the intermediate, C-site position reached after partial slip.

The common dislocation in the FCC structure is the dislocation that causes close packed planes to slip over one another. The close-packed planes are {111} planes, and the Burgers' vectors of the dislocations that induce slip on these planes are the 1 2 <110> vectors that connect atoms to their nearest neighbors in the {111} planes. Figure illustrates the path by which an FCC crystal is slipped by one atom spacing in the (111) plane. The slip vector is $\frac{1}{2}[\bar{1}01]$, and is the slip caused by a dislocation with Burgers' vector $b = \frac{1}{2}[\bar{1}01]$. Assuming that the glide plane of the dislocation lies between (111) planes of A and B-type atom sites, the element of slip carries an atom from one B-site to another, as shown in the figure. For clarity, the figure is drawn twice. The drawing on the left shows the displacement in a stacking of close-packed planes (A, B). The drawing on the right places the A and B planes in an FCC unit cell.

Direct slip from one B position to another is relatively difficult, as one can easily show by trying to slide two close-packed arrays of balls over one another. The atoms (balls) in the B-plane must ride over the A atoms, as illustrated in the figure. It is much easier to accomplish the slip in two sequential steps, as indicated in the figure, that circumvent the A atoms. The B atoms are first slipped into C positions, then moved from C back to B again. The slip can be accomplished by the sequential passage of two dislocations with the Burgers' vectors b_1 and b_2. However, b_1 and b_2 are not lattice vectors; they are examples of partial dislocations. Their sum is the total dislocation, b.

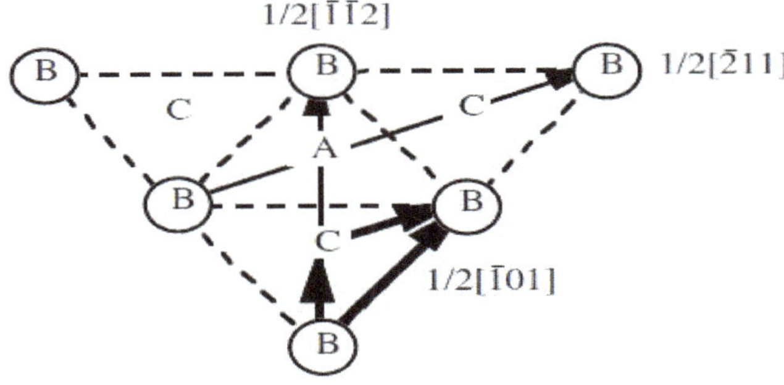

Figure: Configuration of atoms in the (111) plane showing the partial and total slip vectors

To find the Burgers' vectors of the partial dislocations let the close-packed plane be the (111) plane, as shown in the figure. The atom shift is to the shaded atom position, followed by a shift to the final atom position in the face center. The (111) plane of B-sites in the unit cell in figure is redrawn. As shown in this figure, the Burgers' vector of the total dislocation is $b = \frac{1}{2}[\bar{1}01]$. The first partial slip is from the B site at the corner of the unit cell to the nearest C-site in the plane. This slip is 1/3 of the vector connecting B-sites along the height of the triangular configuration of atoms in the plane, which is $\frac{1}{2}[\bar{1}\bar{1}2]$. Hence

$$b_1 = \frac{1}{6}[\bar{1}\bar{1}2]$$

The second partial slip is from the C-site to the final B-site position. As shown in the figure, this vector is parallel to the vector $\frac{1}{2}\{\bar{2}11\}$, and is 1/3 of its length. Hence

$$b_2 = \frac{1}{6}\{\bar{2}11\}$$

The vector sum of the two partial dislocation vectors is

$$b_1 + b_2 = \frac{1}{6}[\bar{1}\bar{1}2] + \frac{1}{6}[\bar{2}11] = \frac{1}{2}[\bar{1}01] = b$$

Splitting the total dislocation, b, into the partials, b_1 and b_2, not only facilitates slip, but also lowers the total energy. The energy per unit length of dislocation line is proportional to $|b|^2$.

$$|b|^2 > |b_1|^2 + |b_2|^2$$

Since the energy of a parallel pair of partial dislocations, b_1 and b_2, is less than that of the total dislocation, $b = b_1 + b_2$.

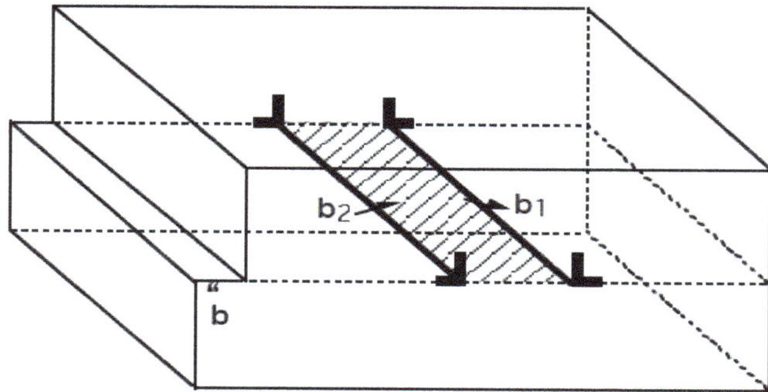

Figure: The total dislocation, b, divided into partial dislocations, b_1 and b_2 separated by a stacking fault.

However, to understand the configuration that appears when the dislocation is split we must recognize that equation $|b|^2 > |b_1|^2 + |b_2|^2$ gives only part of the energy of the pair of partials. Let the total dislocation divide, as illustrated in figure. In the area between the partials, which is shaded in the figure, the region above the slip plane has been slipped by the vector b1, but has not yet experiences the slip b_2. Let the slip plane lie between A and B-planes, as illustrated in figure. The partial slip b_1 carries the atoms in B-sites into C-sites, and causes corresponding changes $C \rightarrow A$

and $A \rightarrow B$ in successively higher planes. The FCC crystal has the stacking ..ABCABC.. But the stacking encountered in a direction perpendicular to the glide plane that passes through the shaded area between the partials is ..ABCA|CABC.., where the slash indicates the glide plane. The partial dislocation perturbs the sequence of close-packed planes; it introduces a planar defect called a stacking fault on the glide plane between the two partials. A stacking fault has a positive energy, σ_s, per unit area. The separation of the two partials, b_1 and b_2, creates a stacking fault whose area increases as the partials move apart. There is, therefore, an equilibrium separation between the partial dislocations. They separate to decrease the energy per length of dislocation, but are attracted by the need to minimize the stacking fault energy. The balance of these two effects determines the equilibrium separation, which is of the order of 5-500Å in typical FCC crystals.

Similar considerations apply to HCP and diamond cubic crystals. Total dislocations in the close-packed planes tend to divide into partials separated by ribbons of stacking fault. In BCC crystals the division of total dislocations into partials is much less pronounced. Separated partials cannot ordinarily be resolved in BCC, even though high resolution electron microscopy. Their presence is inferred from features of the mechanical behavior that are most easily explained by assuming that total dislocations dissociate slightly.

Planar Defect

A planar defect is distortion in a perfect crystal across the plane. The two main types of defects are Stacking & grain boundary.

Stacking Fault

A distortion in the long-range stacking sequence can produce two other types of crystal defects.

1. A stacking fault and
2. A twin region

A change in the stacking sequence over a few atomic planes produces a stacking fault. It arises with the interruption of one or two layer stacking sequence of atom plane while the changes over many atomic spacing give the twin region. It's specially found in closed packed structure like FCC and HCP. Both structures differ only in stacking order and have close packed atomic planes with six fold symmetry. The atoms form equilateral triangles. In both structures, the first two layers arrange themselves identically, and are said to have an AB arrangement. If the third layer is placed, so that, its atoms are directly above the first layer, then the stacking will be ABA which is a hcp arrangement as shown in the figure. So, the arrangements will be ABABAB type.

The another possible arrangement for first layer that the atoms are not situated just above the first layer and they are in the line with the first layer and arranged in ABC type which is a FCC Structure. So, HCP structure switches to FCC structure or ABABAB type changes to ABCABCABC type which shows the presence of stacking fault. In the FCC arrangement the pattern is ABCABCABC but due to stacking fault the pattern would become ABCABCAB_ABCABC.

If it continues over some number of atomic planes, it will produce a next stacking fault which is the twin of the first one. For above example if the stacking pattern is ABABABAB switches to ABCABCABC for a period of time before switching back to prior state, a pair of twin stacking faults is produced. The blue region in the stacking sequence shows the [ABCABCACBACBABCABC] twin plane.

Grain boundary - This is another type of planar defect which is found in poly crystals. Single crystal is found in specially controlled growth conditions. Solids are made of a number of small crystallizes which are also known as Grains. The size range of grains is up to nanometers to millimeters. The orientations of atomic planes rotate with respect to their neighboring grains. All the grains are separated by boundaries which are called Grain boundaries and the atoms in this region are not in perfect arrangement. In the crystallization process of solid, these boundaries give the uneven growth to the solid.

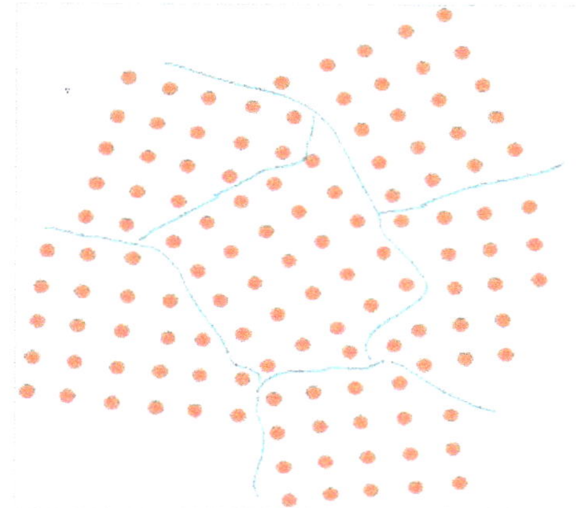

These boundaries are important as they develop the opaque property in the material and also affect the mechanical properties. They limit the lengths and the movement of dislocations. It means more grain boundary surface area or small grain give more strength than a larger grain. Grain's size can be controlled by temperature cooling. Rapid cooling gives the small grains while slow cooling produces larger grains. This is the reason for the low strength, hardness and ductility of large grains at room temperature.

Diffusion

Diffusion is the process by which atoms move in a material. Many reactions in solids and liquids are diffusion dependent. Structural control in a solid to achieve the optimum properties is also dependent on the rate of diffusion.

Atoms are able to move throughout solids because they are not stationary but execute rapid, small-amplitude vibrations about their equilibrium positions. Such vibrations increase with temperature and at any temperature a very small fraction of atoms has sufficient amplitude to move from one atomic position to an adjacent one. The fraction of atoms possessing this amplitude increases markedly with rising temperature. In jumping from one equilibrium position to another, an atom passes through a higher energy state since atomic bonds are distorted and broken, and the increase in energy is supplied by thermal vibrations. As might be expected defects, especially vacancies, are quite instrumental in affecting the diffusion process on the type and number of defects that are present, as well as the thermal vibrations of atoms. Diffusion can be defined as the mass flow process in which atoms change their positions relative to neighbors in a given phase under the influence of thermal and a gradient. The gradient can be a compositional gradient, an electric or magnetic gradient, or stress gradient.

Diffusion Mechanisms, Steady-state and Non-steady-state Diffusion

Diffusion Mechanisms

In pure metals self-diffusion occurs where there is no net mass transport, but atoms migrate in a random manner throughout the crystal. In alloys inter-diffusion takes place where the mass transport almost always occurs so as to minimize compositional differences. Various atomic mechanisms for self-diffusion and inter-diffusion have been proposed. Figure presents schematic view of different atomic diffusion mechanisms.

The most energetically favorable process involves an interchange of places by an atom and a neighboring vacancy – vacancy diffusion. This process demands not only the motion of vacancies, but also the presence of vacancies. The unit step in vacancy diffusion is an atom breaks its bonds and jumps into neighboring vacant site. In interstitial diffusion, solute atoms which are small enough to occupy interstitial sites diffuse by jumping from one interstitial site to another. The unit step here involves jump of the diffusing atom from one interstitial site to a neighboring site. Hydrogen, Carbon, Nitrogen and Oxygen diffuse interstitially in most metals, and the activation energy for diffusion is only that associated with motion since the number of occupied, adjacent interstitial sites usually is large. Substitutional diffusion generally proceeds by the vacancy mechanism. Thus interstitial diffusion is faster than substitutional diffusion by the vacancy mechanism. During self-diffusion or ring mechanism or direct-exchange mechanism, three or four atoms in the form of a ring move simultaneously round the ring, thereby interchanging their positions. This mechanism is untenable because exceptionally high activation energy would be required. A self-interstitial is more mobile than a vacancy as only small activation energy is required for self-interstitial atom to move to an equilibrium atomic position and simultaneously displace the neighboring atom into an interstitial site. However, the equilibrium number of self-interstitial atoms present at any temperature is negligible in comparison to the number of vacancies. This is because the energy to form a self-interstitial is extremely large.

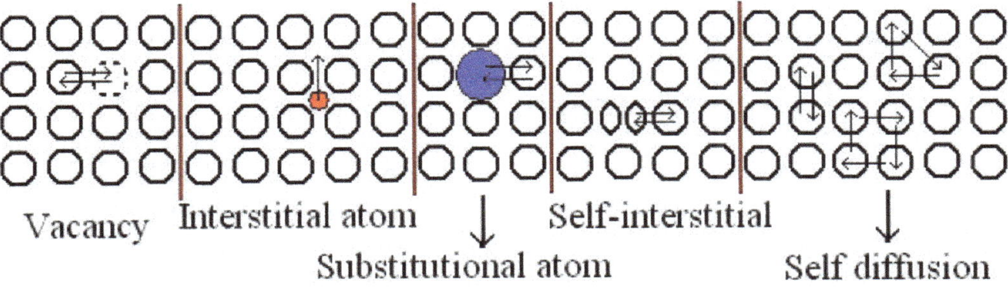

Figure: Diffusion mechanisms.

Diffusion in most ionic solids occurs by a vacancy mechanism. In ionic crystals, Schottky and Frankel defects assist the diffusion process. When Frenkel defects (pair of vacancyinterstial) dominate in an ionic crystal, the cation interstitial of the Frenkel defect carries the diffusion flux. If Schottky defects (pair of vacant sites) dominate, the cation vacancy carries the diffusion flux. In thermal equilibrium, in addition to above defects, ionic crystal may have defects generated by impurities and by deviation from stochiometry. Thus imperfections in ionic materials that influence diffusion arise in two ways:

(1) Intrinsic point defects such as Frenkel and schottky defects whose number depends on temperature, and

(2) Extrinsic point defects whose presence is due to impurity ions of different valance than the host ions.

The former is responsible for temperature dependence of diffusion similar to that for self-diffusion in metals, while the latter result in a temperature dependence of diffusion which is similar to that for interstitial solute diffusion in metals.

For example: Cd^{+2} cation in NaCl crystal will results in a cation vacancy. As Schottky defects form easily in NaCl crystal and thus cation vacancies carry the diffusion flux, even small fraction of Cd^{+2} increases the diffusivity of NaCl by several orders. Excess Zn^{+2} interstitials present in a non-stoichiometric ZnO compound increase the diffusivity of Zn^{+2} ions significantly. It is same with non-stoichiometric FeO.

In addition to diffusion through the bulk of a solid (volume diffusion), atoms may migrate along external or internal paths that afford lower energy barriers to motion. Thus diffusion can occur along dislocations, grain boundaries or external surfaces. The rates of diffusion along such short-circuit paths are significantly higher than for volume diffusion. However, most cases of mass transport are due to volume diffusion because the effective cross-sectional areas available for short-circuit processes are much smaller than those for volume diffusion.

There is a difference between diffusion and net diffusion. In a homogeneous material, atoms also diffuse but this motion is hard to detect. This is because atoms move randomly and there will be an equal number of atoms moving in one direction than in another. In inhomogeneous materials, the effect of diffusion is readily seen by a change in concentration with time. In this case there is a net diffusion. Net diffusion occurs because, although all atoms are moving randomly, there are more atoms moving away from regions where their concentration is higher.

Steady-state Diffusion

Diffusional processes can be either steady-state or non-steady-state. These two types of diffusion processes are distinguished by use of a parameter called flux. It is defined as net number of atoms crossing a unit area perpendicular to a given direction per unit time. For steady-state diffusion, flux is constant with time, whereas for non-steady-state diffusion, flux varies with time. A schematic view of concentration gradient with distance for both steady-state and non-steady-state diffusion processes are shown in figure.

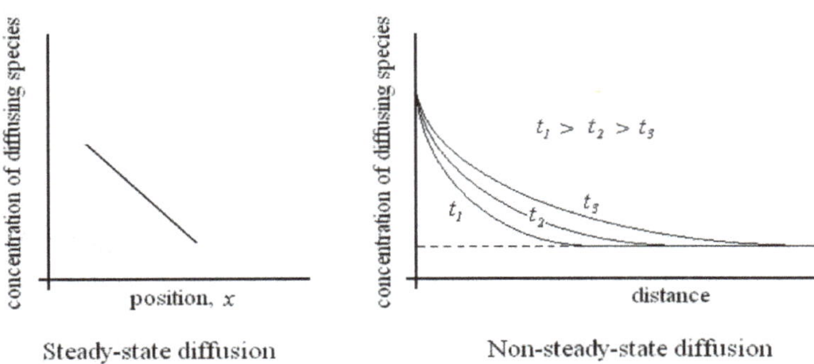

Figure: Steady-state and Non-steady-state diffusion processes.

Steady-state diffusion is described by Fick's first law which states that flux, *J*, is proportional to the concentration gradient. The constant of proportionality is called diffusion coefficient (diffusivity), *D* (cm² /sec). Diffusivity is characteristic of the system and depends on the nature of the diffusing species, the matrix in which it is diffusing, and the temperature at which diffusion occurs. Thus under steady-state flow, the flux is independent of time and remains the same at any cross-sectional plane along the diffusion direction. For the one-dimensional case, Fick's first law is given by

$$J_x = -D\frac{dc}{dx} = \frac{1}{A}\frac{dn}{dt}$$

And

$$J_x \neq f(x,t)$$

where D is the diffusion constant, dc/dx is the gradient of the concentration c, dn/dt is the number atoms crossing per unit time a cross-sectional plane of area A. The minus sign in the equation means that diffusion occurs down the concentration gradient. Although, the concentration gradient is often called the driving force for diffusion (but it is not a force in the mechanistic sense), it is more correct to consider the reduction in total free energy as the driving force.

An example of steady-state diffusion is provided by the permeation of hydrogen atoms through a sheet of palladium with different imposed hydrogen gas pressures on either side the slab. This process has been used to purify the hydrogen gas as other gases like nitrogen, oxygen and water vapor cannot diffuse through palladium.

Non-steady-state Diffusion

Most interesting cases of diffusion are non-steady-state processes since the concentration at a given position changes with time, and thus the flux changes with time. This is the case when the diffusion flux depends on time, which means that a type of atoms accumulates in a region or depleted from a region (which may cause them to accumulate in another region). Fick's second law characterizes these processes, which is expressed as:

$$\frac{dc}{dt} = -\frac{dJ}{dx} = \frac{d}{dx}\left(D\frac{dc}{dx}\right)$$

where dc/dt is the time rate of change of concentration at a particular position, x. If D is assumed to be a constant, then

$$\frac{dc}{dt} = D\frac{d^2c}{dx^2}$$

Solution to the above expression is possible when meaningful boundary conditions are specified. One common set of boundary conditions can be written as

For t = 0, $C = C_0 \ at \ 0 \leq x \leq \infty$

For t > 0, $C = C_s$ at x=0

$$C = C_0 \ at \ x = \infty$$

And the solution is

$$\frac{C_x - C_o}{C_s - C_o} = 1 - erf\left(\frac{x}{2\sqrt{Dt}}\right)$$

where C_x represents the concentration at depth x after time t. The term erf stands for Gaussian error function. Corresponding error function values for a variable are usually found from standard mathematical tables. The above equation demonstrates the relationship between concentration, position, and time. Thus the equation can be used to explain many practical industrial problems like corrosion resistance of duralumin, carburization and de-carburization of steel, doping of semi-conductors, etc.

Factors that Influence Diffusion and Non-equilibrium Transformation & Microstructure

Factors that Influence Diffusion

Ease of a diffusion process is characterized by the parameter D, diffusivity. The value of diffusivity for a particular system depends on many factors as many mechanisms could be operative.

- Diffusing species: If the diffusing species is able to occupy interstitial sites, then it can easily diffuse through the parent matrix. On the other hand if the size of substitutional species is almost equal to that of parent atomic size, substitutional diffusion would be easier. Thus size of diffusing species will have great influence on diffusivity of the system.
- Temperature: Temperature has a most profound influence on the diffusivity and diffusion rates. It is known that there is a barrier to diffusion created by neighboring atoms those need to move to let the diffusing atom pass. Thus, atomic vibrations created by temperature assist diffusion. Empirical analysis of the system resulted in an Arrhenius type of relationship between diffusivity and temperature.

$$D = D_o = \exp\left(-\frac{Q}{RT}\right)$$

where D_0 is a pre-exponential constant, Q is the activation energy for diffusion, R is gas constant (Boltzmann's constant) and T is absolute temperature. From the above equation it can be inferred that large activation energy means relatively small diffusion coefficient. It can also be observed that there exists a linear proportional relation between (lnD) and ($1/T$). Thus by plotting and considering the intercepts, values of Q and Do can be found experimentally.

- Lattice structure: Diffusion is faster in open lattices or in open directions than in closed directions.
- Presence of defects: defects like dislocations, grain boundaries act as short-circuit paths for diffusing species, where the activation energy is diffusion is less. Thus the presence of defects enhances the diffusivity of diffusing species.

Non-equilibrium Transformation & Microstructure

During the processing of metallic materials, they are subjected to different conditions and thus transformation of its structure. During casting process, liquid metal is allowed to cool to become solid component. However, during cooling conditions can be such that they are in equilibrium or non-equilibrium state. Phases and corresponding microstructures, usually shown in a phase diagram, are generated during equilibrium solidification under the conditions that are realized only for extremely slow cooling rates. This is because with change in temperature, there must be readjustments in the compositions of liquids and solid phases in accordance with the phase diagram. These readjustments are accomplished by diffusional processes in both solid and liquid phases and also across the solid-liquid interface. But, it is well understood that diffusion is time dependent phenomenon, and moreover diffusion in solid phases are much lower than in liquid phases, equilibrium solidification requires extremely longer times those are impractical. Thus virtually all practical solidification takes place under non-equilibrium conditions, leading to compositional gradients and formation of meta-stable phases.

As a consequence of compositional gradients during non-equilibrium cooling, segregation (concentration of particular, usually impurity elements, along places like grain boundaries) and coring (gradual compositional changes across individual grains) may occur. Coring is predominantly observed in alloys having a marked difference between liquidus and solidus temperatures. It is often being removed by subsequent annealing (incubation at relatively high temperatures that are close to lower solidus temperature, enhances diffusion in solids) and/or hot-working, and is exploited in zone refining technique to produce high-purity metals. Segregation is also put to good use in zone refining, and also in the production of rimming steel. Micro-segregation is used to describe the differences in composition across a crystal or between neighboring crystals. On the other hand, macro-segregation is used to describe more massive heterogeneities which may result from entrapment of liquid pockets between growing solidifying zones. Micro-segregation can often be removed by prolonged annealing or by hot-working; but macro-segregation persists through normal heating and working operations.

In most situations cooling rates for equilibrium solidification are impractically slow and unnecessary; in fact, on many occasions non-equilibrium conditions are desirable. Two non-equilibrium effects of practical importance are (1) the occurrence of phase changes or transformations at temperatures other than those predicted by phase boundary lines on the phase diagram, and (2) the existence of non-equilibrium phases at room temperature that do.

5

Metals

A metal is any material that is a good conductor of heat and electricity. It is malleable and ductile. Some important materials are iron, aluminium, copper, etc. which have been discussed in this chapter.

The term metal has been applied to a chemical element that has a shiny surface and is a good conductor of heat and electricity. These properties, however, can vary from one metal to the next. More recently, chemists have recognized that the main distinguishing features of a metal are:

(a) The ability of its atoms to lose some of their outermost electrons to form cations, and

(b) The bonding of its atoms by what are called metallic bonds.

Metals form one of three groups of elements—the other two being nonmetals and metalloids. These groups are distinguished by their ionization and bonding properties. On the periodic table, a diagonal line drawn from boron (B) to polonium (Po) separates the metals from the nonmetals. Elements on this line are metalloids, sometimes called semi-metals; elements to the lower left are metals; elements to the upper right are nonmetals. In nature, nonmetals are more abundant than metals, but most elements in the periodic table are metals. Some well-known metals are aluminum, calcium, copper, gold, iron, lead, magnesium, platinum, silver, titanium, uranium, and zinc.

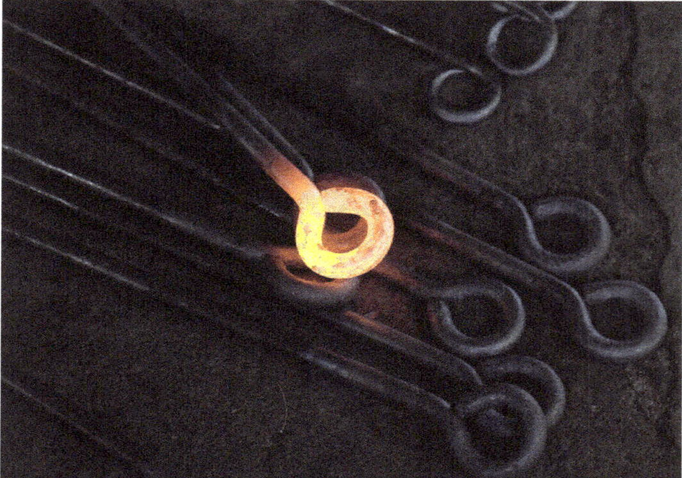

Hot metal work from a blacksmith

Metals and their alloys are extremely useful for both aesthetic and practical purposes. While some are used for jewelry and works of art, many serve as construction materials for buildings, bridges, transportation vehicles, and industrial machinery. Some, such as copper, are used as conductors in electrical wiring; others, such as platinum and palladium, are catalysts for chemical reactions. Tungsten is used in the filaments of incandescent light bulbs; uranium and plutonium are used in nuclear weapons and nuclear power plants. Moreover, metal ions play significant roles in biological processes, such as the functions of enzymes, the transport of oxygen, and the health of teeth and bones.

Properties

All metals, except mercury, are solids at room temperature. In addition, the colors of metals tend to range from silvery white to gray—the two exceptions are copper and gold. As mentioned above, metals tend to be lustrous (shiny) and good conductors of heat and electricity. They also conduct sound well. Furthermore, they are usually ductile and malleable—that is, they can be readily drawn into wires and beaten into sheets. Solid nonmetals, on the other hand, are generally brittle, lack luster, and are insulators.

Metals are usually thought of as being hard substances, high in density and melting point. It should be noted, however, that there are metals that are soft, low in density, and low in melting point, but they are also quite reactive, and we rarely encounter them in their elemental, metallic form. For example, lithium, sodium, and potassium are less dense than water and are also highly reactive. On the other hand, some of the densest metals are osmium, iridium, platinum, and gold. The melting point of cesium is only 28.4 °C, but that of tungsten is as high as 3,422 °C.

The property of electrical conductivity is mainly because each atom exerts only a loose hold on its outermost electrons, called the valence electrons). Thus, a pure metal may be thought of as a lattice of positively charged ions (cations) surrounded by a cloud of delocalized electrons.

Most metals are chemically unstable, reacting with oxygen in the air to form oxides. Depending on the metal, the time scale of the reaction can vary. The alkali metals (in Group 1 of the periodic table) react quickest, followed by the alkaline earth metals (in group 2). The transition metals—such as iron, copper, zinc, and nickel—take much longer to oxidize. For example, potassium burns in seconds, but iron rusts gradually over a long period of time, depending on the prevailing environmental conditions.

Other metals—such as palladium, platinum, and gold—do not react with the atmosphere at all. Some metals form a barrier layer of oxide on their surface, which cannot be penetrated by further oxygen molecules. They thus retain their shiny appearance and good conductivity for many decades. Examples are aluminum and titanium. The oxides of metals are basic (as opposed to those of nonmetals, which are acidic).

Metal cations combine with nonmetal anions to form salts. Some common classes of salts are carbonates, sulfides, sulfates, silicates, chlorides, nitrates, and phosphates. Many of the minerals found in nature are salts.

Base Metals

In alchemy, the term base metal was used as a designation for common and inexpensive metals, to draw a contrast with precious metals such as gold and silver. A long-cherished goal of the alchemists was the transmutation of base metals into precious metals.

In chemistry today, the term base metal is used informally to refer to a metal that oxidizes or corrodes relatively easily and reacts variably with dilute hydrochloric acid (HCl) to form hydrogen. Examples include iron, nickel, lead, and zinc. Copper, too, is considered a base metal because it oxidizes relatively easily, although it does not react with HCl. Metals that resist oxidation or corrosion are called noble metals, which also tend to be precious metals.

In the past, coins used to derive their value primarily from their precious metal content. Most modern currencies, however, are fiat currency—money that is given legal tender status. This approach allows coins to be made of base metals.

Precious Metals

A precious metal is a rare metallic chemical element of high, durable economic value. The best-known precious metals are gold and silver. Although both have industrial uses, they are better known for their uses in art, jewelry, and coinage. Other precious metals include the platinum group metals: ruthenium, rhodium, palladium, osmium, iridium, and platinum, of which platinum is the most widely traded. Plutonium and uranium may also be considered precious metals.

Chemically, the precious metals are less reactive than most elements. They have high luster and higher melting points than other metals. Historically, precious metals were important as currency, but are now regarded mainly as investment and industrial commodities. Investments in gold and silver are often regarded as a hedge against inflation and economic downturn.

A gold nugget

Bullion

Precious metals in bulk form are known as bullion and are traded on commodity markets. Bullion metals may be cast into ingots, or minted into coins. The defining attribute of bullion is that it is valued by its mass and purity rather than by a face value as money.

Many nations mint bullion coins, of which the most famous is probably the gold South African Krugerrand. Although nominally issued as legal tender, the face value of these coins as currency is far below that of their value as bullion. For instance, the United States mints a gold bullion coin (the American Gold Eagle) at a face value of $50 containing 1 troy ounce (31.1035 g) of gold. In January 2006, this coin was worth about $550 as bullion.

The minting of bullion coins by national governments gives them some numismatic value in addition to their bullion value, as well as certifying their purity. The level of purity varies from country to country. Some bullion coins, such as the Canadian Gold Maple Leaf, are available at a purity as high as 99.99%. Note that 100 percent pure bullion is not possible, because the absolute purity of extracted and refined metals can only be approached asymptotically.

One of the largest bullion coins in the world is a 10,000 Australian dollar coin that consists of a full kilogram (kg) of 99.9 percent pure gold. China, however, has produced limited quantities of coins (less than 20 pieces) that exceed 260 troy ounces (8 kg) of gold.

Silver bullion coins have become popular with coin collectors because of their relative affordability. Unlike most gold and platinum issues, which are valued based on prevailing markets, silver issues are more often valued as collectibles—far higher than their actual bullion value.

A 500-gram silver bullion bar

An American Platinum Eagle bullion coin

Precious Metal Status

A given metal is precious if it is rare. If mining or refining processes improve, or new supplies are discovered and exploited, the value of such a metal declines.

An interesting case of a precious metal going common is that of aluminum. It is one of the Earth's most common metals, but when first discovered, it was extremely difficult to separate from its ores. For a while, aluminum was regarded as more valuable than gold, and aluminum bars were exhibited alongside the French crown jewels at the Exposition Universelle (1855). Napoleon Bonaparte is said to have used aluminum plates for his most honored guests at dinner. Its price, however, began to drop and collapsed altogether when an easy extraction method, the Hall-Héroult process, was discovered in 1886.

Gold bars from the Bank of Sweden

Alloys

An alloy is a combination of two or more chemical elements, at least one of which is a metal, and where the resulting material has metallic properties. Examples of alloys are steel (iron and carbon), brass (copper and zinc), bronze (copper and tin), and duralumin (aluminum and copper). The resulting metallic substance generally has properties significantly different from those of its components.

An alloy with two components is called a binary alloy; one with three is a ternary alloy; one with four is a quaternary alloy. Alloys specially designed for highly demanding applications, such as jet engines, may contain more than ten elements. When an alloy contains mercury, it is called an amalgam.

An alloy "inherits" the characteristics of the elements it was made from, but it is usually designed to have properties more desirable than those of its components. For instance, steel is stronger than iron, one of its main components.

Unlike pure metals, most alloys do not have a sharp melting point. Instead, they have a melting range in which the material becomes a mixture of solid and liquid phases. The temperature at which melting begins is called the solidus, and that at which melting is complete is called the liquidus. For most pairs of elements, however, there is a particular ratio that has a single melting point, and this is called a eutectic mixture.

In practice, some alloys are named after their primary constituent. For example, 14 carat (58 percent) gold is an alloy of gold with other elements. Similarly, the silver used in jewelry and the aluminum used as a structural material are alloys.

Astronomy

In the specialized usage of astronomy and astrophysics, the term "metal" is often used in referring to any element other than hydrogen or helium—it includes substances as chemically nonmetallic as neon, fluorine, and oxygen. Given that the temperatures of stellar bodies allow practically no solid or liquid matter, and little normal chemistry can exist at temperatures that break down almost all chemical bonds, the term "metal" refers to materials produced by a type of nuclear reaction called the triple-helium process.

Uses

Based on their properties, metals have a wide range of uses. For example, based on their strength and hardness, many metals and their alloys are used as structural materials for buildings, bridges, industrial machines, railroad tracks, automobiles, trains, ships, aircraft, spacecraft, and so forth. A number of metals, such as platinum and palladium, are good catalysts for chemical reactions. Precious metals, particularly gold and silver, are used in jewelry and decorative items. Amalgams are used in dental fillings. Copper, an abundant metal, is an extremely good conductor of electricity and is widely used for electrical wiring. Tungsten, with its high melting point, is suitable for use in the filaments of incandescent light bulbs. Uranium and plutonium are radioactive materials used for nuclear weapons and nuclear power plants that generate electricity. In biological systems, metal ions play a number of important roles, such as the functions of enzymes, the transport of oxygen by hemoglobin (which contains iron), and the health of bones and teeth (which need calcium ions).

Steel

Steel is an alloy of iron and carbon in which the carbon content ranges up to 2 percent (with a higher carbon content, the material is defined as cast iron). By far the most widely used material for building the world's infrastructure and industries, it is used to fabricate everything from sewing needles to oil tankers. In addition, the tools required to build and manufacture such articles are also made of steel. As an indication of the relative importance of this material, in 2013 the world's raw steel production was about 1.6 billion tons, while production of the next most important engineering metal, aluminum, was about 47 million tons. The main reasons for the popularity of steel are the relatively low cost of making, forming, and processing it, the abundance of its two raw materials (iron ore and scrap), and its unparalleled range of mechanical properties.

Properties Of Steel

The Base Metal: Iron

The major component of steel is iron, a metal that in its pure state is not much harder than copper. Omitting very extreme cases, iron in its solid state is, like all other metals, polycrystalline—that is, it consists of many crystals that join one another on their boundaries. A crystal is a well-ordered arrangement of atoms that can best be pictured as spheres touching one another. They are ordered in planes, called lattices, which penetrate one another in specific ways. For iron, the lattice arrangement can best be visualized by a unit cube with eight iron atoms at its corners. Important for the uniqueness of steel is the allotropy of iron—that is, its existence in two crystalline forms. In the body-centered cubic (bcc) arrangement, there is an additional iron atom in the center of each cube. In the face-centered cubic (fcc) arrangement, there is one additional iron atom at the center of each of the six faces of the unit cube. It is significant that the sides of the face-centered cube, or the distances between neighboring lattices in the fcc arrangement, are about 25 percent larger than in the bcc arrangement; this means that there is more space in the fcc than in the bcc structure to keep foreign (i.e., alloying) atoms in solid solution.

Iron has its bcc allotropy below 912 °C (1,674 °F) and from 1,394 °C (2,541 °F) up to its melting point of 1,538 °C (2,800 °F). Referred to as ferrite, iron in its bcc formation is also called alpha iron in the lower temperature range and delta iron in the higher temperature zone. Between 912° and 1,394 °C iron is in its fcc order, which is called austenite or gamma iron. The allotropic behaviour of iron is retained with few exceptions in steel, even when the alloy contains considerable amounts of other elements.

There is also the term beta iron, which refers not to mechanical properties but rather to the strong magnetic characteristics of iron. Below 770 °C (1,420 °F), iron is ferromagnetic; the temperature above which it loses this property is often called the Curie point.

Effects of Carbon

In its pure form, iron is soft and generally not useful as an engineering material; the principal method of strengthening it and converting it into steel is by adding small amounts of carbon. In solid steel, carbon is generally found in two forms. Either it is in solid solution in austenite and ferrite or it is found as a carbide. The carbide form can be iron carbide (Fe3C, known as cementite), or it can be a carbide of an alloying element such as titanium. (On the other hand, in gray iron, carbon appears as flakes or clusters of graphite, owing to the presence of silicon, which suppresses carbide formation.)

The effects of carbon are best illustrated by an iron-carbon equilibrium diagram. The A-B-C line represents the liquidus points (i.e., the temperatures at which molten iron begins to solidify), and the H-J-E-C line represents the solidus points (at which solidification is completed). The A-B-C line indicates that solidification temperatures decrease as the carbon content of an iron melt is increased. (This explains why gray iron, which contains more than 2 percent carbon, is processed at much lower temperatures than steel.) Molten steel containing, for example, a carbon content of 0.77 percent (shown by the vertical dashed line in the figure) begins to solidify at about 1,475 °C (2,660 °F) and is completely solid at about 1,400 °C (2,550 °F). From this point down, the iron crystals are all in an austenitic—i.e., fcc—arrangement and contain all of the carbon in solid solution. Cooling further, a dramatic change takes place at about 727 °C (1,341 °F) when the austenite crystals transform into a fine lamellar structure consisting of alternating platelets of ferrite and iron carbide. This microstructure is called pearlite, and the change is called the eutectoidic transformation. Pearlite has a diamond pyramid hardness (DPH) of approximately 200 kilograms-force per square millimeter (285,000 pounds per square inch), compared with a DPH of 70 kilograms-force per square millimeter for pure iron. Cooling steel with a lower carbon content (e.g., 0.25 percent) results in a microstructure containing about 50 percent pearlite and 50 percent ferrite; this is softer than pearlite, with a DPH of about 130. Steel with more than 0.77 percent carbon—for instance, 1.05 percent—contains in its microstructure pearlite and cementite; it is harder than pearlite and may have a DPH of 250.

The effects of carbon are best illustrated by an iron-carbon equilibrium diagram. The A-B-C line represents the liquidus points (i.e., the temperatures at which molten iron begins to solidify), and the H-J-E-C line represents the solidus points (at which solidification is completed). The A-B-C line indicates that solidification temperatures decrease as the carbon content of an iron melt is increased. (This explains why gray iron, which contains more than 2 percent carbon, is processed at much lower temperatures than steel.) Molten steel containing, for example, a carbon content of 0.77 percent (shown by the vertical dashed line in the figure) begins to solidify at about 1,475° C (2,660° F) and is completely solid at about 1,400 °C (2,550 °F). From this point down, the iron crystals are all in an austenitic—i.e., fcc—arrangement and contain all of the carbon in solid solution. Cooling further, a dramatic change takes place at about 727 °C (1,341 °F) when the austenite crystals transform into a fine lamellar structure consisting of alternating platelets of ferrite and iron carbide. This microstructure is called pearlite, and the change is called the eutectoidic transformation. Pearlite has a diamond pyramid hardness (DPH) of approximately 200 kilograms-force per square millimeter (285,000 pounds per square inch), compared with a DPH of 70 kilograms-force per square millimeter for pure iron. Cooling steel with a lower carbon content (e.g., 0.25 percent) results in a microstructure containing about 50 percent pearlite and 50 percent ferrite; this is softer than pearlite, with a DPH of about 130. Steel with more than 0.77 percent carbon—for instance, 1.05 percent—contains in its microstructure pearlite and cementite; it is harder than pearlite and may have a DPH of 250.

Effects of Heat-treating

Adjusting the carbon content is the simplest way to change the mechanical properties of steel. Additional changes are made possible by heat-treating—for instance, by accelerating the rate of cooling through the austenite-to-ferrite transformation point, shown by the P-S-K line in the figure. (This transformation is also called the Ar_1 transformation, r standing for refroidissement, or "cooling.") Increasing the cooling rate of pearlitic steel (0.77 percent carbon) to about 200 °C per minute generates a DPH of about 300, and cooling at 400 °C per minute raises the DPH to about 400. The reason for this increasing hardness is the formation of a finer pearlite and ferrite microstructure than can be obtained during slow cooling in ambient air. In principle, when steel cools quickly, there is less time for carbon atoms to move through the lattices and form larger carbides. Cooling even faster—for instance, by quenching the steel at about 1,000° C per minute—results in a complete depression of carbide formation and forces the undercooled ferrite to hold a large amount of carbon atoms in solution for which it actually has no room. This generates a new microstructure, martensite. The DPH of martensite is about 1,000; it is the hardest and most brittle form of steel. Tempering martensitic steel—i.e., raising its temperature to a point such as 400 °C and holding it for a time—decreases the hardness and brittleness and produces a strong and tough steel. Quench-and-temper heat treatments are applied at many different cooling rates, holding times, and temperatures; they constitute a very important means of controlling steel's properties.

Effects of Alloying

A third way to change the properties of steel is by adding alloying elements other than carbon that produce characteristics not achievable in plain carbon steel. Each of the approximately 20 elements used for alloying steel has a distinct influence on microstructure and on the temperature, holding time, and cooling rates at which microstructures change. They alter the transformation points between ferrite and austenite, modify solution and diffusion rates, and compete with other elements in forming intermetallic compounds such as carbides and nitrides. There is a huge amount of empirical information on how alloying affects heat-treatment conditions, microstructures, and properties. In addition, there is a good theoretical understanding of principles, which, with the help of computers, enables engineers to predict the microstructures and properties of steel when alloying, hot-rolling, heat-treating, and cold-forming in any way.

A good example of the effects of alloying is the making of a high-strength steel with good weldability. This cannot be done by using only carbon as a strengthener, because carbon creates brittle zones around the weld, but it can be done by keeping carbon low and adding small amounts of other strengthening elements, such as nickel or manganese. In principle, the strengthening of metals is accomplished by increasing the resistance of lattice structures to the motion of dislocations. Dislocations are failures in the lattices of crystals that make it possible for metals to be formed. When elements such as nickel are kept in solid solution in ferrite, their atoms become embedded in the iron lattices and block the movements of dislocations. This phenomenon is called solution hardening. An even greater increase in strength is achieved by precipitation hardening, in which certain elements (e.g., titanium, niobium, and vanadium) do not stay in solid solution in ferrite during the cooling of steel but instead form finely dispersed, extremely small carbide or nitride crystals, which also effectively restrict the flow of dislocations. In addition, most of these strong carbide or nitride formers generate a small grain size, because their precipitates have a nucleation effect and slow down crystal growth during recrystallization of the cooling metal. Producing a small grain size is another method of strengthening steel, since grain boundaries also restrain the flow of dislocations.

Alloying elements have a strong influence on heat-treating, because they tend to slow the diffusion of atoms through the iron lattices and thereby delay the allotropic transformations. This means, for example, that the extremely hard martensite, which is normally produced by fast quenching, can be produced at lower cooling rates. This results in less internal stress and, most important, a deeper hardened zone in the work piece. Improved hardenability is achieved by adding such elements as manganese, molybdenum, chromium, nickel, and boron. These alloying agents also permit tempering at higher temperatures, which generates better ductility at the same hardness and strength.

Types of Steel

Steel can be categorized into four basic groups based on the chemical compositions:

1. Carbon Steel
2. Alloy Steel
3. Stainless Steel
4. Tool Steel

There are many different grades of steel that encompass varied properties. These properties can be physical, chemical and environmental.

All steel is composed of iron and carbon. It is the amount of carbon, and the additional alloys that determine the properties of each grade.

Carbon Steel

Carbon Steel can be segregated into three main categories: Low carbon steel (sometimes known as mild steel); Medium carbon steel; and High carbon steel:

- Low Carbon Steel (Mild Steel): Typically contain 0.04% to 0.30% carbon content. This is one of the largest groups of Carbon Steel. It covers a great diversity of shapes; from Flat Sheet to Structural Beam. Depending on the desired properties needed, other elements are added or increased. For example: Drawing Quality (DQ) – The carbon level is kept low and Aluminum is added, and for Structural Steel the carbon level is higher and the manganese content is increased.
- Medium Carbon Steel: Typically has a carbon range of 0.31% to 0.60%, and a manganese content ranging from .060% to 1.65%. This product is stronger than low carbon steel, and it is more difficult to form, weld and cut. Medium carbon steels are quite often hardened and tempered using heat treatment.

- High Carbon Steel: Commonly known as "carbon tool steel" it typically has a carbon range between 0.61% and 1.50%. High carbon steel is very difficult to cut, bend and weld. Once heat treated it becomes extremely hard and brittle.

Alloy Steel

Alloy steel is steel that has had small amounts of one or more alloying elements (other than carbon) such as such as manganese, silicon, nickel, titanium, copper, chromium and aluminum added. This produces specific properties that are not found in regular carbon steel. Alloy steels are workhorses of industry because of their economical cost, wide availability, ease of processing, and good mechanical properties. Alloy steels are generally more responsive to heat and mechanical treatments than carbon steels.

The heat-treated type is available in both Annealed and Normalized.

Stainless Steel

Stainless steel is a steel alloy with increased corrosion resistance compared to carbon/alloy steel. Common alloying ingredients include chromium (usually at least 11%), nickel, or molybdenum. Alloy content often is on the order of 15-30%.

Common applications include food handling/processing, medical instruments, hardware, appliances, and structural/architectural uses.

Tool Steel

Tool steel is a term used for a variety of high-hardness, abrasion resistant steels. Specific tool applications are dies (stamping or extrusion), cutting, mold making, or impact applications like hammers (personal or industrial). It is also a common material used to make knives.

Tool Steels are extremely hard and are quite often used to form other metal products.

Tool Steel is available in a wide variety of shapes including round bar, flat bar, square bar and more.

Iron

Iron is a chemical element with Fe as its symbol. It belongs to group 8, periodic number 4 of the periodic table. Its atomic number is 26.

Iron makes up 5% of the Earth's crust and is one of the most abundantly available metals. It is primarily obtained from the minerals hematite and magnetite. It can also be obtained from taconite, limonite, and siderite. Iron is the most used of all the metals.

Iron is also found in meat, potatoes and vegetables and is essential for animals and humans. It is an essential part of hemoglobin.

Iron metal is greyish in appearance, and is very ductile and malleable. It begins to rust in damp air and at elevated temperatures, but not in dry air. It dissolves readily in dilute acids, and is chemically active.

Characteristics

Pure iron is a silver-colored metal that conducts heat and electricity well. Iron is too reactive to exist alone so it only occurs naturally in the earth's crust as iron ores, such as hematite, magnetite, and siderite.

One of iron's identifying characteristics is that it is strongly magnetic. Exposed to a strong magnetic field, any piece of iron can be magnetized. Scientists believe that the earth's core is made up of about 90% iron. The magnetic force produced by this iron is what creates the magnetic North and South poles.

Rust

Iron's most troublesome characteristic is its tendency to form rust. Rust (or ferric oxide) is a brown, crumbly compound that is produced when iron is exposed to oxygen. The oxygen gas that is contained in water speeds up the process of corrosion. The rate of rust - how quickly iron turns into ferric oxide - is determined by the oxygen content of the water and the surface area of the iron. Salt water contains more oxygen than fresh water, which is why salt water rusts iron faster than fresh water.

Rust can be prevented by coating iron with other metals that are chemically more attractive to oxygen, such as zinc (the process of coating iron with zinc is referred to as 'galvanizing'). However, the most effective method of protecting against rust is the use of steel.

Production

Most iron is produced from ores found near the earth's surface. Modern extraction techniques use blast furnaces, which are characterized by their tall stacks (chimney-like structures). The iron is poured into the stacks along with coke (carbon-rich coal) and limestone (calcium carbonate). Nowadays, the iron ore normally goes through a process of sintering before entering the stack. This process forms pieces of ore that are between 10-25mm, which are then mixed with coke and limestone.

The sintered ore, coke and limestone are then poured into the stack where it burns at temperatures of 1800°C. Coke burns as a source of heat and, along with oxygen that is shot into the furnace, helps to form the reducing gas carbon monoxide. The limestone mixes with impurities in the iron to form slag. Slag is lighter than molten iron ore, so it rises to the surface and can easily be removed. The hot iron is then poured into molds to produce pig iron or directly prepared for steel production.

Pig iron still contains between 3.5-4.5% carbon, along with other impurities, and is brittle and difficult to work with. Various processes are used in order to lower the phosphorus and sulfur impurities in pig iron in order to produce cast iron. Wrought iron, which contains less than 0.25% carbon, is tough, malleable and easily welded, but is much more laborious and costly to produce than low carbon steel.

In 2010, global iron ore production was around 2.4 billion tones. China, the largest producer, accounted for about 37.5% of all production, while other major producing countries include Australia, Brazil, India, and Russia.

Applications

Iron was once the primary structural material, but it has long been replaced by steel in most applications. Nevertheless, cast iron is still used in pipes and to make automotive parts, such as cylinder heads, cylinder blocks and gearbox cases. Wrought iron is still used to produce home décor items, such as wine racks, candle holders, and curtain rods.

Aluminium

Aluminium is a silvery-white metal, the 13 element in the periodic table. One surprising fact about aluminium is that it's the most widespread metal on Earth, making up more than 8% of the Earth's core mass. It's also the third most common chemical element on our planet after oxygen and silicon.

At the same time, because it easily binds with other elements, pure aluminium does not occur in nature. This is the reason that people learned about it relatively recently. Formally aluminium was produced for the first time in 1824 and it took people another fifty years to learn to produce it on an industrial scale.

The most common form of aluminium found in nature is aluminium sulphates. These are minerals that combine two sulphuric acids: one based on an alkaline metal (lithium, sodium, potassium rubidium or caesium) and one based on a metal from the third group of the periodic table, primarily aluminium.

Aluminium sulphates are used to this day to clean water, for cooking, in medicine, in cosmetology, in the chemical industry and in other sectors. By the way, aluminium got its name from aluminium sulphates which in Latin were called alumen.

Properties of Aluminium

Aluminium offers a rare combination of valuable properties. It is one of the lightest metals in the world: it's almost three times lighter than iron but it's also very strong, extremely flexible and corrosion resistant because its surface is always covered in an extremely thin and yet very strong layer of oxide film. It doesn't magnetise, it's a great electricity conductor and forms alloys with practically all other metals.

Aluminium can be easily processed using pressure both when it's hot and when it's cold. It can be rolled, pulled and stamped. Aluminium doesn't catch fire, it doesn't need special paint and unlike plastics it's not toxic. It's also very pliable so sheets just 4 microns thick can be made from it, as well as extra thin wire. The extra-thin foil that can be made from aluminium is three times thinner than a human hair. In addition, aluminium is more cost effective than other metals and materials.

Since aluminium easily forms compounds with other chemical elements, a huge variety of aluminium alloys have been developed. Even a very small amount of admixtures can drastically change the properties of the metal, making it possible to use it in new areas. For example, in ordinary life you can find aluminium mixed with silicon and magnesium literally on the road, i.e. in the aluminium alloy wheels, in the engines, chassis and other parts of modern automobiles. As for aluminium zinc alloy, chances are you might be holding it in your hands right now as it's this alloy that's widely used in the production of mobile phones and tablet PCs. In the meantime, scientists keep developing new aluminium alloys.

The modern construction, automotive, aviation, energy, food and other industries would be impossible without aluminium. In addition, aluminium has become a symbol of progress: all cutting edge devices and vehicles are made from aluminium.

Copper

Copper is a transition metal element with the chemical symbol Cu and atomic number 29. It is a reddish-orange malleable metal with high thermal and electrical conductivity.

Metallic copper is well known to undergo corrosion in air and ultimately produce a characteristic green color when converted to a copper salt such as copper(II) carbonate, basic copper chloride or copper(II) acetate.

Copper is commonly used to produce wires, heat sinks, electromagnets and electric motors due to its excellent electrical and thermal conductivity, but it is also used as structural components and piping. Because the air oxidation of copper produces protection against further corrosion, copper can have advantages over other metals in these applications.

Biological processes are also highly dependent on copper atoms to facilitate important electron transfer and oxygen transfer reactions.

Copper from the Earth is comprised of 69.15% of the ^{63}Cu isotope and 30.85% of the ^{65}Cu isotope, with 34 and 36 neutrons respectively. Its appearance is characterized both by its unoxidized shiny reddish-orange coloring and by the oxidized and modified greenish coloring. Copper is prone to oxidation and is commonly found in the Cu^{+} and Cu^{2+} oxidation states. The corresponding copper salts are often colorful.

Physical Properties

- Strength: Copper is a weak metal with a tensile strength about half that of mild carbon steel. This explains why copper is easily formed by hand but is not a good choice for structural applications.
- Toughness: Copper may not be strong, but it is not easy to break due to its high toughness. This property comes in handy for piping and tube applications, where a rupture can be dangerous and expensive.
- Ductility: Copper is very ductile and also very malleable. The electrical and jewelry industries benefit from the ductility of copper.
- Conductivity: Second only to silver, copper is not only an excellent conductor of electricity but also of heat. As a result, copper serves well in applications such as cookware, where it quickly draws heat to the food inside.

References

- Steel, technology: britannica.com, Retrieved 11 June 2018
- Types-of-steel: metalsupermarkets.com, Retrieved 10 July 2018
- Metal, entry: newworldencyclopedia.org, Retrieved 20 April 2018
- Metal-profile-iron-2340139: thebalance.com, Retrieved 30 June 2018
- What-is-aluminum: aluminiumleader.com, Retrieved 16 May 2018
- What-is-copper-2340037: thebalance.com, Retrieved 25 April 2018

6

Polymer Structure and Properties

A polymer is a large molecule that is composed of repeating subunits. The study of the structure of a polymer is important for an understanding of polymers. This chapter closely examines the structures of polymers through the study of the molecular structure, polymer crystal, polymer classification based on its structure, etc.

Polymer

Polymers are modern materials produced in megaton quantities. The raw materials for their synthesis or biosynthesis are low-molecular-weight substances. In the synthesis, hundreds to many thousands of small monomer molecules undergo a polymerization reaction resulting in large molecules, termed macromolecules, mostly of chain-like structure. Polymers are substances consisting of macromolecules and the high molecular weight and chain-like structure of the macromolecules are responsible for the unique properties of polymers. The spectrum of applications of polymers is immense. For each application, the requirements for properties are specific and there are a number of ways to meet the requirements, such as mixing polymers with various types of additives, blending two or more polymers, combining polymers with particulate or fibrous materials, developing a new type of monomer, polymerizing two or more monomers together, or modifying chemically existing polymers. Mechanical properties of polymers can be varied over very broad ranges from hard to soft and from brittle to tough. To fabricate polymers into useful articles, the most frequently used techniques are molding, blowing, calendering, casting, extrusion, foaming, spinning of fibers, etc. Polymer waste is a serious burden for the environment because common organisms existing in nature are incapable of metabolizing them. Specialty polymers are materials having useful properties which cannot be found with any other material, produced in small quantities and applied in cases when the price is not too important. Main fields of application of those polymers are special electronics, space technology and medicine.

Classification of Polymers

Since Polymers are numerous in number with different behaviours and can be naturally found or synthetically created, they can be classified in various ways. The following below are some basic ways in which we classify polymers:

Classification Based on Source

The first classification of polymers is based on their source of origin.

- Natural polymers

The easiest way to classify polymers is their source of origin. Natural polymers are polymers which occur in nature and are existing in natural sources like plants and animals. Some common examples are Proteins (which are found in humans and animals alike), Cellulose and Starch (which are found in plants) or Rubber (which we harvest from the latex of a tropical plant).

- Synthetic polymers

Synthetic polymers are polymers which humans can artificially create/synthesize in a lab. These are commercially produced by industries for human necessities. Some commonly produced polymers which we use day to day are Polyethylene (a mass-produced plastic which we use in packaging) or Nylon Fibers (commonly used in our clothes, fishing nets etc.)

- Semi-Synthetic polymers

Semi-Synthetic polymers are polymers obtained by making modification in natural polymers artificially in a lab. These polymers formed by chemical reaction (in a controlled environment) and are of commercial importance. Example: Vulcanized Rubber (Sulphur is used in cross bonding the polymer chains found in natural rubber) Cellulose acetate (rayon) etc.

Classification Based on Structure of Polymers

Classification of polymers based on their structure can be of three types:

- Linear polymers

These polymers are similar in structure to a long straight chain which identical links connected to each other. The monomers in these are linked together to form a long chain. These polymers have high melting points and are of higher density. A common example of this is PVC (Poly-vinyl chloride). This polymer is largely used for making electric cables and pipes.

- Branch chain polymers

As the title describes, the structure of these polymers is like branches originating at random points from a single linear chain. Monomers join together to form a long straight chain with some branched chains of different lengths. As a result of these branches, the polymers are not closely packed together. They are of low density having low melting points. Low-density polyethene (LDPE) used in plastic bags and general purpose containers is a common example.

- Crosslinked or Network polymers

In this type of polymers, monomers are linked together to form a three-dimensional network. The monomers contain strong covalent bonds as they are composed of bi-functional and tri-functional in nature. These polymers are brittle and hard. Ex:- Bakelite (used in electrical insulators), Melamine etc.

Based on Mode of Polymerization

Polymerization is the process by which monomer molecules are reacted together in a chemical reaction to form a polymer chain (or three-dimensional networks). Based on the type of polymerization, polymers can be classified as:

- Addition polymers

These type of polymers are formed by the repeated addition of monomer molecules. The polymer is formed by polymerization of monomers with double or triple bonds (unsaturated compounds). Note, in this process, there is no elimination of small molecules like water or alcohol etc (no by-product of the process). Addition polymers always have their empirical formulas same as their monomers. Example: ethene n(CH_2=CH_2) to polyethene -(CH_2-CH_2)n^-.

- Condensation polymers

These polymers are formed by the combination of monomers, with the elimination of small molecules like water, alcohol etc. The monomers in these types of condensation reactions are bi-functional or tri-functional in nature. A common example is the polymerization of Hexamethylenediamine and adipic acid to give Nylon – 66, where molecules of water are eliminated in the process.

Classification Based on Molecular Forces

Intramolecular forces are the forces that hold atoms together within a molecule. In Polymers, strong covalent bonds join atoms to each other in individual polymer molecules. Intermolecular forces (between the molecules) attract polymer molecules towards each other.

Note that the properties exhibited by solid materials like polymers depend largely on the strength of the forces between these molecules. Using this, Polymers can be classified into 4 types:

- Elastomers

Elastomers are rubber-like solid polymers, that are elastic in nature. When we say elastic, we basically mean that the polymer can be easily stretched by applying a little force.

The most common example of this can be seen in rubber bands(or hair bands). Applying a little stress elongates the band. The polymer chains are held by the weakest intermolecular forces, hence allowing the polymer to be stretched. But as you notice removing that stress also results in the rubber band taking up its original form. This happens as we introduce crosslinks between the polymer chains which help it in retracting to its original position, and taking its original form. Our car tyres are made of Vulcanized rubber. This is when we introduce sulphur to cross bond the polymer chains.

- Thermoplastics

Thermoplastic polymers are long-chain polymers in which inter-molecules forces (Van der Waal's forces) hold the polymer chains together. These polymers when heated are softened (thick fluid like) and hardened when they are allowed to cool down, forming a hard mass. They do not contain any cross bond and can easily be shaped by heating and using molds. A common example is Polystyrene or PVC (which is used in making pipes).

- Thermosetting

Thermosetting plastics are polymers which are semi-fluid in nature with low molecular masses. When heated, they start cross-linking between polymer chairs, hence becoming hard and infusible.

They form a three-dimensional structure on the application of heat. This reaction is irreversible in nature. The most common example of a thermosetting polymer is that of Bakelite, which is used in making electrical insulation.

- Fibres

In the classification of polymers, these are a class of polymers which are a thread like in nature, and can easily be woven. They have strong inter-molecules forces between the chains giving them less elasticity and high tensile strength. The intermolecular forces may be hydrogen bonds or dipole-dipole interaction. Fibres have sharp and high melting points. A common example is that of Nylon-66, which is used in carpets and apparels.

Physical Properties

Physical properties of polymers include molecular weight, molar volume, density, degree of polymerization, crystallinity of material, and so on. Some of these are discussed herewith in the following sections.

Degree of Polymerization and Molecular Weight

The degree of polymerization (DP)-n in a polymer molecule is defined as the number of repeating units in the polymer chain. For example,

$$-(-CH - CH -)-_n$$

The molecular weight of a polymer molecule is the product of the degree of polymerization and the molecular weight of the repeating unit. The polymer molecules are not identical but are a mixture of many species with different degrees of polymerization, that is, with different molecular weights. Therefore, in the case of polymers we talk about the average values of molecular weights.

Molecular Weight Averages

Suppose we have a set of values $\{x_1, x_2, \dots, x_n\}$ and the corresponding probability of occurrence is given by $\{P_1, P_2, \dots, P_n\}$, then the average value is defined as follows:

$$\sum_{i=0}^{\infty} Pixi$$

Number-Average Molecular Weight

If N_i is the number of polymer molecules having the molecular weight M_i, then the "number-average" probability of the given mass is given below:

$$P_i = \frac{N_i}{\sum_{j=0}^{\infty} N_j}$$

The number-average molecular weight is given by:

$$M_n = \sum_{i=0}^{\infty}\left[\frac{N_i}{\sum_{j=0}^{\infty} N_j}\right] M_i = \frac{\sum_{i=0}^{\infty} M_i N_i}{\sum_{j=0}^{\infty} N_j}$$

The physical properties (such as transition temperature, viscosity, etc.) and mechanical properties (such as strength, stiffness, and toughness) depend on molecular weight of polymer. The lower the molecular weight, lower the transition temperature, viscosity, and the mechanical properties. Due to increased entanglement of chains with increased molecular weight, the polymer gets higher viscosity in molten state, which makes the processing of polymer difficult.

Weight-Average Molecular Weight

The weight-average probability is given by:

$$P_i = \frac{N_i M_i}{\sum_{j=0}^{\infty} N_j M_j}$$

The weight- average molecular weight is given by:

$$M_w = \sum_{i=0}^{\infty}\left[\frac{N_i M_i}{\sum_{j=0}^{\infty} N_j M_j}\right] M_i = \frac{\sum_{i=0}^{\infty} N_i M_i^2}{\sum_{j=0}^{\infty} N_j M_j}$$

A typical plot showing the number-average and weight-average molecular weight is shown in following figure. The number-average molecular weight is less than the weight-average molecular weight. The degree of polymerization can be calculated using the number-average molecular weight.

$$\textit{Degree of polymerization} = \frac{\text{Numberaverage molecular weight}}{\text{molecular weight of the repeat unit}}$$

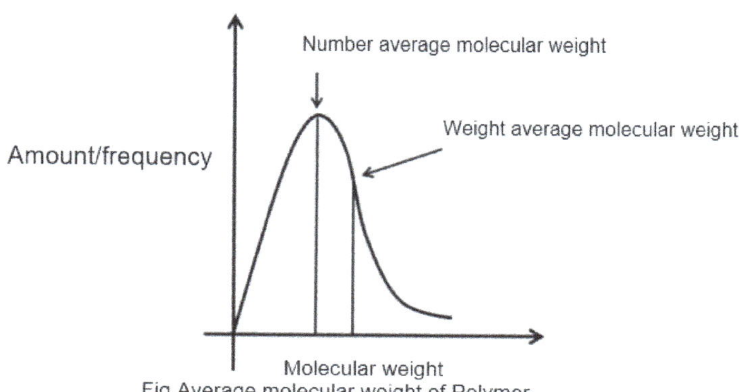

Fig.Average molecular weight of Polymer

Polydispersity Index or Heterogeneity Index

The ratio of the weight-average molecular weights to the number-average molecular weights is called polydispersity index (PDI) or heterogeneity index, which measures the polydispersity of the polymer mixture.

$$PDI = \frac{M_W}{M_n}$$

The dispersity measures heterogeneity of sizes of molecules or particles in the mixture. The mixture is called monodisperse if the molecules have the same size, shape, or mass. If the molecules in the mixture have an inconsistent size, shape and mass distribution, the mixture is called polydisperse.

The natural polymers are generally monodisperse as all synthetic polymers are poly-disperse with some exceptions. The PDI is equal to or greater than 1 where, as the polymer chains approach uniform chain length, the PDI tends to unity.

Crystallinity

The state where the polymer chains exist in parallel position that a polymer can be achieve at a particular temperature. It depends on chemical nature, geometrical structure, molecular weight and molecular weight distribution etc. Maximum crystallinity become possible if the polymer is annealed just at its Tm (Melting Temperature) for sufficiently long time.

Crystallinity is expressed as degree of crystallinity, given by,

$$Degree\ of\ Crystallinity = \frac{c - c_{am}}{C_{cr} - c_{am}}$$

Where,

C = Refractive index/density of whole polymer

C_{am} = The value of property when the polymer is amorphous.

C_{cr} = The value of property when the polymer is crystalline.

Crystallinity Increases with Density.

Amorphous:

The state of polymer where the chains are palced in haphazard direction is termed as the amorphousness of polymer. The density of polymer decreases of this area.

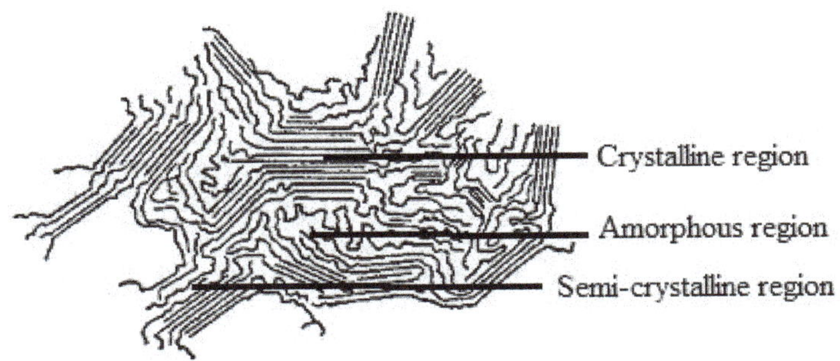

Semi-crystalline:

The state between crystalline and amorphous of a polymer is called semi crystalline.

Comparison of properties of crystalline and amorphous polymer:

Properties	Crystalline	Amorphous
1. Strength	High	Less
2. Melting point	high	Less
3. Elongation	Less	high
4. Dye takeup	Less	high
5. Light deflection	Regular	Irregular
6. Density	high	Less
7.Phase	Crystalline	Liquid
8. Structure	Linear	Branch or random
9. Polarity	high	Less
10. Flexural modulus	high	Less
11. Hardness	high	Less
12. Permeability	Less	high
13. Acid hydrolysis	Difficult	Attack rapidly
14. Young modulus	high	Less
15. Chain mobolity	easy	high

Factor that Control Crystallinity of Polymer

1. Regularity of the molecular structure of the polymer chain: In general, crystalline. substance is of regular molecular.
2. Polarity: Polarity increases the crystallinty of a polymer.
3. The mobility of polymer chain: The mobility of polymer chain is another factor that control crystallinity of polymer. If the polymer chain is mobile, the polymer will be.

Extended Chain Crystals

1. Less common, often take a needle.
2. Usually formed with low molecular weight polymer by slow crystallization or under pressure.

Nucleation – onset of crystallinity

1. Homogeneous nucleation – occur randomly throughout the.
2. Heterogeneous nucleation –occur at the interface of a foreign impurity 9e.g. a finely divided silica).

Mechanical Properties

It is of great importance to be familiar with some basic mechanical properties of the material before its application in any field, such as how much it can be stretched, how much it can be bent, how hard or soft it is, how it behaves on the application of repeated load and so on.

a. Strength: In simple words, the strength is the stress required to break the sample. There are several types of the strength, namely tensile (stretching of the polymer), compressional (compressing the polymer), flexural (bending of the polymer), torsional (twisting of the polymer), impact (hammering) and so on. The polymers follow the following order of increasing strength: linear < branched < cross-linked < network.

Factors Affecting the Strength of Polymers

1. Molecular Weight: The tensile strength of the polymer rises with increase in molecular weight and reaches the saturation level at some value of the molecular weight. The tensile strength is related to molecular weight by the following equation.

$$\sigma = \sigma_{\infty} - \frac{A}{M}$$

$\sigma\infty$ is the tensile strength of the polymer with molecular weight of infinity. A is some constant, and M is the molecular weight. At lower molecular weight, the polymer chains are loosely bonded by weak van der Waals forces and the chains can move easily, responsible for low strength, although crystallinity is present. In case of large molecular weight polymer, the chains become large and hence are entangled, giving strength to the polymer.

2. Cross-linking: The cross-linking restricts the motion of the chains and increases the strength of the polymer.

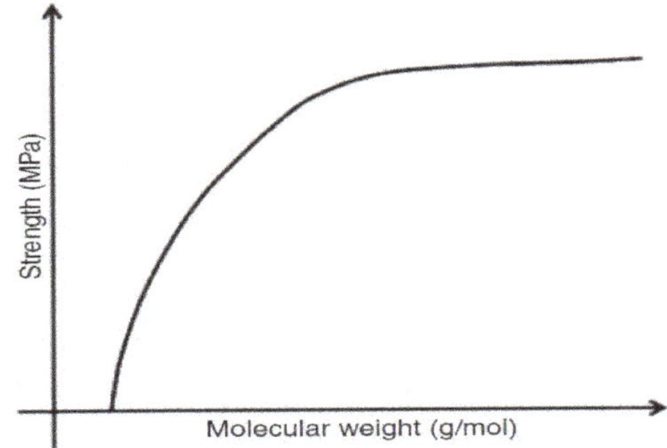

Figure: Variation of tensile strength with molecular weight of the polymer.

3. Crystallinity: The crystallinity of the polymer increases strength, because in the crystalline phase, the intermolecular bonding is more significant. Hence, the polymer deformation can result in the higher strength leading to oriented chains.

b. Percent Elongation to Break (Ultimate Elongation): It is the strain in the material on its breakage. It measures the percentage change in the length of the material before fracture. It is a measure of ductility. Ceramics have very low (<1%), metals have moderate (1-50%)) and thermoplastic (>100%),thermosets (<5%) value of elongation to break.

c. Young's Modulus (Modulus of Elasticity or Tensile Modulus): Young's Modulus is the ratio of stress to the strain in the linearly elastic region. Elastic modulus is a measure of the stiffness of the material.

$$E = \frac{Tensile\ Stress(\sigma)}{Tensile\ Strain(\varepsilon)}$$

d. Toughness: The toughness of a material is given by the area under a stress–strain curve.

$$Toughness = \int \sigma \, d\varepsilon$$

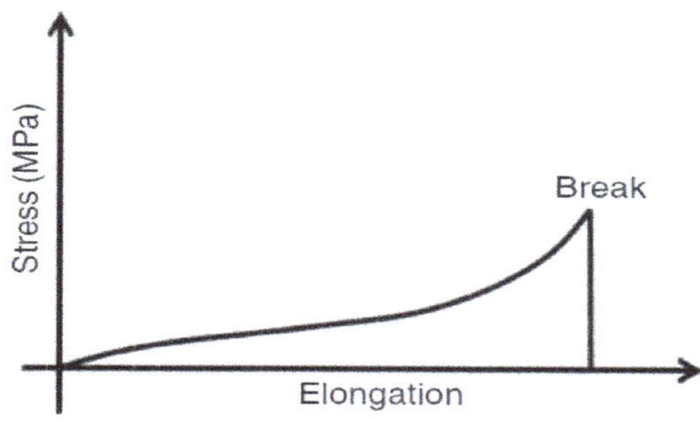

Figure: Elongation to break of the polymer.

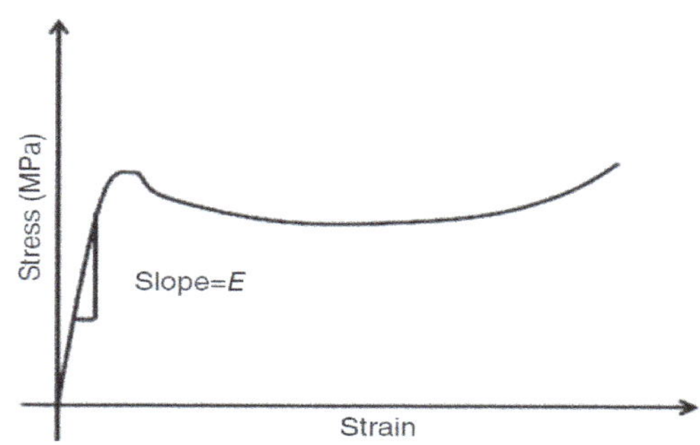

Figure: Young's modulus of the polymer.

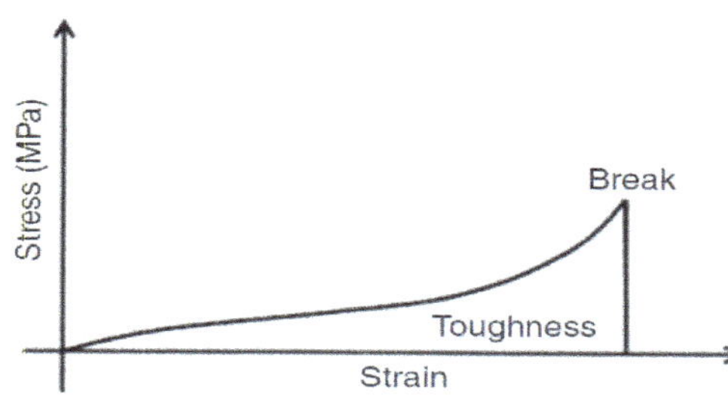

Figure: The toughness of polymer material.

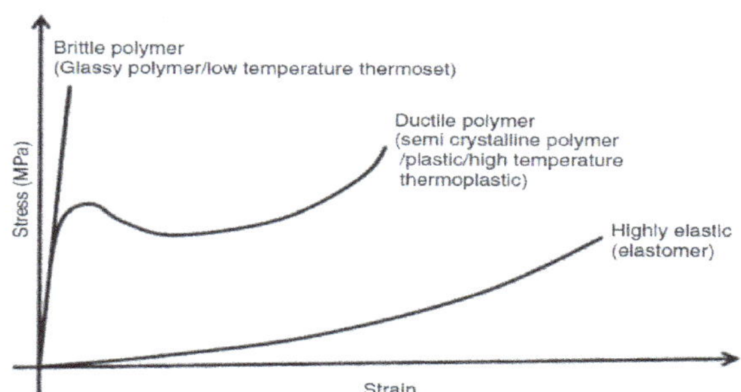

Figure: Stress–strain behavior of different types of materials.

The toughness measures the energy absorbed by the material before it breaks. A typical stress–strain curve is shown in figure, which compares the stress–strain behavior of different types of materials. The rigid materials possess high Young's modulus (such as brittle polymers), and ductile polymers also possess similar elastic modulus, but with higher fracture toughness. However, elastomers have low values of Young's modulus and are rubbery in nature. The yield strength of the plastic polymer is the corresponding stress where the elastic region (linear portion of the curve) ends. The tensile strength is the stress corresponding to the fracture of the polymer. The tensile strength may be higher or lower than the yield strength. The mechanical properties of the polymer are strongly affected by the temperature. A typical plot of stress versus strain is shown in figure. From the plot, it is clear that with increase in the temperature, the elastic modulus and tensile strength are decreased, but the ductility is enhanced.

e. Viscoelasticity: There are two types of deformations: elastic and viscous. Consider the constant stress level applied to a material as shown in the figure. In the elastic deformation, the strain is generated at the moment the constant load (or stress) is applied, and this strain is maintained until the stress is not released.

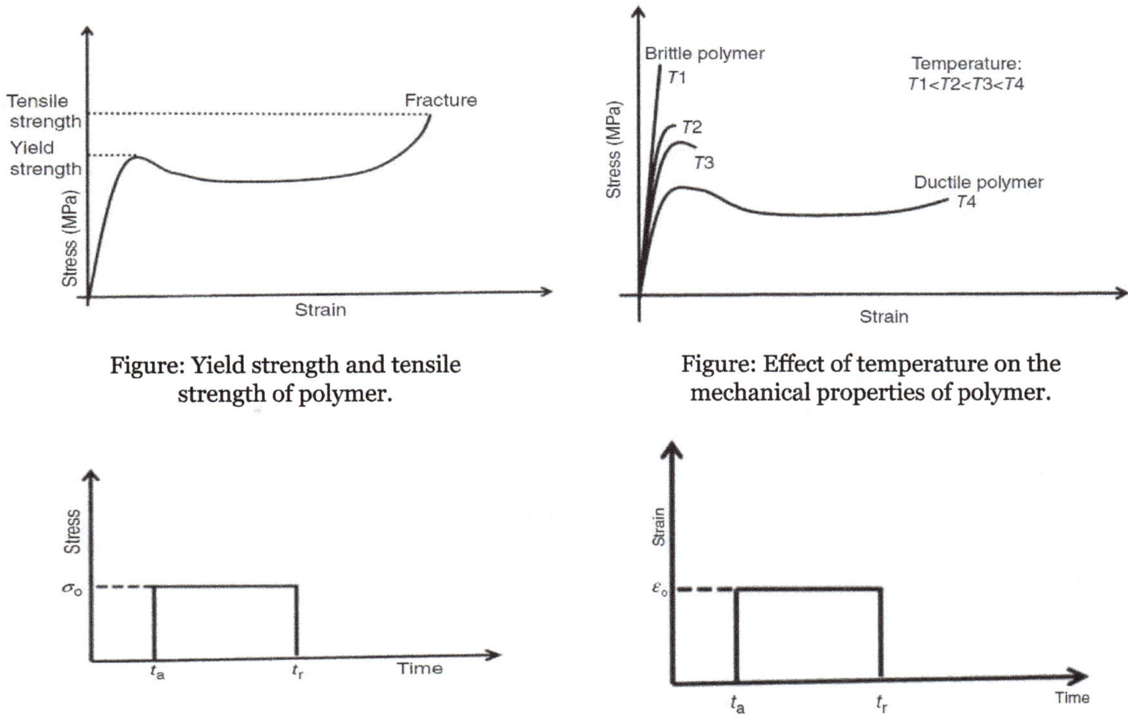

Figure: Yield strength and tensile strength of polymer.

Figure: Effect of temperature on the mechanical properties of polymer.

Figure: Constant stress applied to a polymer.

Figure: Elastic deformation.

On removal of the stress, the material recovers its original dimensions completely, that is the deformation is reversible, that is:

$$\sigma = E\varepsilon$$

where E is the elastic modulus, σ is applied stress, and ϵ is the strain developed.

However, in viscous deformation, the strain generated is not instantaneous and it is time dependent. The strain keeps on increasing with time on application of the constant load, that is, the recovery process is delayed. When the load is removed, the material does not return to its original dimensions completely, that is, this deformation is irreversible.

$$\sigma = \gamma \frac{d\varepsilon}{dt}$$

where

γ =viscosity, and

$d\epsilon/dt$=strain rate

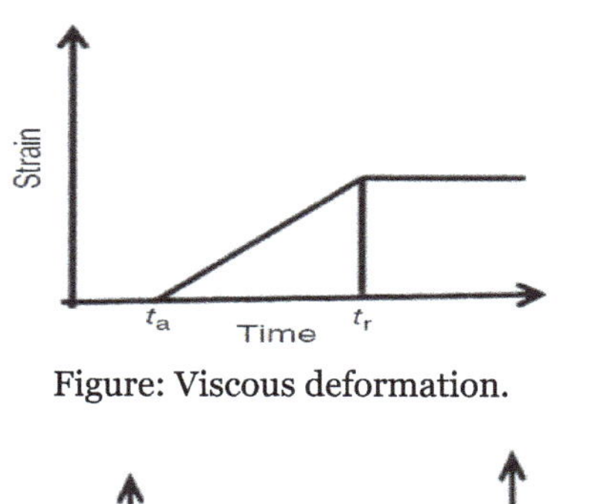

Figure: Viscous deformation.

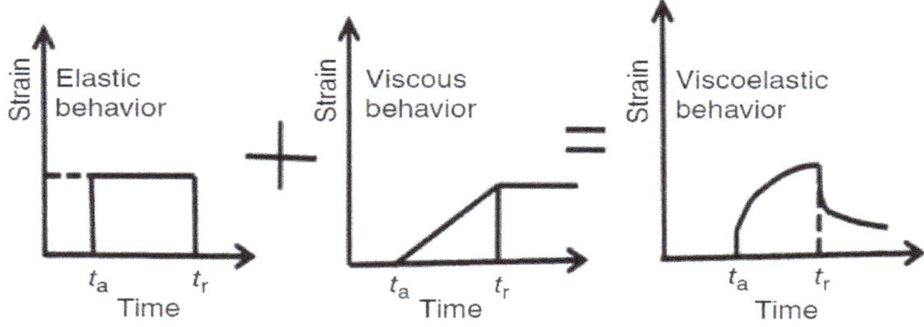

Figure: Viscoelastic deformation: the combined behavior of viscosity and elasticity.

Usually, polymers show a combined behavior of elastic and plastic deformation) depending on the temperature and strain rate. At low temperature and high strain rate, elastic behavior is observed, and at high temperature but low strain rate, the viscous behavior is observed. The combined behavior of viscosity and elasticity is observed at intermediate temperature and strain rate values. This behavior is termed as viscoelasticity, and the polymer is termed as viscoelastic.

Viscoelastic Relaxation Modulus

At a given temperature, when the polymer is strained to a given value, then the stress required to maintain this strain is found to decrease with time. This is called stress relaxation. The stress required to maintain the constant strain value is decreased with time, because the molecules of polymer get relaxed with time, and to maintain the level of strain, somewhat lower value of stress is sufficient.

$$E_{rel}(t) = \frac{\sigma(t)}{\varepsilon_o}$$

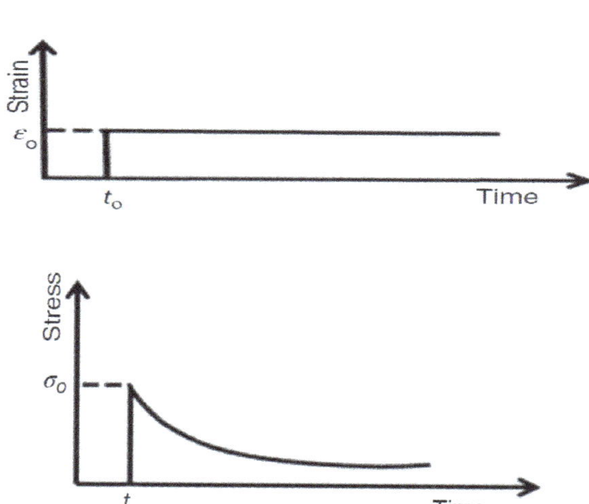

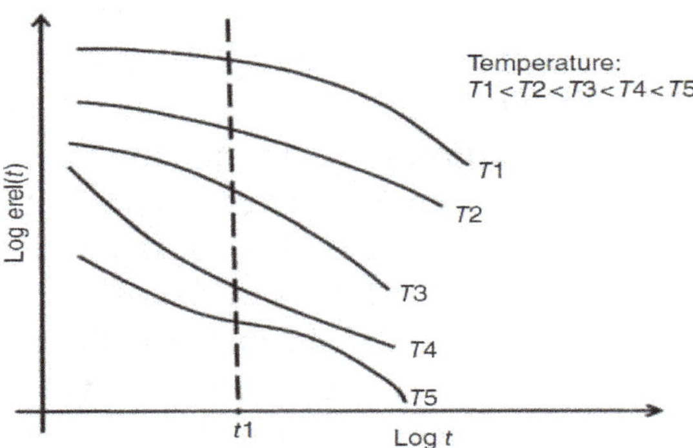

Figure: Variation of relaxation modulus with temperature and time.

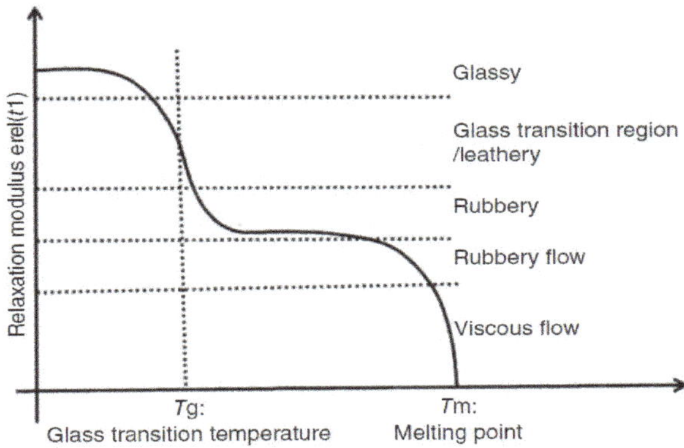

Figure: Variation of relaxation modulus with temperature after a given time t.

The decrease in stress follows the exponential decay:

$$\sigma = \sigma_0 e^{-t/r}$$

where

σ =stress at time t,

σo =peak stress level, and

τ =relaxation time constant.

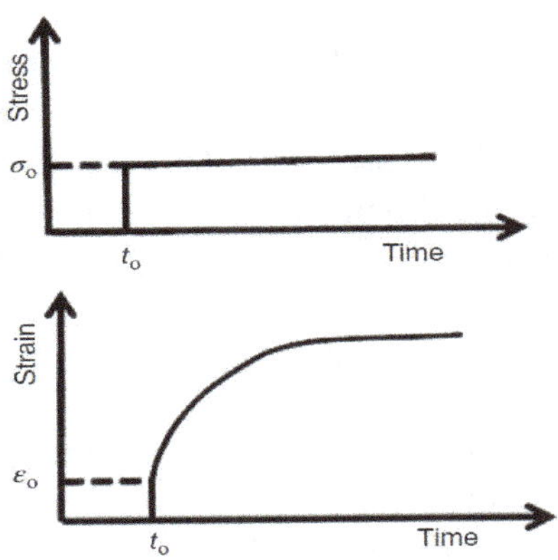

Figure: Viscoelastic creep.

The relaxation modulus is found to decrease with increase in temperature and time as shown in figure. Now, consider the time t1. Measure the values of relaxation modulus at time t1 at different temperatures, say T1, T2, T3 ... for the viscoelastic polymer and plot the relaxation modulus versus temperature. A typical plot is shown in figure. The glass transition temperature lies near the upper temperature extremity. Viscoelastic Creep: creep can be considered as the opposite of stress relaxation (lowering of the stress with time maintaining the constant strain level) where the polymer suffers time-dependent deformation (increasing strain with time) at constant stress level. At a given temperature, when a constant load is applied to the material, there is a time-dependent increase of strain in the material. This behavior of the material is called viscoelastic creep. The increase in molecular weight and stiffness of the chains results in better creep resistance of the material. We can define the time-dependent creep modulus as follows:

$$E_{Creep}(t) = \frac{\sigma_o}{\varepsilon(t)}$$

Or we can define the creep compliance as follows:

$$J_{Creep}(t) = \frac{\varepsilon(t)}{\sigma_o}$$

where

σ_o =constant applied stress,

$\varepsilon(t)$ =strain at time t.

Thermal Properties

In the amorphous region of the polymer, at lower temperature, the molecules of the polymer are in, say, frozen state, where the molecules can vibrate slightly but are not able to move significantly. This state is referred as the glassy state. In this state, the polymer is brittle, hard and rigid analogous to glass. Hence, the name glassy state. The glassy state is similar to a super cooled liquid where the molecular motion is in the frozen state. The glassy state shows hard, rigid, and brittle nature analogous to a crystalline solid with molecular disorder as a liquid. Now, when the polymer is heated, the polymer chains are able to wiggle around each other, and the polymer becomes soft and flexible similar to rubber. This state is called the rubbery state. The temperature at which the glassy state makes a transition to rubbery state is called the glass transition temperature Tg. Note that the glass transition occurs only in the amorphous region, and the crystalline region remains unaffected during the glass transition in the semi-crystalline polymer.

Glass Transition Point (Secondary Transition Point)

The glass transition point of polycarbonate obtained from the inflection point of refractive index is 141~149 °C, as shown in figure. In addition, the glass transition point obtained by the measurement of expansion coefficient, specific heat, differential thermal analysis, and viscoelasticity, etc. is in the range of 130~155 °C.

When the glass transition point is studied in further detail, it differs in accordance with the molecular weight, as shown in figure (Tg was obtained by the differential thermal analysis). Also, the glass transition point is known to have the pressure dependence and as for polycarbonate, it is as follows:

$$\delta T_g / \partial p = 0.044\ ^\circ C / atm$$

A comparison with other resins is shown in table.

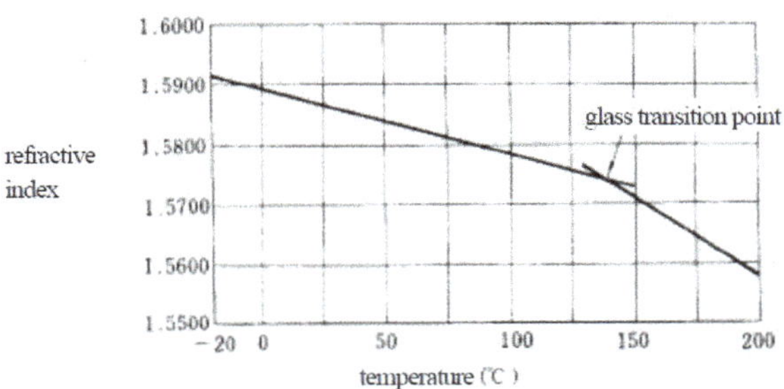

Figure: Temperature characteristics of refractive index of polycarbonate

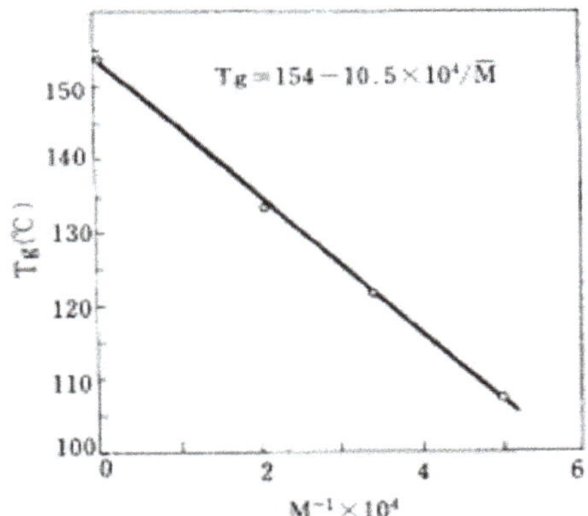

Figure: Relation between Tg (glass transition point) and M (molecular weight)

Table: Tm, Tg and dispersion temperature of other resins

Name	Tm melting point	Tg glass transition point (C)	Dispersion temperature				
			Crystalline dispersion	Primary dispersion	Secondary dispersion	Secondary dispersion	Secondary dispersion
Polyethylene	107-138	-53--23	60 72	-18 - 8 23	-126 -111 - 65		
Polypropylene	168-170	-35	82	- 2 22	- 83 - 40	-215 -173	
Polyvinyl chloride (PVC)	217	77		91 117 127	- 38 12		

Polyvinylidene chloride	190	-18	77	33	- 23		
Polytetrafluoroeth-ylene	327	-73--63	127 150	-33	- 93 - 66 - 31		
Polystyrol	230	80-90		117 131 148	40	-153 - 53 87	
Polymethyl methacrylate (PMMA)		82-102		127 143 167	27 103	-115 - 69 17	-183 -143 - 53
Polyvinyl acetate		7-27		30 90	- 47	-113 -17	
Polyethylene· terephthalate (PETP)	265	63-83		57 127	-10 7		
Polyoxymethylene	177	-40--60	127	-13 45 127	- 73 - 58 - 33		
Nylon 6	223	33-53		57 82	- 61 - 23 32	-128 -105 30	
Nylon 66	275	33-53		67 82	- 53 - 23	-125 -103	

Melting Point

The melting point of Iupilon / NoVAREX pellet is 220~230 °C. The melting point of crystallized polycarbonate is about 230~260 °C. The melting heat of the crystalline is 134J/g (32cal/g). The melting point of other resins is shown in table.

Dispersion Temperature

As for the thermoplastic resin, it is known that there are the primary dispersion zone where the macro-Brownian motion of molecular chain occurs, and the secondary dispersion zone where the local thermal motion (for example, thermal motion of methyl group of a part of main chain or side chain) occurs. Polycarbonate is not exception, too and its existence is recognized in many literatures. The dispersion temperature of polycarbonate is summarized in table.

Dispersion type	Primary dispersion α	Secondary dispersion β	Secondary dispersion γ	Secondary dispersion δ
Dispersion temperature (C)	157(1) 175(10^3) 195(10^5)	0-100	-100(1) -80(10^3) -3(10^6)	-200 ~ -100

Thermal motion type	Macro-Brownian motion of main chain		Free rotation motion of the C=o group which accompanies the restricted motion of the phenyl group	Free rotation motion of the Me group (by MMR)

The results of these dispersions are shown in figures. The primary dispersion, theγ dispersion and the δ dispersion are shown in figures respectively. As for the β dispersion, its existence is recognized in many literatures but there is not definite one. Figure is an example.

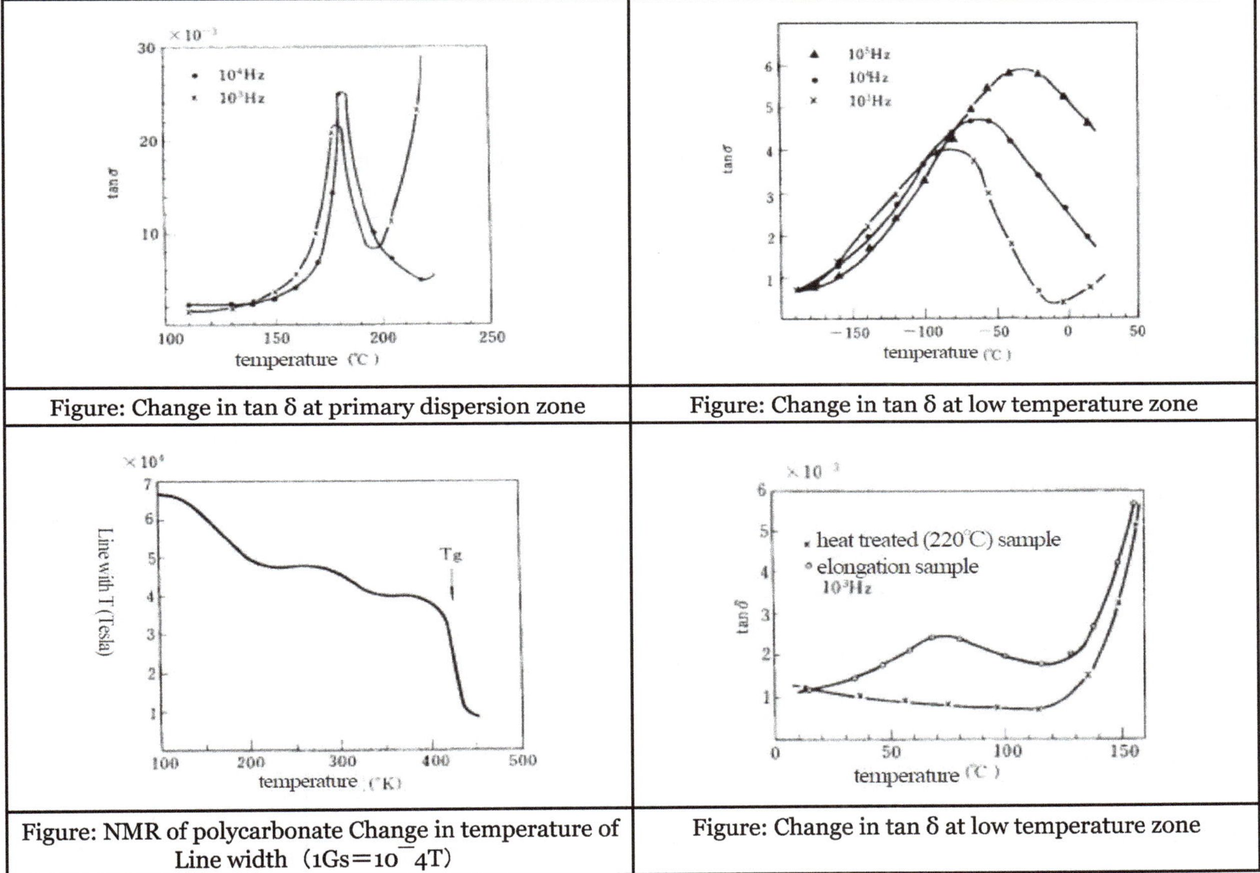

Figure: Change in tan δ at primary dispersion zone

Figure: Change in tan δ at low temperature zone

Figure: NMR of polycarbonate Change in temperature of Line width (1Gs=10¯4T)

Figure: Change in tan δ at low temperature zone

Thermal Conductivity and Specific Heat

As shown in figure, the specific heat of polycarbonate changes with temperature but it can be considered that this is 1.09~1.17J/(g.k) (0.26~0.28cal/g.°C) for the practical temperature range. This value does not differ very much from the common synthetic resins and corresponds to about 3 times of iron and copper etc.

The thermal conductivity of polycarbonate is

$$0.19W/(m.k)(4.6x10^{-4}\ cal/cm.sec\ °C)$$

This value does not differ very much from the common synthetic resins and is very small when compared with those of metals as it is 1/400 of iron, 1/1000 of aluminum and 1/2000 of copper.

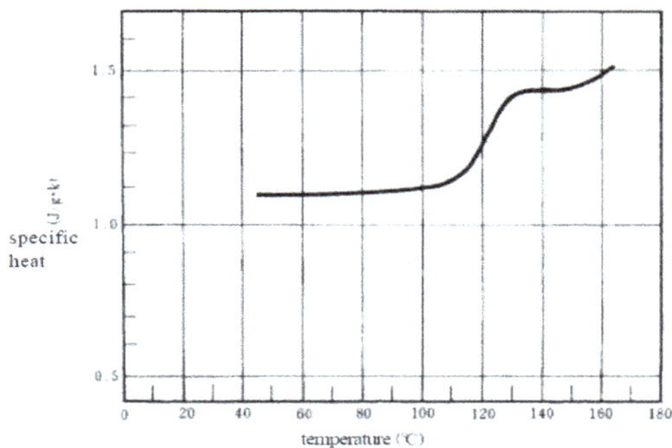

Figure: Specific heat – temperature characteristic of polycarbonate

Table: Comparison of thermal properties

Name	Thermal conductivity W/(m·k) (caI/cm.secC) x 10^{-4}	Specific heat kJ/(kg·k (caI/g.·°C	Linear expansion coefficient 10^{-5}.k^{-1}	Brittle temperature °C	Deflection temperature under load(C)	
					1820kpa (18.6kgf/cm^2)	455kpa (4.6kgf/cm^2)
Low density polyethylene	0.33 (8)	2.3 (0.55)	16-18	-85--55	32-41	38-49
High density polyethylene	0.46-0.52 (11-12.4)	2.3 (0.55)	11-13	-140	43-54	60-88
Polypropylene	0.14 (3.3)	1.9 (0.46)	6-10	-10--35	57-63	93-110
Acrylate (PMMA)	0.17-0.25 (4-6)	1.5 (0.35)	5-9	90	70-100	74-110
Polystyrene	0.10-0.14 (2.4-3.3)	1.4 (0.33)	6-8		80-90	
Polyvinyl chloride	0.13-0.17 (3-4)	1.0 (0.24)	5-6	81	54-79	57-82
Polyvinylidene chloride	0.13 (3)	1.3 (0.32)	19	0--30	55-65	
Polytetrafluoroethylene	0.25 (6)	1.0 (0.25)	10			121
Polyvinyl acetate	0.16 (3.8)	1.6 (0.39)	8.6		38	
Acryl nitrile· Styrene(AS)	0.12 (2.9)	1.4 (0.33)	6		88-102	
66 Nylon	0.22-0.24 (5.2-5.8)	1.7 (0.40)	10-15		66	
6 Nylon	0.21 (5.0)	1.6 (0.38)	8-13	-85-60	62	150
Polyethylene· terephthalate(PETP)			27			
Polyether	0.23	1.5	8.2	-76--120	110	170

Triacetyl cellulose	(5.5)	(0.35)	8-16		44-91	49-98
	0.17-0.33	1.5				
	(4-8)	(0.35)	5.6		117	128
Modified PPE	0.22	1.3		<-40		
Polysulfone	(5.2)	(0.32)	5.6		174	181
		1.3				
		(0.3)				
Polycarbonate	0.19	1.1	6	-135	130-136	136-142
	(4.6)	(0.27)				

Coefficient of Thermal Expansion

The coefficient of linear expansion of Iupilon / NOVAREX at 20~120 °C is

$6\sim7 \times 10^{-5}$ /K

The coefficient of volume expansion of Iupilon / NOVAREX at 30~130 °C is

$(20\pm5) \times 10^{-5}$/K

The change in length and volume weight ratio is shown in figures, respectively. The relation between temperature and volume expansion coefficient is shown in figure. The coefficient of linear expansion has the refraction point near the room temperature and becomes small in the low temperature region.

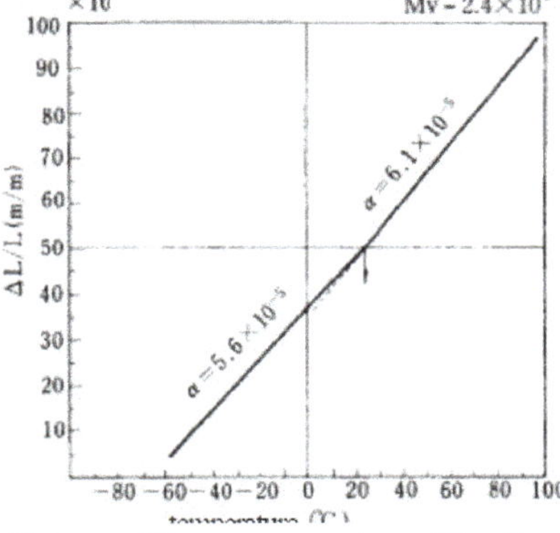

Figure: Coefficient of linear expansion of Iupilon / NOVAREX

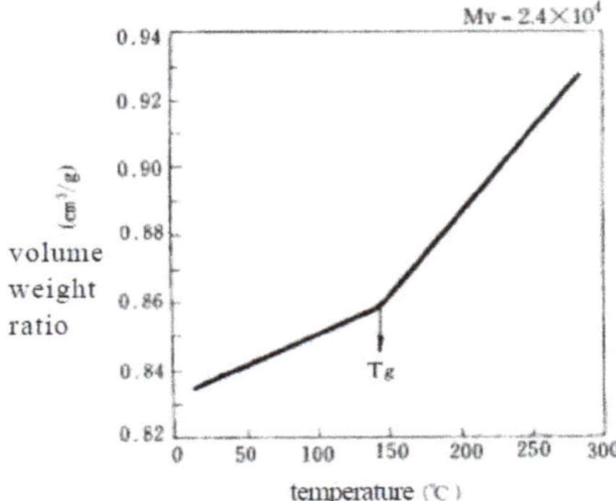

Figure: Relation between temperature and volume weight ratio

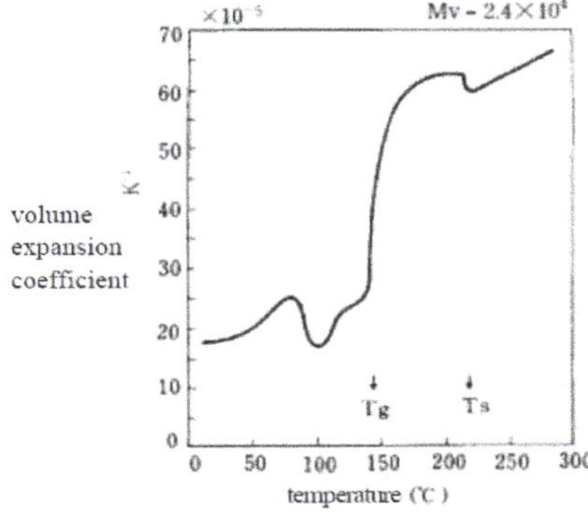

Figure: Relation between temperature and volume expansion coefficient

Deflection Temperature

The deflection temperature (ASTM - D648 - 56) of Iupilon / NOVAREX is

Stress 1.82MPa (18.6kgf/cm^2) 132~138 °C

Stress 0.45MPa (4.6kgf/cm2) 138~144 °C

The deflection temperature changes in accordance with the added load and is shown in figure in case of Iupilon / NOVAREX. Also, the deflection temperature is influenced by the molecular weight in the same way as Tg (glass transition point) as shown in figure. When Iupilon / NOVAREX is heat-treated, as indicated in other physical properties, the heat hardening is shown and the deflection temperature changes rapidly as shown in figure. A comparison with other resins is shown in table and figure.

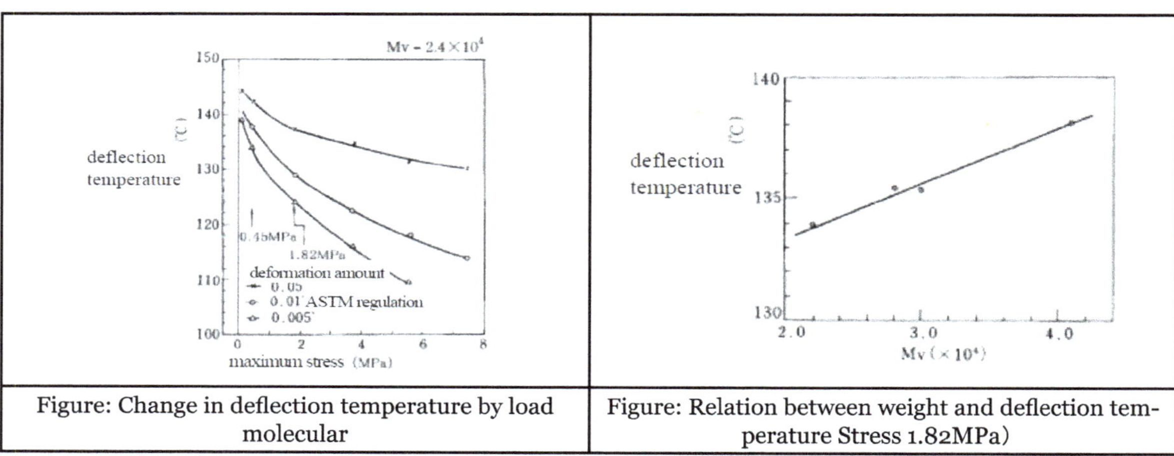

Figure: Change in deflection temperature by load molecular	Figure: Relation between weight and deflection temperature Stress 1.82MPa)

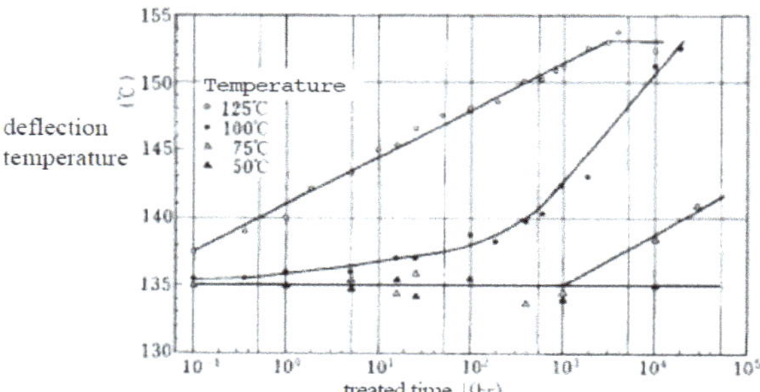

Figure: Change in deflection temperature by heat-treatment (Stress 1.82MPa) (Mv=2. 8×10^4)

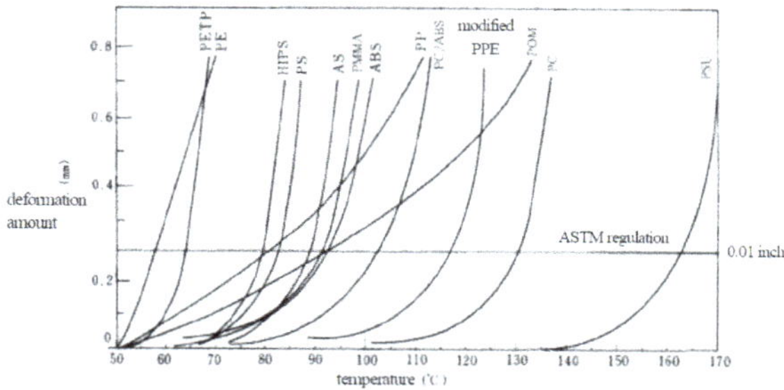

Figure: Deflection temperature of other resins (Stress 1.82MPa)

Thermal Stability and Pyrolysis

It is possible to know the excellent heat resistance as shown in figure where the differential thermal analysis result of Iupilon / NOVAREX is indicated. However, when examining it more in detail, the different aspect of every change in various temperature regions (practical temperature region, processing temperature region, decomposition combustion region), in the environment (in oxygen, in air, in nitrogen, in vacuum, in steam) is recognized.

Low temperature region When Iupilon / NOVAREX is heat-treated at the temperature below Tg, the fact that the change in physical properties occurs due to the hardening phenomenon has already been known and there are a lot of researches to look for the cause in the change of the solid structure. However, when heating it in air at this temperature region for long time, it is observed that the chemical changes (oxidation, decomposition), discoloration, decrease in molecular weight etc. take place. The result of arranging the yellowed degree of this temperature region is shown in figure. Curve (1) shows the influence of temperature on the yellowed speed, but the aspect of change is different at the up-and-down region of Tg. This might be due to the difference of the thermal effect of the molecular chain, that is, the difference of oxygen diffusion speed. Curve (2) shows the relation between temperature and treated time that the yellowed degree becomes equivalent.

Figure shows the CO_2 generation speed at this temperature region and the breakage of carbon bond, namely the decrease in molecular weight. Figures show the comparison of the oxidation with other resins and the antioxidative property of polycarbonate

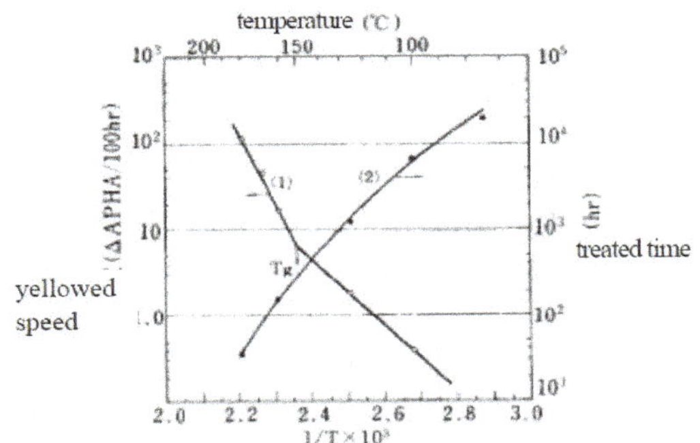

Figure: Yellowed speed of Iupilon / NOVAREX (Mv=2.4×10⁴

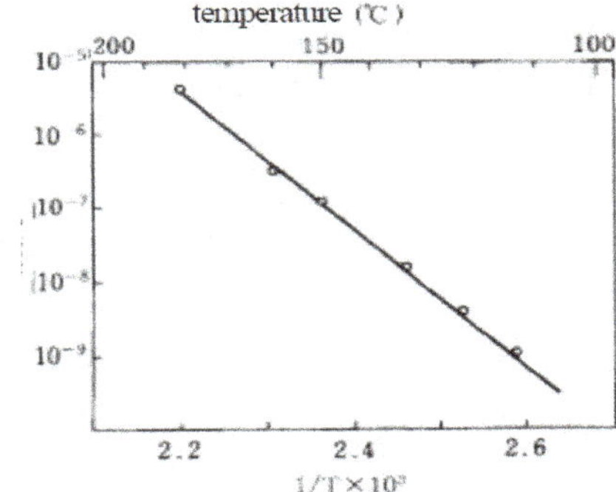

Figure: CO_2 generation at 110~180 °C Mv=2.4×10⁴)

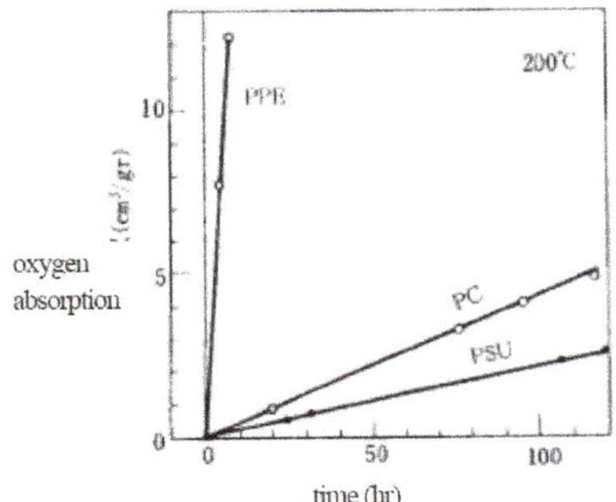

Figure: Oxygen absorption speed at 200 °C

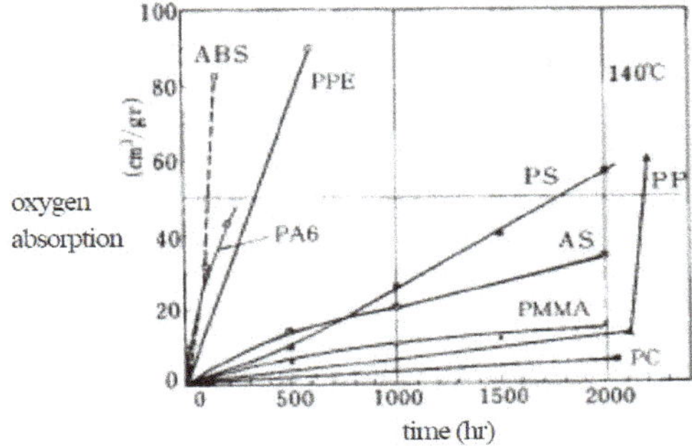

Figure: Oxygen absorption speed of various resins

High Temperature Region

The heat stability of Iupilon / NOVAREX is excellent. As shown in figure, the change in heat stability is observed at the temperature above 450 °C, and the influence of atmosphere, impurities, and additives is large in such a high temperature region, especially oxygen and moisture promote the heat degradation considerably.

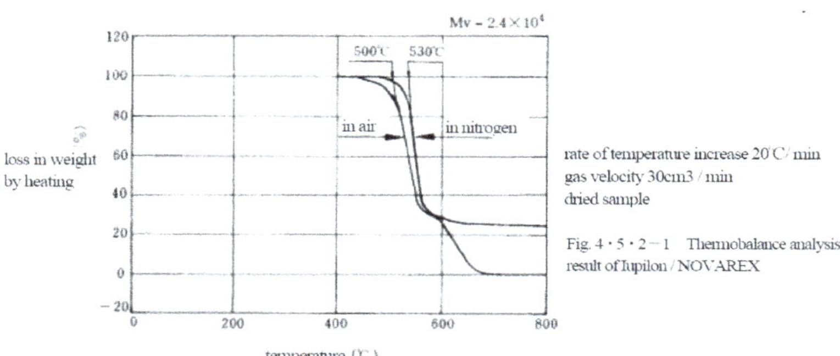

Fig. 4・5・2－1 Thermobalance analysis result of Iupilon / NOVAREX

The pyrolysis of Iupilon / NOVAREX、 consists of one exothermic region and two endothermic regions as shown from the result of the differential thermal analysis in figure. The exothermic region is the first stage of pyrolysis, the oxidation reaction is observed as an exothermic peak that starts at about 340 °Cand is highest at about 470 °C. The first endothermic region is based on the depolymerization and peaks at 500 °C. The second one is the region where the bond energy becomes equivalent to thermal energy, and dissociation of all molecular bonds takes place. The exothermic peak based on thermal oxidation and endothermic peak also decrease considerably in nitrogen and show that the influence of oxygen is remarkable. The decomposition gases generated by pyrolysis change by atmosphere as shown in table, but CO, CO_2 by the decomposition of carbonic acid group, CH_4 by the dissociation of methyl group, and various phenols by the decomposition of BPA (bisphenol A) are main decomposition products. The generation of CO_2, CH_4, and various phenols are shown in figures. The generation becomes active at around 300 °C in air, but shifts to high temperature side about 50 °Cin nitrogen atmosphere. The decrease in molecular weight of Iupilon when heated for 2 hours in nitrogen and in air is shown in figures. Also, the result of heating for a long time in the sealed tube in vacuum was shown in figure. The relation between temperature and pyrolysis kinetics is shown in figures, and the influence of oxygen and moisture is extremely big.

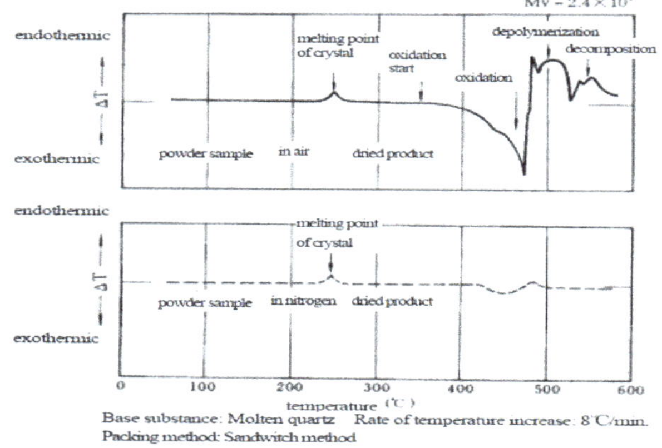

Figure: Result of differential thermal analysis of Iupilon / NOVAREX

Table: Decomposition products of polycarbonate

Decom-position products	In oxy-gen	In air	In vacuum sealed	In vacuum continuous	Decomposition products	In oxy-gen	In air	In vacu-um sealed	In vacuum continu-ous
CO2	+	+	+	+	Benzene	+	+		
CO	+	+	+	+	Toluene	+	+	+	+
CH4	+	+	+	+	Ethyl benzene	+	+	+	+
H2 H20	+ +	+ 	+ 		Phenol Cresol	+ +	+ +		
HCHO	+				Ethyl phenol		+	+	+
CH3CHO	+								
Acetone Methanol	+	+			Isopropyl phenol		+	+	+
Diphenyl					Isopropenyl phenol		+	+	+
			+	+	Bisphenol A	+	+	+	+

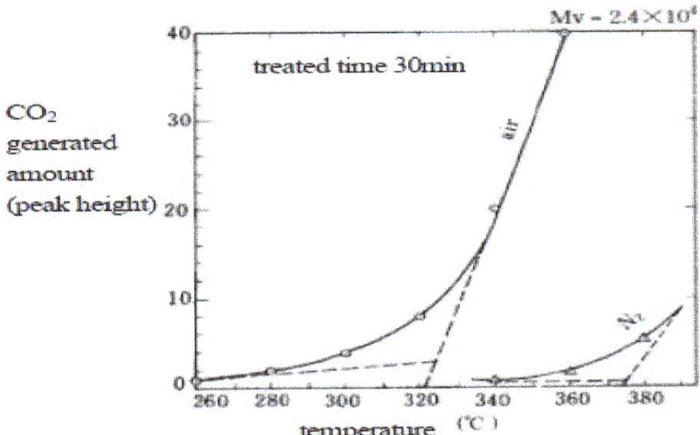

Figure: Influence of atmosphere on CO_2 generated amount

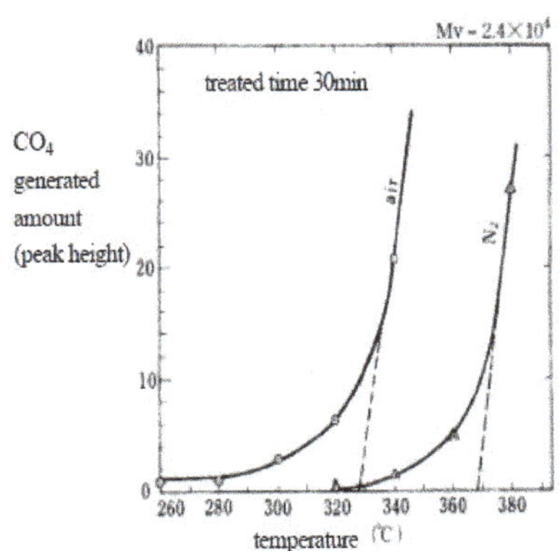

Figure: Influence of atmosphere

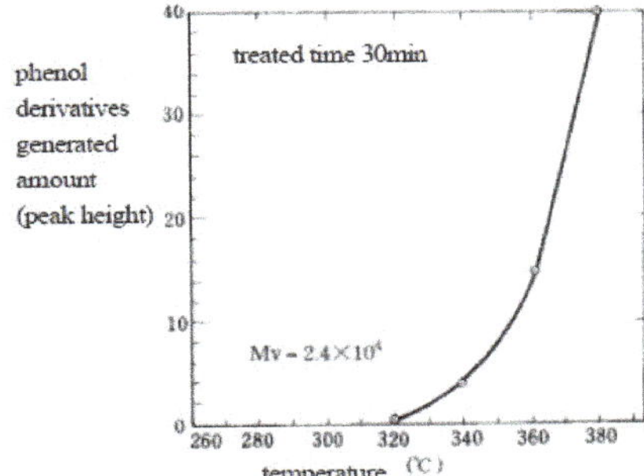

Figure: Relation between temperature and generated amount of phenol derivatives by decomposition when heated in air

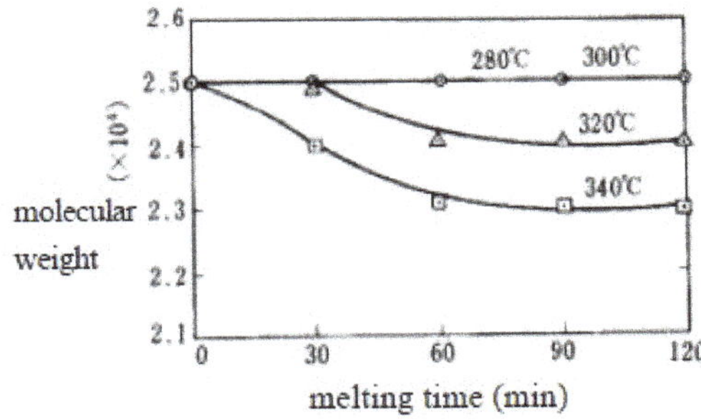

Figure: Decrease in molecular weight of Iupilon / NOVAREX by meltin (Completely dry, melted in nitrogen stream)

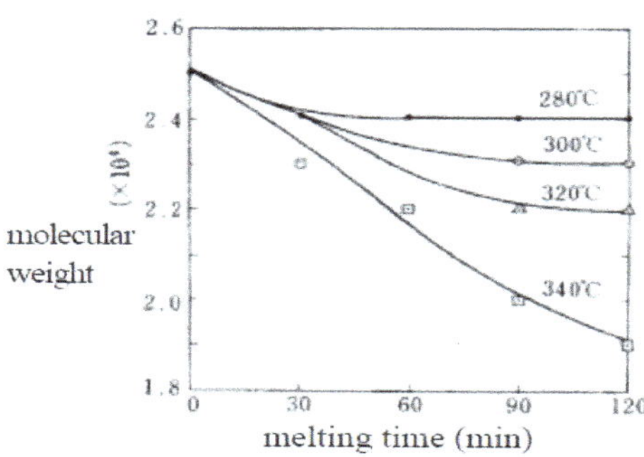

Figure: Decrease in molecular weight of Iupilon / NOVAREX by melting (Undried, melted in air)

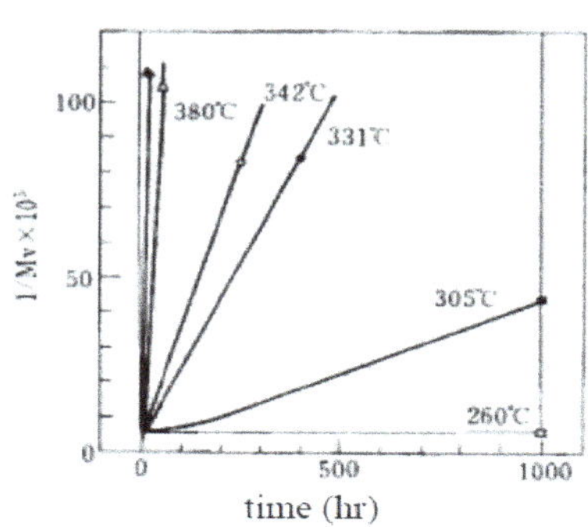

Figure: Decrease in molecular weight (Mv) of sealed tube in vacuum

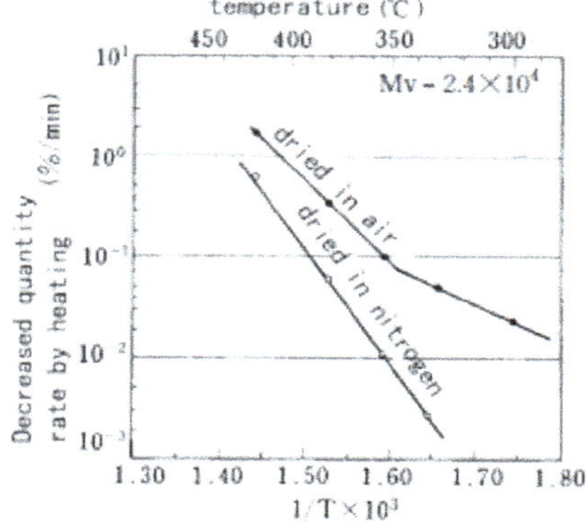

Figure: Decreased quantity rate of Iupilon / NOVAREX by heating

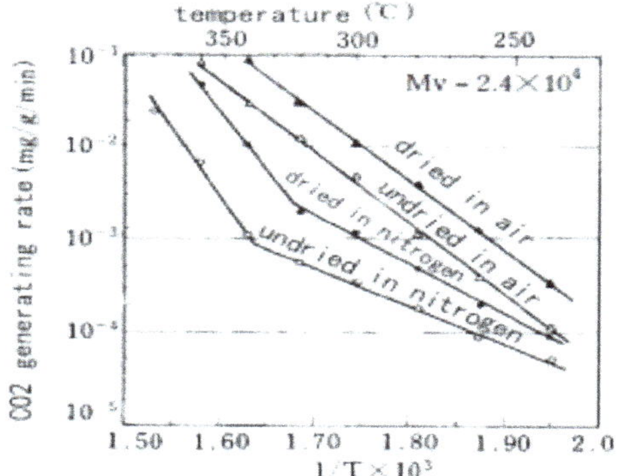

Figure: CO_2 generating rate of

Iupilon / NOVAREX

When the inorganic filling agent is added to Iupilon / NOVAREX, the influence on the pyrolysis is big. For ezample, decreased quantity rate in case of adding an iorganic filling agent in figure and CO_2 generated amount indicate the value which is bigger than the material in any case. Also, the influence of metal salts is shown in table. The influence of carbonates is extremely big, and the others also have influence to some degree.

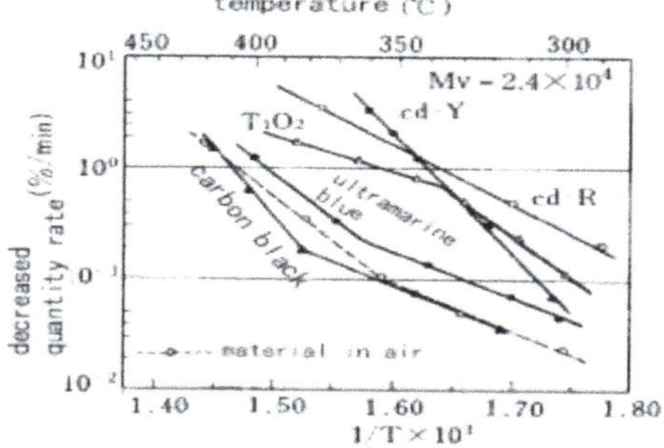

Figure: Influence on decreased quantity rate of the pigment (pigment additive amount: 1.0%)

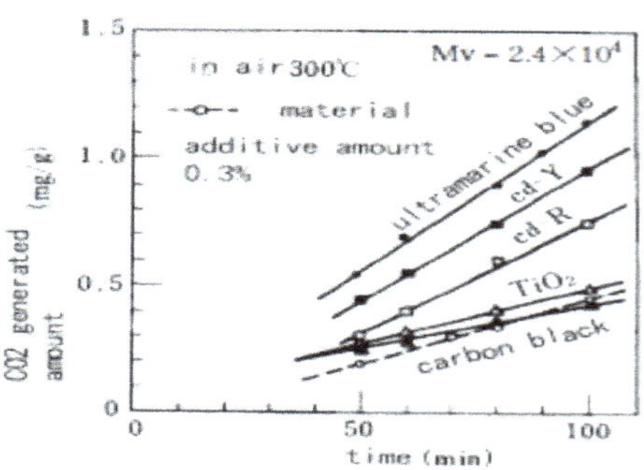

Figure: Influence on CO_2 generation of the pigment

Table: Influence of metal oxides on heat stability of Iupilon / NOVAREX

Metal oxides	Chemical Composition	pH	Molecular weight Mv x 10^4		Start temperature of decreased quantity (C)
			0%	1%	
Stannic oxide	SnO_2	4.2	2.8	2.4	340
Lead sulfate	$PbSO_4$	4.5	2.8	2.6	320
Lead chromate	$PbCr_2O_3$	5.4	2.8	2.4	306
Lead oxide	Pb_3O_4	7.8	2.8	2.2	210
Lead monoxide	PbO	10.2	2.8	2.5	363
Zinc sulfide	ZnS	2.4	2.8	2.4	250
Zinc oxide	ZnO	7.2	2.8	2.7	352
Zinc carbonate	$ZnCO_3$	7.1	2.8	1.9	315
Cadmium sulfate	$CdSO_4$	6.3	2.8	2.7	335
Cadmium sulfite	CdS	6.3	2.8	2.4	340
Cadmium oxide	CdO	9.4	2.8	2.6	315
Cadmium carbonate	$CdCO_3$	7.0	2.8	2.0	280
Aluminum oxide	$A1_2CO_3$	9.0	2.8	2.7	320
Cobalt oxide	CoO	8.2	2.8	2.7	330
Barium sulfate	$BaSO_4$	7.2	2.8	2.7	340
Titanic oxide	TiO_2	6.8	2.8	2.7	343
Copper oxide	CuO	6.9	2.8	2.5	340
Manganese dioxide	MnO_2	6.6	2.8	2.6	350
Ferric oxide	Fe_2O_3	6.4	2.8	2.7	320
Chromic oxide	Cr_2O_3	5.5	2.8	2.7	358
Cadmium selenide	CdSe	6.0	2.8	2.8	345

As for the molecular weight, the sample melted in nitrogen for 1 hour is used for measurement

As for the start temperature of decreased quantity, the sample added with 1% of the pigment is used for measurement in air.

When adding an organic additive (for example, ultraviolet absorber, stabilizer, antistatic agent, blowing agent, and plasticizer, etc.) to Iupilon / NOVAREX, the one that causes the chemical reaction with polycarbonate cannot be used. Also, it is necessary to consider sufficiently not only the reactivity but also the heat stability of the additive to be used because the processing temperature of polycarbonate is high, close to the decomposition temperature of the organic substances at the temperature range above 300° C.

When polycarbonate is heated in vacuum system, it is known that if the decomposition product is removed continuously, the peculiar phenomenon to cause rapidly the gel generation is observed.

The state of gel generation, and the change of soluble part [η] of methyl chloride are shown in figures, respectively. The generation rate of the decomposition product in this system is indicated in figure. It is known that this gel phenomenon is recognized not only in polycarbonate but also in polysulfone, PPE(polyphenylene ether) etc. as shown in figure.

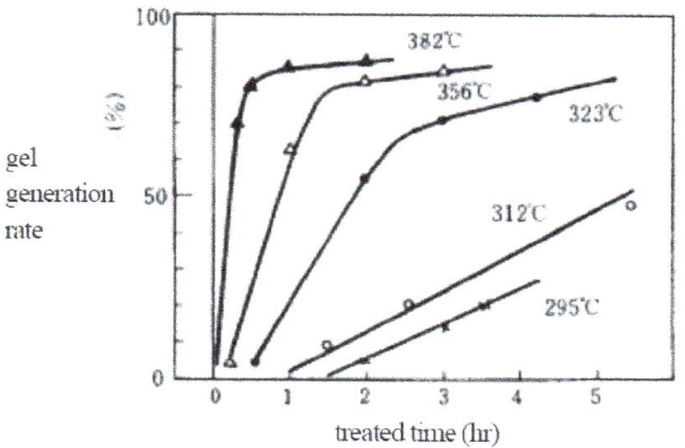

Figure: Gel generation by heating in continuous vacuum system

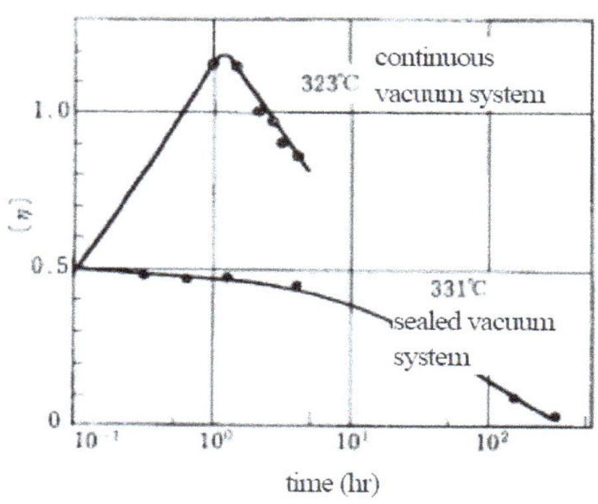

Figure: Change in [η] by heating in vacuum system

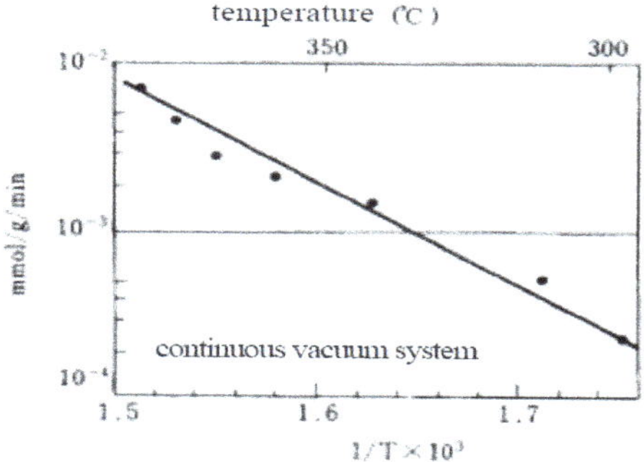

Figure: Generation rate of decomposed gas in continuous vacuum system

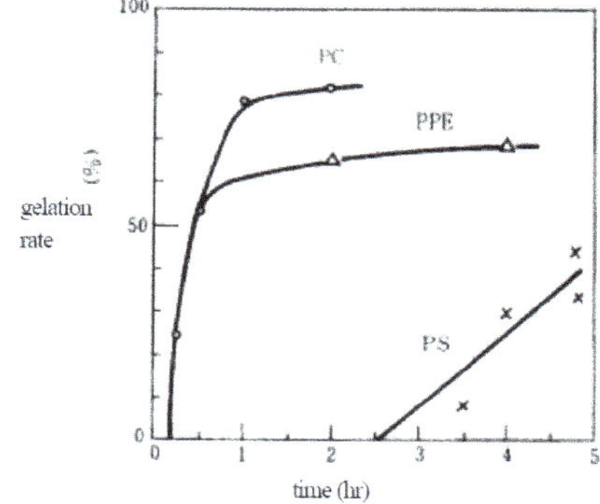

Figure: Gel generation of other resins

The generation of decomposed gas when heating polycarbonate at 700-1200 °C is shown in figure. CO_2 and CH_4 generation show a constant value regardless of temperature.

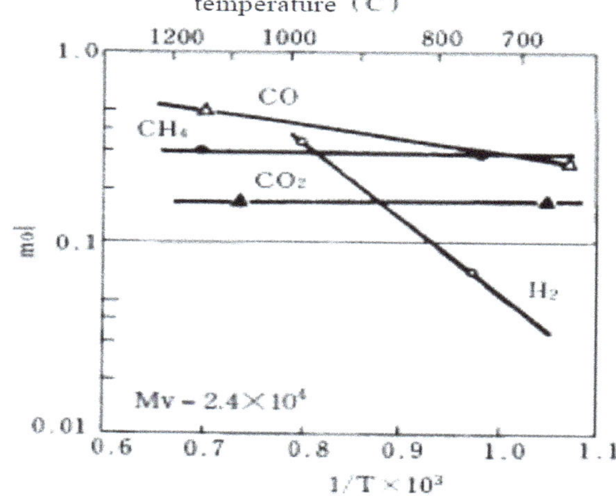

Figure: Generation of decomposed gas at high temperature

Hot Water Resisting Property

As the bond of the main chain of Iupilon / NOVAREX is an ester bond, hydrolysis takes place gradually and molecular weight decreases when it comes in contact with hot water and steam. At the same time, cracks form with the decrease in mechanical strength after a long time.

The decrease in molecular weight of Iupilon / NOVAREX by treating with hot water is shown in figure. The decrease in molecular weight occurs rapidly by treating at high temperature. Also, the decrease in case of only one surface of the molding contacts with hot water is gentle than the case of the immersion, for example, at 75 °C, the treated time when the molecular weight becomes 2.0×104 is 3~4 times.

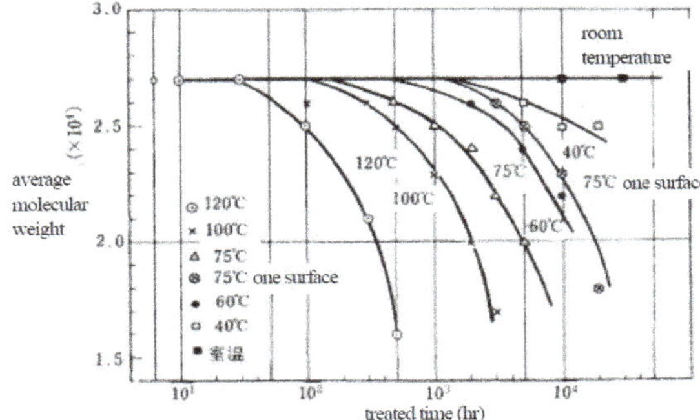

Figure: Decrease in molecular weight by treating with hot water

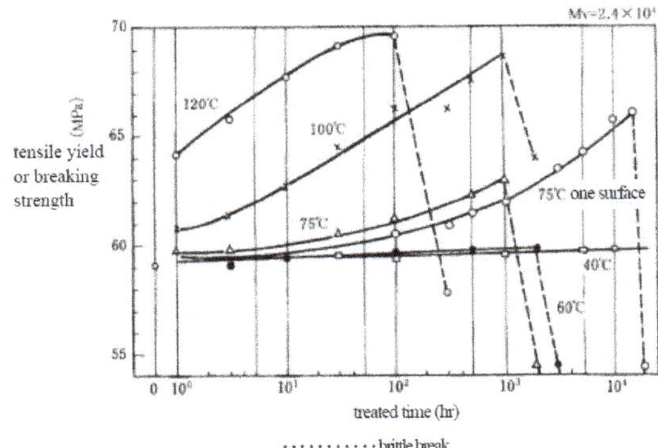

Figure: Change in tensile yield or breaking strength by treating with hot water

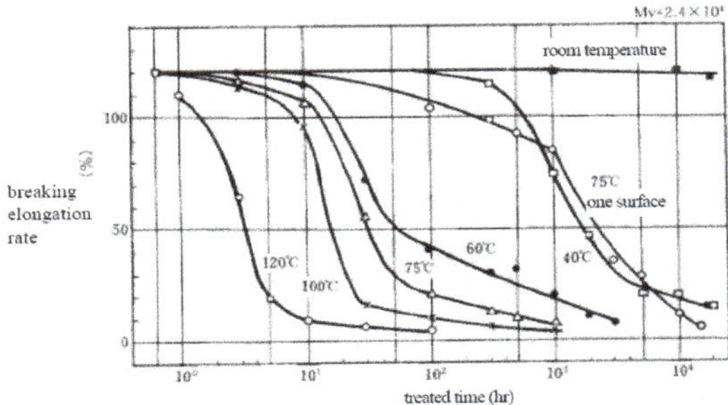

Figure: Tensile breaking elongation rate by treating with hot water

The tensile properties of Iupilon / NOVAREX treated with hot water show the deterioration by crack generation with decrease in molecular weight as indicated in figures.

The time when ductile breaking moves to brittle breaking is as follows : 100~200 hours at 120 °C (in steam of 98kPa , 1kgf/cm2), 1000~2000 hours at 100 °C and 75 °C, 2000~3000 hours at 60 °C, 20000 hours at 75 °C one surface, above 20000 hours at 40 °C. Although the decrease in molecular weight at 75 °C and 60 °C is small but the tensile property is deteriorated. This is due to crack generation.

The deterioration of Izod impact strength is shown in figure. The deterioration rate becomes fast compared with the case of dry-heat treatment. For example, as for dry-heat treatment at 100 °C, 1000 hours is needed, but only 30~50 hours in case of treating with hot water.

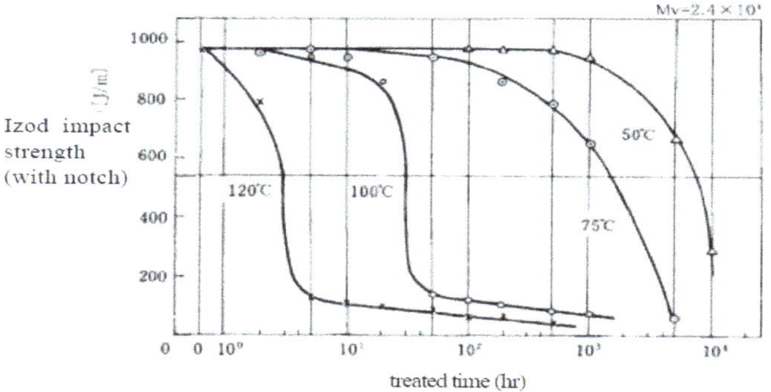

Figure: Change in Izod impact strength by treating with hot water

Flammability

A comparison with other resins is shown in Table.

Table: Comparison of flammability with other resins

Polymers	Combustion heat kJ/g(cal/g	Heat value kJ/g(cal/g	Generated moisture wt%	Flammability cm/ min (in/min)	Oxygen index (%)
Polyethylene Polypropylene Polyvinyl chloride Tetrafluoro ethylene Polymethylmethacrylate (PMMA) Polystyrene Acrylnitryl· stylene(AS) ABS Polyether Ethylcellulose Polyamide(nylon) Polyphenylether(PPE) Polysulphone Polycarbonate Copolymerization Polycarbonate (Iupilon N-3)	45.9 (10965) 44.0 (10506) 18.1 (4315) 4.2 (1004) 26.2 (6265) 40.2 (9604) 35.3 (8424) 16.9 (4046) 23.7 (5659) 30.9 (7371) 30.5 (7294)	42.8 (10225) 41.1 (9828) 16.8 (4015) 24.6 (5869) 38.4 (9182) 33.8 (8066) 15.9 (3790) 28.7 (6863) 29.4 (7020)	126.6 115.9 51.3 67.8 72.1 61.2 43.7 86.8 46.8	2.5 (1.0) 2.5 (1.0) self-extinguishing nonflammable 2.8 (1.1) 2.5-5.1 (1.0-2.0 2.5 (1.0) 3.3 (1.3) 2.8 (1.1) 2.0-3.6 (0.8-1.4 self-extinguishing self-extinguishing self-extinguishing self-extinguishing	17.4 17.4 47.0 95.0 17.3 18.3 18.1 16.2 28.0 30.2 30.4 25.0 31.0

The problem should be considered when making plastic flame-resistant does not include only the improvement of its flammability but also the composition, quantity and fuming property of generated gas. In case of Iupilon / NOVAREX, because the composition element is C · H · O, the generation of toxic gas such as HCI (PVC, Polyvinylidene chloride etc.), NH_3, cyanide (polyamide, ABS, AS etc.), SO_2 (polysulfone etc.) does not occur. Also, as for the fuming property, results of various resins are shown in table. Polycarbonate has a moderate fuming property in case of ignition combustion, but shows a characteristic with extremely low fuming property in case of burning combustion.

Table: Fuming property of plastics

Plastics	Thickness mm	Ignition combustion			Burning combustion		
		Dm	Rm	T16(min	Dm	Rm	T16(min
Polyvinyl chloride	6.4	660	134	0.8	300	12	3.9
Polyvinylidene chloride	2.8	125					
Polydifluoride vinyl chloride	0.04	0					
Polyfluoro vinyl	0.05	4					
Polystyrol	6.4	660	243	1.3	322	24	7.3
ABS	1.2	660	400	0.6	71	4	4.8
Polymethyl methacrylate (PMMA)	5.6	660	23	2.6	156	60	9.2
Cellulose acetate butyrate	6.4	49	12	5.0	434	45	2.7
Polycarbonate	3.2	174	43	2.1	12	1	
Polyphenylene ether(PPE)	2.0	183					
Polysulphone	1.5	40					
Nylon fiber	7.6	269	105	1.8	320	45	2.8
Acryl fiber	7.6	159	29	0.6	319	49	1.5
Polypropylene fiber	4.6	110	50	1.7	456	60	2.3
Oak	6.4	155	18	3.9	350	34	4.8

Dm: fuming quantity per unit area
Rm: fuming rate
T 16: t ime when Dm becomes 16

Other Thermal Properties

Brittleness Temperature

The brittleness temperature of Iupilon / NOVAREX is -135C. A comparison with other resins is shown in table. The low temperature resisting property of Iupilon / NOVAREX is the best among plastics.

Heat Shrinkage

The change of heat shrinkage of Iupilon / NOVAREX when treated in hot air atmosphere is shown in figure. The heat shrinkage takes place even at low temperature and shows a 0.1~0.2% change. However, such a heat shrinkage also changes in accordance with the molding conditions etc.

The frozen orientation strain is released by macro-Brownian motion of the molecular chain in an atmosphere over 150 °Cand the shrinkage becomes 5~10%.

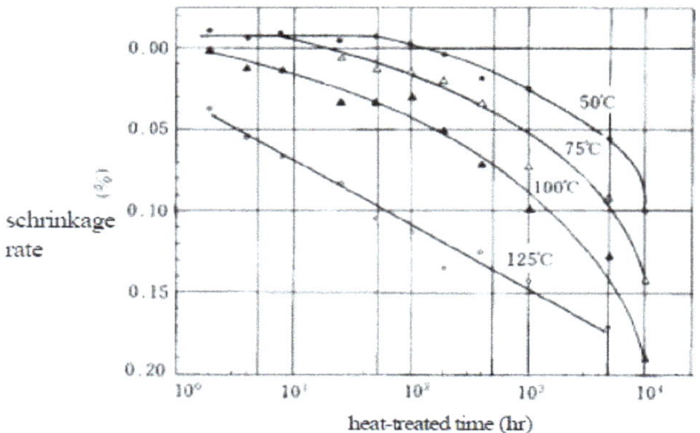

(Sample dimension 6. 4×12. 7×152mm, Mv=2. 2×104)
Figure: Change in dimension by heat treating

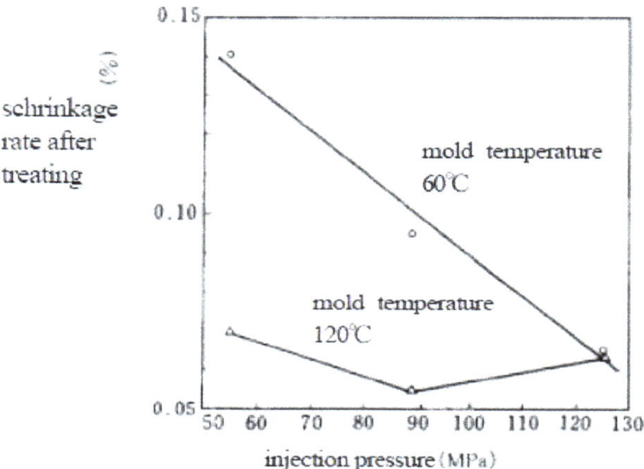

Figure: Relation between injection molding conditions and heat shrinkage

Polymer Crystallization

The crystallization of polymers can be broadly classified under three groups:

(A) Crystallization during polymerization.

(B) Crystallization induced by orientation.

(C) Crystallization under quiescent condition.

Crystallization During Polymerization

A special attribute of this kind of polymerization is the formation of macroscopic single polymer crystals. During such a process the monomers forming a crystal can be joined up into chains by solid state polymerization, while the original "monomer" crystals are preserved. The final polymer crystal is obtained due to chemical reactions at the gas/solid or liquid/solid interface and not just as a consequence of change in physical state of the material as is observed in normal crystallization processes. The final properties of crystals formed by such methods can be very interesting, for e.g. poly (sulfur nitride) crystals formed by such methods conduct electricity like metals along the crystal axis (corresponding to the chain direction) and can even become superconducting at sufficiently low temperatures. The mechanism of such a process can be (a) the simultaneous polymerization and crystallization and (b) successive polymerization and crystallization. In (a) the primary and secondary bonds are set at the same time, and in (b) the polymerization and crystallization sites can be separated and thus the nature of the polymer segments as yet uncrystallized becomes important. While macromolecular crystallization can occur from only the melt or solution state, the crystallization during polymerization can occur from the monomer being in either gaseous or condensed state. It is thus also possible to get chain folded crystals below the glass transition temperature of the final polymer (e.g. 100 °C below Tg for poly (p-xylylenes)).

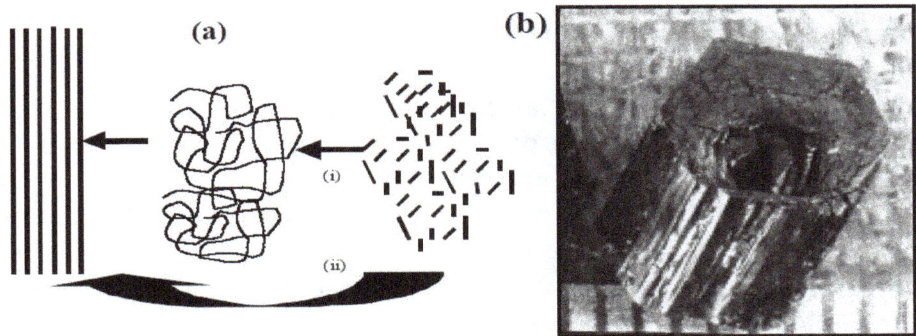

Figure: (a) Crystallization of macromolecules (i) polymerization followed by crystallization [(i.e.) separate polymerization and crystallization] (ii) crystallization during polymerization4 and (b) example of macroscopic single crystal obtained by simultaneous polymerization and crystallization-poly (sulfur nitride) 1 division = 0.5 mm

Crystallization Induced by Orientation

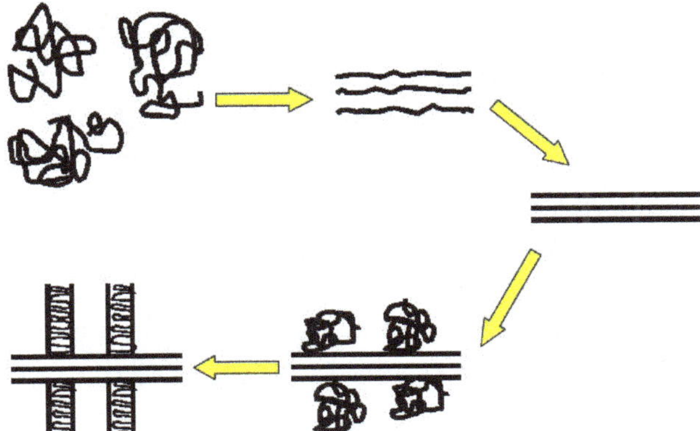

Figure: Schematic representation of orientation induced crystallization. The first three drawings illustrate the orientation and crystallization of random coils while the last two drawings show the growth of f olded chain kebabs around the central shish

The schematic of orientation induced crystallization is illustrated in figure. The process can be described as stretching of long chains to form fibrous crystals. In fact this is the underlying process governing the formation of fibers though any perfectly smooth and completely elongated chain morphology as illustrated in the schematic is difficult to attain under the most perfect of circumstances. During stretching, the distortion of chains from their most probable conformation results and hence a decrease in the conformational entropy takes place. If this deformation is maintained in this lower conformational entropy state then less conformational entropy needs to be sacrificed by transforming to a crystalline state. This decrease in total entropy of fusion allows the crystallization to occur at higher temperatures than will take place under quiescent conditions. Natural rubber and polyisobutylene are excellent examples of such an effect as they show great propensity to crystallize under stretched conditions whereas they crystallize slowly under quiescent conditions. Also, crystallization in an already oriented polymer results in reduc-tion retractive force (with respect to oriented state).

This can be explained on the crystallization induced by orientation. The schematic of orientation induced crystallization is illustrated in figure. The process can be described as stretching of long chains to form fibrous crystals. In fact this is the underlying process governing the formation of fibers though any perfectly smooth and completely elongated chain morphology as illustrated in the schematic is difficult to attain under the most perfect of circumstances. During stretching, the distortion of chains from their most probable conformation results and hence a decrease in the conformational entropy takes place. If this deformation is maintained in this lower conformational entropy state then less conformational entropy needs to be sacrificed by transforming to a crystalline state. This decrease in total entropy of fusion allows the crystallization to occur at higher temperatures than will take place under quiescent conditions. Natural rubber and polyisobutylene are excellent examples of such an effect as they show great propensity to crystallize under stretched conditions whereas they crystallize slowly under quiescent conditions. Also, crystallization in an already oriented polymer results in reduction retractive force (with respect to oriented state). This can be explained on the basis of rubbery elasticity theory according to which the force exerted by fixed chain ends is inversely proportional to number of statistical elements and the magnitude of end to end distance. The reduction in force results, due to lesser number of statistical units available in the amorphous regions and also because the end to end distance of the basis of rubbery elasticity theory according to which the force exerted by fixed chain ends is inversely proportional to number of statistical elements and the magnitude of end to end distance. The reduction in force results, due to lesser number of statistical units available in the amorphous regions and also because the end to end distance of the amorphous units is smaller than the end to end distance in the crystal. Melting of such elongated crystals lead to contraction and crystallization leads to elongation. Thus macroscopic dimensional changes and changes in retractive force can be related to the crystal-liquid phase transformation. Normally, the formation of such fibrous morphology is accompanied by formation of an epitaxial layer over and around the inner fiber giving rise to the socalled 'shish-kebab' kind of morphology. It is well documented that the outside 'kebab' like regions are essentially folded chain regions comprised of chains which did not crystallize during the orientation process. Thus, while the inner 'shish' regions form first, the formation of folded chain discs occurs due to nucleation events taking place on the extended chain surface. It is interesting that though the nature of the nucleating surface is a partially extended chain, and thus a great propensity to crystallize into the thermodynamically more favorable extended chain form exists, sub-sequent crystallization is still of the folded chain type. This has been used as a strong argument in favor of kinetic theories that argue for the chain folded model of crystallization. The lamellar kebabs are usually spaced at distances of ca. 200 to 1000 Å along the chain extended shish. Some researchers have studied the rate of growth perpendicular to the stretch direction and found it to be independent of the percentage extension (though the rate of extension was not a factor in these studies). The unaligned chains, which give rise to this undetachable plate like growth, can be the uncrystallized part of the main chain or totally separate chains. Some researchers found that the central shish was of higher molecular weight than the kebabs, while some have demonstrated that a minimum chain length was required for the chain extension process. It has also been argued that the growth of chain folded structures is aided in large part due to the dangling cilia which mostly result along the central fiber like morphology. These cilia, it has been proposed, then act as nucleation sites for the chain folded region to develop.

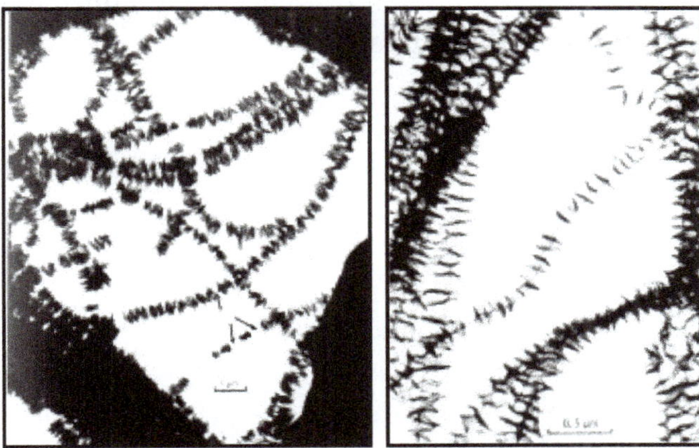

Figure: (a) Shish kebab morphology of polyethylene from solution. (b) Shish kebabs of cellulose formed by recrystallizing cellulose II onto microfibrils of high molecular weight

Crystallization under Quiescent Condition

Crystallization of long-chain flexible molecules of sufficient structural regularity is widely observed under quiescent conditions for a large number of macromolecules of both synthetic and natural origin. While it has been long established that similar to low molecular weight compounds, polymers can exhibit considerable long range order in the crystalline regions, the exact nature and morphological form of these crystalline regions (specifically at the molecular level) has been a matter of considerable debate. In this regard it is important to classify the quiescent polymer crystallization into two general types:

(1) Crystallization from dilute solutions; and

(2) Crystallization from the melt.

Crystallization from dilute solutions often provides a more fundamental avenue for structural analysis of polymer crystals as these entities can be isolated and precisely studied. Crystallization from the melt is often closer to pragmatic use of the polymer of interest though it adds an additional degree of difficulty to the fundamental structural studies. While this discussion will refer to results and attributes of dilute solution crystallization intermittently, it is crystallization from the melt that is of direct relevance to the present study. The nucleation, growth and kinetics of development of these crystalline regions are of both profound fundamental and practical interest. These characteristics are however directly linked to understanding of the morphological detail of these crystalline regions. On this account, there have been various models proposed over the past five decades- each involving considerable amount of controversy and debate, much of that debate persisting even to date. These models are elucidated below, some of which will be elaborated in more detail in the ensuing discussion. The type of morphology can, however, be first classified into two broad classes:

(1) The fringed micelle model and

(2) Lamellar type of morphology.

The models of lamellar morphology themselves differ on the basis of the nature of the fold surface, type of reentry of the chains and on accounts of presence of an intermediate region for the chain traveling from the crystal to the amorphous phase.

7

Ceramics and Composites

A ceramic is a non-metallic material that is comprised of an inorganic compound of metal, non-metal or metalloid held together through ionic or covalent bonds. A material formed from two or more materials having widely differing physical and chemical properties, such that the combined material has unique properties from its constituents is known as a composite. The topics included in this chapter cover the diverse aspects of ceramic and composite materials such as their properties, fabrication, processing, etc.

Ceramic

The term covers inorganic, nonmetallic materials that have been hardened by baking at a high temperature. Up until the 1950s or so, the most important of these were the traditional clays, made into pottery, dinnerware, bricks, tiles, and the like. Since then, new materials called advanced ceramics have been prepared and are being used for a wide range of applications, including components used by the aerospace, automotive, defense, environmental, fiber-optic, and medical technologies.

Ceramic parts are also used in cellular phones and personal computers. Each of NASA's space shuttles has a coating of roughly 34,000 ceramic tiles, which protect it from the searing heat (up to 2,300 °F) produced during reentry into the atmosphere. Thus, advanced ceramics are an integral part of the modern technological revolution.

The space shuttle has a coat of ceramic tiles that protect it from the searing heat produced during reentry into the atmosphere.

Composition and Classification

Traditional ceramics have been mainly silicate-based. Advanced ceramics are made from various other types of materials as well. Depending on their composition, they are classified as oxides, non-oxides, and composites.

- Oxides: alumina, zirconia.
- Non-oxides: carbides, borides, nitrides, silicides, and silicates.
- Composites: particulate reinforced, combinations of oxides and non-oxides.

The materials in each of these classes can have unique properties.

Ceramic Materials and their uses

A small sample of the high-temperature superconductor BSCCO-2223 (bismuth strontium calcium copper oxide); the two lines in the background are 1 mm apart.

- Barium titanate (often mixed with strontium titanate), which has properties called ferroelectricity and piezoelectricity, is widely used in electromechanical devices known as transducers, as well as in ceramic capacitors and data storage elements.
- Bismuth strontium calcium copper oxide (BSCCO) is a high-temperature superconductor.
- Boron carbide (B_4C) is used in some types of personal, helicopter and tank armor.
- Boron nitride takes on physical forms that are similar to those of carbon: a graphite-like form used as a lubricant, and a diamond-like one used as an abrasive.
- Bricks, which are mostly aluminum silicates, are used for construction.
- "Cermet," the name given to a composite of ceramic and metal, is used to produce capacitors, resistors, and other electronic parts for high-temperature applications.
- Earthenware is often made from clay, quartz, and feldspar.
- Ferrite (Fe_3O_4), which is ferrimagnetic, is used in the core of electrical transformers and in magnetic core memory.
- Lead zirconate titanate (also known as PZT) is a ferroelectric and piezoelectric material and has similar uses as barium titanate mentioned above.

- Magnesium diboride (MgB_2) is an unconventional superconductor.
- Porcelain, which usually contains the clay mineral kaolinite, is used to make decorative and household items.
- Silicon carbide (SiC) is used as an abrasive, a refractory material, and a "susceptor" that helps cook food in microwave furnaces.
- Silicon nitride (Si_3N_4) is used as an abrasive powder.
- Steatite (a type of soapstone) is used as an electrical insulator.
- Uranium oxide (UO_2) is used as fuel in nuclear reactors.
- Yttrium barium copper oxide ($YBa_2Cu_3O_{7-x}$) is another high-temperature superconductor.
- Zinc oxide (ZnO) is a semiconductor and is used in the construction of varistors (a class of electrical resistors).
- Zirconia, which in the pure form undergoes many phase changes when heated, can be chemically "stabilized" in several different forms. Most ceramic knife blades are made of this material. Also, as it is a good conductor of oxygen ions, it could be useful in fuel cells.

Other Applications

In the early 1980s, Toyota researched production of a ceramic engine that could run at a temperature above 6,000 °F (3,300 °C). Ceramic engines do not require a cooling system and hence permit major weight reduction and higher fuel efficiency. In a conventional metallic engine, much of the energy released by combustion of the fuel must be dissipated as waste heat, to prevent the metallic parts from melting. Despite these desirable properties, such engines are not being mass-produced because it is difficult to manufacture ceramic parts with the requisite precision and durability. Imperfections in the ceramic material can lead to cracks and potentially dangerous equipment failure.

Efforts are being made to develop ceramic parts for gas turbine engines. Currently, even blades made of advanced metal alloys for the hot section of an engine require cooling and careful limiting of operating temperatures. Turbine engines made with ceramics could operate more efficiently, giving aircraft greater range and payload for a set amount of fuel.

Since the late 1990s, highly specialized ceramics, usually based on boron carbide, have been used in ballistic armored vests to repel large-caliber rifle fire. Such plates are commonly known as "small-arms protective inserts" (SAPI). Similar technology is used to armor the cockpits of some military airplanes because of the lightness of the material.

Recent advances in ceramics include bio-ceramics such as dental implants and synthetic bones. Hydroxyapatite, the natural mineral component of bone, has been synthesized from a number of biological and chemical sources and can be formed into ceramic materials. Orthopedic implants made from these materials bond readily to bone and other tissues in the body without rejection or inflammatory reactions.

Hydroxyapatite ceramics, however, are usually porous and lack mechanical strength. They are therefore used to coat metal orthopedic devices, to aid in forming a bond to bone, or as bone fillers.

They are also used as fillers for orthopedic plastic screws to help reduce inflammation and increase absorption of the plastic materials. Work is being done to make strong, dense, nano-crystalline hydroxyapatite ceramics for orthopedic weight-bearing devices, replacing metal and plastic materials. Ultimately, these ceramic materials, with the incorporation of proteins called collagens, may be used to make synthetic bones.

Processing of Ceramic Materials

Non-crystalline ceramics, being glasses, are usually formed from melts. The glass is shaped when fully molten or when in a state of toffee-like viscosity, by methods such as casting or blowing to a mold. If the material becomes partly crystalline by later heat treatments, the resulting material is known as a "glass ceramic."

Crystalline ceramic materials are generally processed by one of two approaches: (a) the ceramic is made in the desired shape by reaction in situ, or (b) powders are "formed" into the desired shape and then heated ("sintered") until the particles adhere to one another and produce a solid body. Some methods use a hybrid of the two approaches.

In Situ Manufacturing

This method is most commonly used for producing cement and concrete. In this case, the dehydrated powders are mixed with water, which starts what are called hydration reactions. As a result, long, interlocking crystals begin to form around the aggregates. Over time, a solid ceramic is produced.

The biggest problem with this method is that most reactions are so fast that good mixing is not possible, preventing large-scale construction. On the other hand, small-scale systems can be made by "deposition" techniques—various materials (reactants) are introduced above a substrate, and made to react to form the ceramic on the substrate. This process borrows techniques from the semiconductor industry and is very useful for making ceramic coatings.

Sintering-based Approach

The term sintering refers to the process of heating a material at a temperature below its melting point, so that the particles adhere to one another. In a sintering-based approach, the object (called a "green body") that is prepared from a powder is initially held together loosely, but it is hardened by baking in a kiln. The pores in the object close up, so that the body shrinks and is converted to a denser, stronger product. There is virtually always some porosity left, but the real advantage of this method is that the green body can be produced in a wide variety of ways and then sintered.

Slurry can be used in place of a powder, then cast into a desired shape, dried, and sintered. This approach is used for traditional pottery, in which a plastic mixture is worked by hand.

When a mixture of different materials is used to form a ceramic, the sintering temperature is sometimes above the melting point of a minor component, leading to "liquid phase" sintering. In this case, the sintering time is shorter than for solid phase sintering.

Types and Applications of Ceramics

Ceramics greatly differ in their basic composition. The properties of ceramic materials also vary greatly due to differences in bonding, and thus found a wide range of engineering applications. Classification of ceramics based on their specific applications and composition are two most important ways among many. Based on their composition, ceramics are classified as:

- Oxides,
- Carbides,
- Nitrides,
- Sulfides,
- Fluorides, etc.

The other important classification of ceramics is based on their application, such as:

- Glasses,
- Clay products,
- Refractories,
- Abrasives,
- Cements, and
- Advanced ceramics.

In general, ceramic materials used for engineering applications can be divided into two groups: traditional ceramics, and the engineering ceramics. Typically, traditional ceramics are made from three basic components: clay, silica (flint) and feldspar. For example bricks, tiles and porcelain articles. However, engineering ceramics consist of highly pure compounds of aluminium oxide (Al_2O_3), silicon carbide (SiC) and silicon nitride (Si_3N_4).

- Glasses: glasses are a familiar group of ceramics – containers, windows, mirrors, lenses, etc. They are non-crystalline silicates containing other oxides, usually CaO, Na_2O, $K_2O$and Al_2O_3 which influence the glass properties and its color. Typical property of glasses that is important in engineering applications is its response to heating. There is no definite temperature at which the liquid transforms to a solid as with crystalline materials.

A specific temperature, known as glass transition temperature or fictive temperature is defined based on viscosity above which material is named as super cooled liquid or liquid, and below it is termed as glass.

- Clay products: clay is the one of most widely used ceramic raw material. It is found in great abundance and popular because of ease with which products are made. Clay products are mainly two kinds – structural products (bricks, tiles, sewer pipes) and white wares (porcelain, chinaware, pottery, etc.).
- Refractories: these are described by their capacity to withstand high temperatures without melting or decomposing; and their inertness in severe environments. Thermal insulation is also an important functionality of refractories.
- Abrasive ceramics: these are used to grind, wear, or cut away other material. Thus the

prime requisite for this group of materials is hardness or wear resistance in addition to high toughness. As they may also exposed to high temperatures, they need to exhibit some refractoriness. Diamond, silicon carbide, tungsten carbide, silica sand, aluminium oxide / corundum are some typical examples of abrasive ceramic materials.

- Cements: cement, plaster of paris and lime come under this group of ceramics. The characteristic property of these materials is that when they are mixed with water, they form slurry which sets subsequently and hardens finally. Thus it is possible to form virtually any shape. They are also used as bonding phase, for example between construction bricks.
- Advanced ceramics: these are newly developed and manufactured in limited range for specific applications. Usually their electrical, magnetic and optical properties and combination of properties are exploited. Typical applications: heat engines, ceramic armors, electronic packaging, etc.

Some typical ceramics and respective applications are as follows:

- Aluminium oxide / Alumina (Al_2O_3): it is one of most commonly used ceramic material. It is used in many applications such as to contain molten metal, where material is operated at very high temperatures under heavy loads, as insulators in spark plugs, and in some unique applications such as dental and medical use. Chromium doped alumina is used for making lasers.
- Aluminium nitride (AlN): because of its typical properties such as good electrical insulation but high thermal conductivity, it is used in many electronic applications such as in electrical circuits operating at a high frequency. It is also suitable for integrated circuits. Other electronic ceramics include – barium titanate ($BaTiO_3$) and Cordierite ($2MgO$-$2Al_2O_3$-$5SiO_2$).
- Diamond (C): it is the hardest material known to available in nature. It has many applications such as industrial abrasives, cutting tools, abrasion resistant coatings, etc. it is, of course, also used in jewelry.
- Lead zirconium titanate (PZT): it is the most widely used piezoelectric material, and is used as gas igniters, ultrasound imaging, in underwater detectors.
- Silica (SiO_2): is an essential ingredient in many engineering ceramics, thus is the most widely used ceramic material. Silica-based materials are used in thermal insulation, abrasives, laboratory glassware, etc. it also found application in communications media as integral part of optical fibers. Fine particles of silica are used in tires, paints, etc.
- Silicon carbide (SiC): it is known as one of best ceramic material for very high temperature applications. It is used as coatings on other material for protection from extreme temperatures. It is also used as abrasive material. It is used as reinforcement in many metallic and ceramic based composites. It is a semiconductor and often used in high temperature electronics. Silicon nitride (Si_3N_4) has properties similar to those of SiC but is somewhat lower, and found applications in such as automotive and gas turbine engines.
- Titanium oxide (TiO_2): it is mostly found as pigment in paints. It also forms part of certain glass ceramics. It is used to making other ceramics like $BaTiO_3$.

- Titanium boride (TiB_2): it exhibits great toughness properties and hence found applications in armor production. It is also a good conductor of both electricity and heat.
- Uranium oxide (UO_2): it is mainly used as nuclear reactor fuel. It has exceptional dimensional stability because its crystal structure can accommodate the products of fission process.
- Yttrium aluminium garnet (YAG, $Y_3Al_5O_{12}$): it has main application in lasers (Nd-YAG lasers)
- Zirconia (ZrO_2): it is also used in producing many other ceramic materials. It is also used in making oxygen gas sensors, as additive in many electronic ceramics. Its single crystals are part of jewelry.

Properties of Ceramic

Mechanical Properties

Ceramic materials are somewhat limited in applicability by their mechanical properties, which in many respects are inferior to those of metals.

Brittle Fracture of Ceramics

The brittle fracture process consists of the formation and propagation of cracks through the cross section of material in a direction perpendicular to the applied load. Crack growth in crystalline ceramics may be either trans granular (i.e., through the grains) or intergranular (i.e., along grain boundaries); for trans granular fracture, cracks propagate along specific crystallographic (or cleavage) planes, planes of high atomic density.

Stress–Strain Behavior

At room temperature, virtually all ceramics are brittle. Micro cracks, the presence of which is very difficult to control, result in amplification of applied tensile stresses and account for relatively low fracture strengths (flexural strengths). This amplification does not occur with compressive loads, and, consequently, ceramics are stronger in compression Engineering Materials.

The stress at fracture using this flexure test is known as the flexural strength, modulus of rupture, fracture strength, or the bend strength, an important mechanical parameter for brittle ceramics.

Hardness

One beneficial mechanical property of ceramics is their hardness, which is often utilized when an abrasive or grinding action is required; in fact, the hardest known materials are ceramics.

Creep

Often ceramic materials experience creep deformation as a result of exposure to stresses (usually compressive) at elevated temperatures. In general, the time deformation creep behavior of ceramics is similar to that of metals; however, creep occurs at higher temperatures in ceramics. High temperature compressive creep tests are conducted on ceramic materials to ascertain creep deformation as a function of temperature and stress level.

Electrical Properties

Semiconductivity

A number of ceramics are semiconductors. Most of these are oxides of transition metals, such as zinc oxide.

One common use of these semiconductors is for varistors. These are electrical resistors with the unusual property of "negative resistance." Once the voltage across the device reaches a certain threshold, a change in the electrical structure of the material causes its electrical resistance to drop from several megaohms down to a few hundred ohms. As a result, these materials can dissipate a lot of energy. In addition, they self-reset after the voltage across the device drops below a threshold, its resistance returns to being high.

This property makes them ideal for surge-protection applications. The best demonstration of their ability can be found in electrical substations, where they are employed to protect the infrastructure from lightning strikes. They have rapid response, require low maintenance, and do not appreciably degrade from use.

When various gases are passed over a polycrystalline ceramic, its electrical resistance changes. Based on this property, semiconducting ceramics are used to make inexpensive gas sensors.

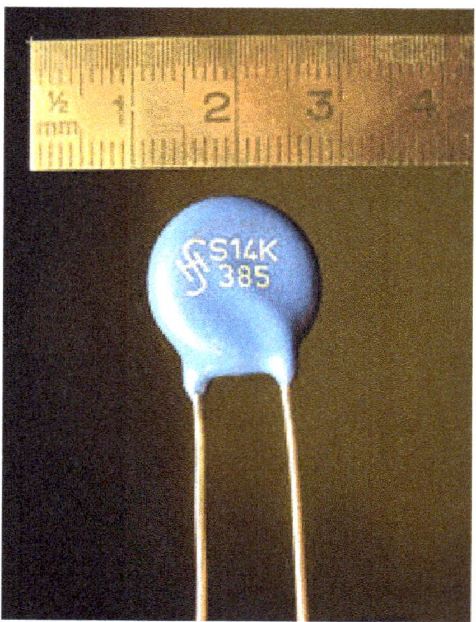

A 385-volt metal oxide varistor

Superconductivity

Under some conditions, such as extremely low temperatures, some ceramics exhibit superconductivity. The exact reason for this property is not known, but there are two major families of superconducting ceramics.

Piezoelectricity, Pyroelectricitya and Ferroelectricity

Many ceramic materials exhibit the property of piezoelectricity. A piezoelectric material develops a voltage difference between two faces when compressed or made to vibrate. This property links electrical and mechanical responses. Such ceramics are used in digital watches and other electronics that rely on quartz resonators. In these devices, electricity is used to produce a mechanical motion (powering the device) and the mechanical motion is in turn used to generate an electrical signal.

The piezoelectric effect is generally stronger in materials that also exhibit pyroelectricity. Such materials generate an electrical potential when heated or cooled. All pyroelectric materials are also piezoelectric. Thus, pyroelectric materials can be used to interconvert between thermal, mechanical, and electrical forms of energy. Such materials are used in motion sensors, where the tiny rise in temperature from a warm body entering a room is enough to produce a measurable voltage in the crystal.

Pyroelectricity, in turn, is observed most strongly in materials that also display the ferroelectric effect. Ferroelectric materials have a spontaneous polarization (formation of an electric dipole) that can be reversed by applying an electric field. Pyroelectricity is a necessary consequence of ferroelectricity.

Barium titanate and lead zirconate titanate have both piezoelectric and ferroelectric properties. They are used in the manufacture of capacitors, high-frequency loudspeakers, transducers for sonar, and actuators for atomic force and scanning tunneling microscopes.

Change of Electrical Properties with Temperature

Some crystalline semiconducting ceramics (mostly mixtures of heavy metal titanates) can conduct electricity as long as they are below a certain "transition" temperature. When heated above that temperature, their "grain boundaries" (boundaries between the little crystals that make up the material) suddenly become insulating, breaking the circuit. Such ceramics are used as self-controlled heating elements in, for example, the rear-window defrost circuits of most automobiles.

Fabrication and Processing of Ceramics

Ceramics melt at high temperatures and they exhibit a brittle behavior under tension. As a result, the conventional melting, casting and thermo-mechanical processing routes are not suitable to process the polycrystalline ceramics. Inorganic glasses, though, make use of lower melting temperatures due to formation of eutectics. Hence, most ceramic products are made from ceramic powders through powder processing starting with ceramic powders. The powder processing of ceramics is very close to that of metals, powder metallurgy. However there is an important consideration in ceramic-forming that is more prominent than in metal forming: it is dimensional tolerance. Post forming shrinkage is much higher in ceramics processing because of the large differential between the final density and the as-formed density.

Glasses, however, are produced by heating the raw materials to an elevated temperature above which melting occurs. Most commercial glasses are of the silica-soda-lime variety, where silica is supplied in form of common quartz sand, soda (Na_2O) in form of soda ash (Na_2CO_3) while the lime

(CaO) is supplied in form of limestone ($CaCO_3$). Different forming methods- pressing, blowing, drawing and fiber forming- are widely in practice to fabricate glass products. Thick glass objects such as plates and dishes are produced by pressing, while the blowing is used to produce objects like jars, bottles and light bulbs. Drawing is used to form long objects like tubes, rods, fibers, whiskers etc. The pressing and blowing process is shown in figure below.

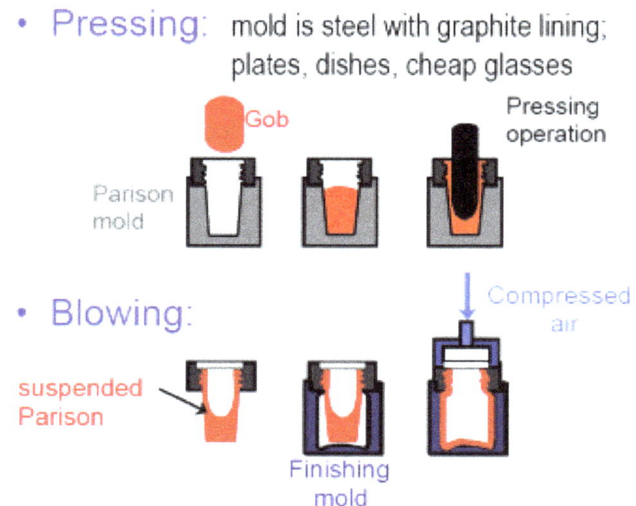

Figure: Schematic diagram of pressing and blowing processes.

Ceramic powder processing consists of powder production by milling/grinding, followed by fabrication of green product, which is then consolidated to obtain the final product. A powder is a collection of fine particles. Synthesis of powder involves getting it ready for shaping by crushing, grinding, separating impurities, blending different powders, drying to form soft agglomerates. Different techniques such as compaction, tape casting, slip casting, injection molding and extrusion are then used to convert processed powders into a desired shape to form what is known as green ceramic. The green ceramic is then consolidated further using a high-temperature treatment known as sintering or firing.

As-mined raw materials are put through a milling or grinding operation in which particle size is reduced to and physically 'liberate' the minerals of interest from the rest of the 'gangue' material. Wet milling is much more common with ceramic materials than with metals. The combination of dry powders with a dispersant such as water is called slurry. Ball- and vibratory- milling is employed to further reduce the size of minerals and to blend different powders.

Ceramic powders prepared are shaped using number of techniques, such as casting, compaction, extrusion/hydro-plastic forming, injection molding. Tape casting, also known as doctor blade process, is used for the production of thin ceramic tapes. The schematic diagram of tape casting process is shown in figure. In this technique slurry containing ceramic particles, solvent, plasticizers, and binders is then made to flow under a blade and onto a plastic substrate. The shear thinning slurry spreads under the blade. The tape is then dried using clean hot air. Later-on the tape is subjected to binder burnout and sintering operations. Tape thickness normally range between 0.1 and 2 mm. Commercially important electronic packages based on alumina substrates and barium titanate capacitors are made using this technique. A schematic diagram of doctor blade process is shown in the figure.

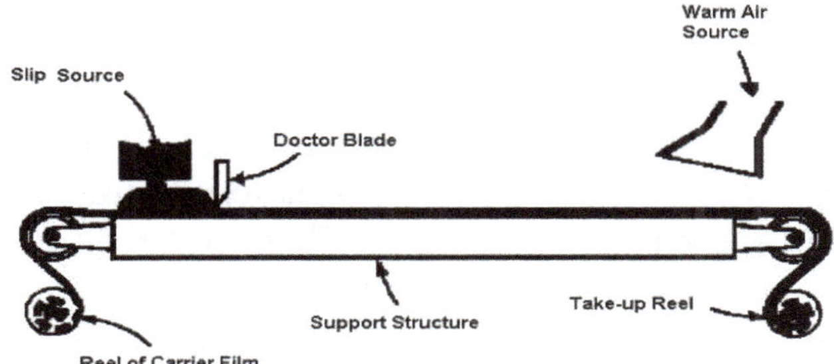

Figure: Schematic diagram of tape casting process.

Slip casting is another casting technique widely used. This technique uses aqueous slurry, also known as slip, of ceramic powder. The slip is poured into a plaster of Paris ($CaSO_4$:$2H_2O$) mold. As the water from slurry begins to move out by capillary action, a thick mass builds along the mold wall. When sufficient product thickness is built, the rest of the slurry is poured out (drain casting). It is also possible to continue to pour more slurry in to form a solid piece (solid casting). The schematic diagram of slip casting process is shown in figure below.

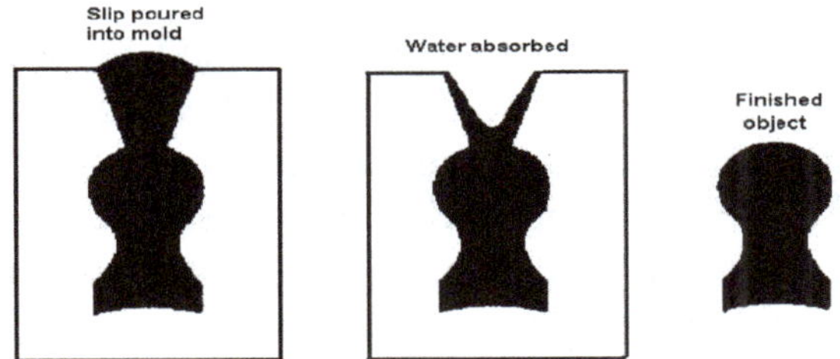

Figure 10.3: Schematic diagram of slip casting process

Extrusion and injection molding techniques are used to make products like tubes, bricks, tiles etc. The basis for extrusion process is a viscous mixture of ceramic particles, binder and other additives, which is fed through an extruder where a continuous shape of green ceramic is produced. The product is cut to required lengths and then dried and sintered. Injection molding of ceramics is similar to that of polymers. Ceramic powder is mixed with a plasticizer, a thermoplastic polymer, and additives. Then the mixture is injected into a die with use of an extruder. The polymer is then burnt off and the rest of the ceramic shape is sintered at suitable high temperatures. Ceramic injection molding is suitable for producing complex shapes. Figure shows schematically the injection molding process.

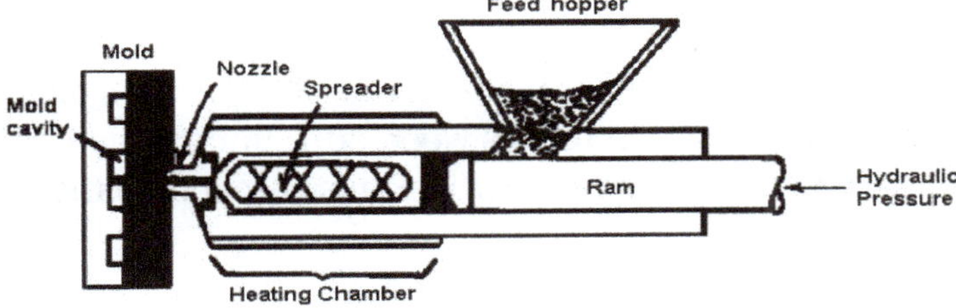

Figure: Schematic diagram of Injection molding.

Most popular technique to produce relatively simple shapes of ceramic products in large numbers is combination of compaction and sintering. For example: electronic ceramics, magnetic ceramics, cutting tools, etc. Compaction process is used to make green ceramics that have respectable strength and can be handled and machined. Time for compaction process varies from within a minute to hours depending on the complexity and size of the product. Basically compaction process involves applying equal pressure in all directions to a mixture ceramic powder to increase its density. In some cases, compaction involves application of pressure using oil/fluid at room temperatures, called cold iso-static pressing (CIP). Then the green ceramic is sintered with or without pressure. CIP is used to achieve higher ceramic density or where the compaction of more complex shapes is required. In some instances, parts may be produced under conditions in which compaction and sintering are conducted under pressure at elevated temperatures. This technique is known as hot iso-static pressing (HIP), and is used for refractory and covalently bonded ceramics that do not show good bonding characteristics under CIP. HIP is also used when close to none porosity is the requirement. Another characteristic feature of HIP is high densities can be achieved without appreciable grain growth.

Sintering is the firing process applied to green ceramics to increase its strength. Sintering is carried out below the melting temperature thus no liquid phase presents during sintering. However, for sintering to take place, the temperature must generally be maintained above one-half the absolute melting point of the material. During sintering, the green ceramic product shrinks and experiences a reduction in porosity. This leads to an improvement in its mechanical integrity. These changes involve different mass transport mechanisms that cause coalescence of powder particles into a more dense mass. With sintering, the grain boundary and bulk atomic diffusion contribute to densification, surface diffusion and evaporation condensation can cause grain growth, but do not cause densification. After pressing, ceramic particles touch one another. During initial stages of sintering, necks form along the contact regions between adjacent particles thus every interstice between particles becomes a pore. The pore channels in the compact grow in size, resulting in a significant increase in strength. With increase in sintering time, pores become smaller in size. The driving force for sintering process is the reduction in total particle surface area, and thus the reduction in total surface energy. During sintering, composition, impurity control and oxidation protection are provided by the use of vacuum conditions or inert gas atmospheres.

Composite Material

A composite material (or just composite) is a mixture of two or more materials with properties superior to the materials of which it is made. Many common examples of composite materials can be found in the world around us. Wood and bone are examples of natural composites. Wood consists of cellulose fibers embedded in a compound called lignin. The cellulose fibers give wood its ability to bend without breaking, while the lignin makes wood stiff. Bone is a combination of a soft form of protein known as collagen and a strong but brittle mineral called apatite.

Traditional Composites

Humans have been using composite materials for centuries, long before they fully understood the structures of such composites. The important building material concrete, for example, is a mixture of rocks, sand, and Portland cement. Concrete is a valuable building material because it is much stronger than any one of the individual components of which it is made. Interestingly enough, two of those components are themselves natural composites. Rock is a mixture of stony materials of various sizes, and sand is a composite of small-grained materials.

Reinforced concrete is a composite developed to further improve the strength of concrete. Steel rods embedded in concrete add both strength and flexibility to the concrete.

Cutting wheels designed for use with very hard materials are also composites. They are made by combining fine particles of tungsten carbide with cobalt powder. Tungsten carbide is one of the hardest materials known, so the composite formed by this method can be used to cut through almost any natural or synthetic material.

Classification of Composites

Composite materials are commonly classified at following two distinct levels:

- The first level of classification is usually made with respect to the matrix constituent. The major composite classes include Organic Matrix Composites (OMCs), Metal Matrix Composites (MMCs) and Ceramic Matrix Composites (CMCs). The term organic matrix composite is generally assumed to include two classes of composites, namely Polymer Matrix Composites (PMCs) and carbon matrix composites commonly referred to as carbon-carbon composites.
- The second level of classification refers to the reinforcement form - fibre reinforced composites, laminar composites and particulate composites. Fibre Reinforced composites (FRP) can be further divided into those containing discontinuous or continuous fibres.
- Fibre Reinforced Composites are composed of fibres embedded in matrix material. Such a composite is considered to be a discontinuous fibre or short fibre composite if its properties vary with fibre length. On the other hand, when the length of the fibre is such that any further increase in length does not further increase, the elastic modulus of the composite, the composite is considered to be continuous fibre reinforced. Fibres are small in diameter and when pushed axially, they bend easily although they have very good tensile properties. These fibres must be supported to keep individual fibres from bending and buckling.

- Laminar Composites are composed of layers of materials held together by matrix. Sandwich structures fall under this category.
- Particulate Composites are composed of particles distributed or embedded in a matrix body. The particles may be flakes or in powder form. Concrete and wood particle boards are examples of this category.

Organic Matrix Composites

Polymer Matrix Composites (PMC)/Carbon Matrix Composites or Carbon-Carbon Composites

Polymers make ideal materials as they can be processed easily, possess lightweight, and desirable mechanical properties. It follows, therefore, that high temperature resins are extensively used in aeronautical applications.

Two main kinds of polymers are thermosets and thermoplastics. Thermosets have qualities such as a well-bonded three-dimensional molecular structure after curing. They decompose instead of melting on hardening. Merely changing the basic composition of the resin is enough to alter the conditions suitably for curing and determine its other characteristics. They can be retained in a partially cured condition too over prolonged periods of time, rendering Thermosets very flexible. Thus, they are most suited as matrix bases for advanced conditions fiber reinforced composites. Thermosets find wide ranging applications in the chopped fiber composites form particularly when a premixed or moulding compound with fibers of specific quality and aspect ratio happens to be starting material as in epoxy, polymer and phenolic polyamide resins.

Thermoplastics have one- or two-dimensional molecular structure and they tend to at an elevated temperature and show exaggerated melting point. Another advantage is that the process of softening at elevated temperatures can reversed to regain its properties during cooling, facilitating applications of conventional compress techniques to mould the compounds.

Resins reinforced with thermoplastics now comprised an emerging group of composites. The theme of most experiments in this area to improve the base properties of the resins and extract the greatest functional advantages from them in new avenues, including attempts to replace metals in die-casting processes. In crystalline thermoplastics, the reinforcement affects the morphology to a considerable extent, prompting the reinforcement to empower nucleation. Whenever crystalline or amorphous, these resins possess the facility to alter their creep over an extensive range of temperature. But this range includes the point at which the usage of resins is constrained, and the reinforcement in such systems can increase the failure load as well as creep resistance. Figure below shows kinds of thermoplastics.

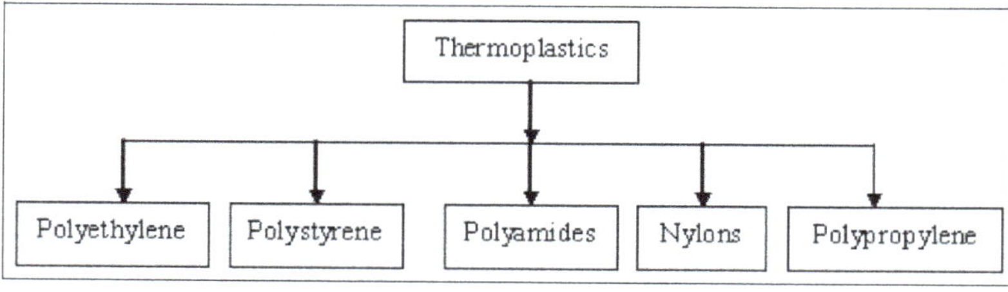

Figure: Thermoplastics

A small quantum of shrinkage and the tendency of the shape to retain its original form are also to be accounted for. But reinforcements can change this condition too. The advantage of thermoplastics systems over thermosets are that there are no chemical reactions involved, which often result in the release of gases or heat. Manufacturing is limited by the time required for heating, shaping and cooling the structures.

Thermoplastics resins are sold as moulding compounds. Fiber reinforcement is apt for these resins. Since the fibers are randomly dispersed, the reinforcement will be almost isotropic. However, when subjected to moulding processes, they can be aligned directionally.

There are a few options to increase heat resistance in thermoplastics. Addition of fillers raises the heat resistance. But all thermoplastic composites tend loose their strength at elevated temperatures. However, their redeeming qualities like rigidity, toughness and ability to repudiate creep, place thermoplastics in the important composite materials bracket. They are used in automotive control panels, electronic products encasement etc.

Newer developments augur the broadening of the scope of applications of thermoplastics. Huge sheets of reinforced thermoplastics are now available and they only require sampling and heating to be moulded into the required shapes. This has facilitated easy fabrication of bulky components, doing away with the more cumbersome moulding compounds.

Thermosets are the most popular of the fiber composite matrices without which, research and development in structural engineering field could get truncated. Aerospace components, automobile parts, defense systems etc., use a great deal of this type of fiber composites. Epoxy matrix materials are used in printed circuit boards and similar areas. Figure below shows some kinds of thermosets.

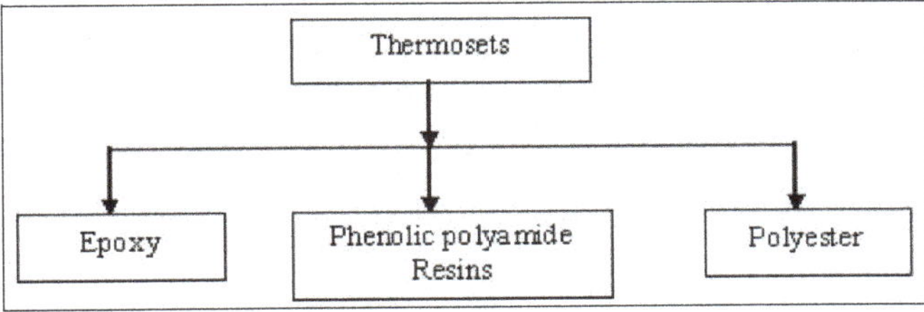

Figure: Thermoset

Materials Direct condensation polymerization followed by rearrangement reactions to form heterocyclic entities is the method generally used to produce thermoset resins. Water, a product of the reaction, in both methods, hinders production of void-free composites. These voids have a negative effect on properties of the composites in terms of strength and dielectric properties. Polyesters phenolic and Epoxies are the two important classes of thermoset resins.

Epoxy resins are widely used in filament-wound composites and are suitable for moulding prepress. They are reasonably stable to chemical attacks and are excellent adherents having slow shrinkage during curing and no emission of volatile gases. These advantages, however, make the use of epoxies rather expensive. Also, they cannot be expected beyond a temperature of 140°C. Their use in high technology areas where service temperatures are higher, as a result, is ruled out.

Polyester resins on the other hand are quite easily accessible, cheap and find use in a wide range of fields. Liquid polyesters are stored at room temperature for months, sometimes for years and the mere addition of a catalyst can cure the matrix material within a short time. They are used in automobile and structural applications.

The cured polyester is usually rigid or flexible as the case may be and transparent. Polyesters withstand the variations of environment and stable against chemicals. Depending on the formulation of the resin or service requirement of application, they can be used up to about 75°C or higher. Other advantages of polyesters include easy compatibility with few glass fibers and can be used with verify of reinforced plastic accoutrey.

Aromatic Polyamides are the most sought after candidates as the matrices of advanced fiber composites for structural applications demanding long duration exposure for continuous service at around 200-250°C .

Metal Matrix Composites (MMC)

Metal matrix composites, at present though generating a wide interest in research fraternity, are not as widely in use as their plastic counterparts. High strength, fracture toughness and stiffness are offered by metal matrices than those offered by their polymer counterparts. They can withstand elevated temperature in corrosive environment than polymer composites. Most metals and alloys could be used as matrices and they require reinforcement materials which need to be stable over a range of temperature and non-reactive too. However the guiding aspect for the choice depends essentially on the matrix material. Light metals form the matrix for temperature application and the reinforcements in addition to the aforementioned reasons are characterized by high moduli.

Most metals and alloys make good matrices. However, practically, the choices for low temperature applications are not many. Only light metals are responsive, with their low density proving an advantage. Titanium, Aluminium and magnesium are the popular matrix metals currently in vogue, which are particularly useful for aircraft applications. If metallic matrix materials have to offer high strength, they require high modulus reinforcements. The strength-to-weight ratios of resulting composites can be higher than most alloys.

The melting point, physical and mechanical properties of the composite at various temperatures determine the service temperature of composites. Most metals, ceramics and compounds can be used with matrices of low melting point alloys. The choice of reinforcements becomes more stunted with increase in the melting temperature of matrix materials.

Ceramic Matrix Materials (CMM)

Ceramics can be described as solid materials which exhibit very strong ionic bonding in general and in few cases covalent bonding. High melting points, good corrosion resistance, stability at elevated temperatures and high compressive strength, render ceramic-based matrix materials a favorite for applications requiring a structural material that doesn't give way at temperatures above 1500°C. Naturally, ceramic matrices are the obvious choice for high temperature applications.

High modulus of elasticity and low tensile strain, which most ceramics posses, have combined to cause the failure of attempts to add reinforcements to obtain strength improvement. This is because at the stress levels at which ceramics rupture, there is insufficient elongation of the matrix which keeps composite from transferring an effective quantum of load to the reinforcement and the composite may fail unless the percentage of fiber volume is high enough. A material is reinforcement to utilize the higher tensile strength of the fiber, to produce an increase in load bearing capacity of the matrix. Addition of high-strength fiber to a weaker ceramic has not always been successful and often the resultant composite has proved to be weaker.

The use of reinforcement with high modulus of elasticity may take care of the problem to some extent and presents pre-stressing of the fiber in the ceramic matrix is being increasingly resorted to as an option.

When ceramics have a higher thermal expansion coefficient than reinforcement materials, the resultant composite is unlikely to have a superior level of strength. In that case, the composite will develop strength within ceramic at the time of cooling resulting in micro cracks extending from fiber to fiber within the matrix. Micro cracking can result in a composite with tensile strength lower than that of the matrix.

Fiber Reinforced Composites/Fibre Reinforced Polymer (FRP) Composites

Fibers are the important class of reinforcements, as they satisfy the desired conditions and transfer strength to the matrix constituent influencing and enhancing their properties as desired.

Glass fibers are the earliest known fibers used to reinforce materials. Ceramic and metal fibers were subsequently found out and put to extensive use, to render composites stiffer more resistant to heat.

Fibers fall short of ideal performance due to several factors. The performance of a fiber composite is judged by its length, shape, orientation, and composition of the fibers and the mechanical properties of the matrix.

The orientation of the fiber in the matrix is an indication of the strength of the composite and the strength is greatest along the longitudinal directional of fiber. This doesn't mean the longitudinal fibers can take the same quantum of load irrespective of the direction in which it is applied. Optimum performance from longitudinal fibers can be obtained if the load is applied along its direction. The slightest shift in the angle of loading may drastically reduce the strength of the composite.

Unidirectional loading is found in few structures and hence it is prudent to give a mix of orientations for fibers in composites particularly where the load is expected to be the heaviest.

Monolayer tapes consisting of continuous or discontinuous fibers can be oriented unidirectional stacked into plies containing layers of filaments also oriented in the same direction. More complicated orientations are possible too and nowadays, computers are used to make projections of such variations to suit specific needs. In short, in planar composites, strength can be changed from unidirectional fiber oriented composites that result in composites with nearly isotropic properties.

Properties of angle-plied composites which are not quasi-isotropic may vary with the number of plies and their orientations. Composite variables in such composites are assumed to have a constant ratio and the matrices are considered relatively weaker than the fibers. The strength of the fiber in any one of the three axes would, therefore be one-third the unidirectional fiber composite, assuming that the volume percentage is equal in all three axes.

However, orientation of short fibers by different methods is also possible like random orientations by sprinkling on to given plane or addition of matrix in liquid or solid state before or after the fiber deposition. Even three-dimensional orientations can achieve in this way. There are several methods of random fiber orientations, which in a two-dimensional one, yield composites with one-third the strength of a unidirectional fiber-stressed composite, in the direction of fibers. In a 3-dimension, it would result in a composite with a comparable ratio, about less than one-fifth.

In very strong matrices, moduli and strengths have not been observed. Application of the strength of the composites with such matrices and several orientations is also possible. The longitudinal strength can be calculated on the basis of the assumption that fibers have been reduced to their effective strength on approximation value in composites with strong matrices and non-longitudinally orientated fibers.

It goes without saying that fiber composites may be constructed with either continuous or short fibers. Experience has shown that continuous fibers (or filaments) exhibit better orientation, although it does not reflect in their performance. Fibers have a high aspect ratio, i.e., their lengths being several times greater than their effective diameters. This is the reason why filaments are manufactured using continuous process. This finished filaments.

Mass production of filaments is well known and they match with several matrices in different ways like winding, twisting, weaving and knitting, which exhibit the characteristics of a fabric. Since they have low densities and high strengths, the fiber lengths in filaments or other fibers yield considerable influence on the mechanical properties as well as the response of composites to processing and procedures. Shorter fibers with proper orientation composites that use glass, ceramic or multi-purpose fibers can be endowed with considerably higher strength than those that use continuous fibers. Short fibers are also known to their theoretical strength. The continuous fiber constituent of a composite is often joined by the filament winding process in which the matrix impregnated fiber wrapped around a mandrel shaped like the part over which the composite is to be placed, and equitable load distribution and favorable orientation of the fiber is possible in the finished product. However, winding is mostly confined to fabrication of bodies of revolution and the occasional irregular, flat surface.

Short-length fibers incorporated by the open- or close-mould process are found to be less efficient, although the input costs are considerably lower than filament winding.

Most fibers in use currently are solids which are easy to produce and handle, having a circular cross-section, although a few non-conventional shaped and hollow fibers show signs of capabilities that can improve the mechanical qualities of the composites.

Given the fact that the vast difference in length and effective diameter of the fiber are assets to a fiber composite, it follows that greater strength in the fiber can be achieved by smaller diameters due to minimization or total elimination of surface of surface defects.

After flat-thin filaments came into vogue, fibers rectangular cross sections have provided new options for applications in high strength structures. Owing to their shapes, these fibers provide perfect packing, while hollow fibers show better structural efficiency in composites that are desired for their stiffness and compressive strengths. In hollow fibers, the transverse compressive strength is lower than that of a solid fiber composite whenever the hollow portion is more than half the total fiber diameter. However, they are not easy to handle and fabricate.

Laminar Composites

Laminar composites are found in as many combinations as the number of materials. They can be described as materials comprising of layers of materials bonded together. These may be of several layers of two or more metal materials occurring alternately or in a determined order more than once, and in as many numbers as required for a specific purpose.

Clad and sandwich laminates have many areas as it ought to be, although they are known to follow the rule of mixtures from the modulus and strength point of view. Other intrinsic values pertaining to metal-matrix, metal-reinforced composites are also fairly well known.

Powder metallurgical processes like roll bonding, hot pressing, diffusion bonding, brazing and so on can be employed for the fabrication of different alloys of sheet, foil, powder or sprayed materials. It is not possible to achieve high strength materials unlike the fiber version. But sheets and foils can be made isotropic in two dimensions more easily than fibers. Foils and sheets are also made to exhibit high percentages of which they are put. For instance, a strong sheet may use over 92% in laminar structure, while it is difficult to make fibers of such compositions. Fiber laminates cannot over 75% strong fibers.

The main functional types of metal-metal laminates that do not posses high strength or stiffness are single layered ones that endow the composites with special properties, apart from being cost-effective. They are usually made by pre-coating or cladding methods.

Pre-coated metals are formed by forming by forming a layer on a substrate, in the form of a thin continuous film. This is achieved by hot dipping and occasionally by chemical plating and electroplating. Clad metals are found to be suitable for more intensive environments where denser faces are required.

There are many combinations of sheet and foil which function as adhesives at low temperatures. Such materials, plastics or metals, may be clubbed together with a third constituent. Pre-painted or pre-finished metal whose primary advantage is elimination of final finishing by the user is the best known metal-organic laminate. Several combinations of metal-plastic, vinyl-metal laminates, organic films and metals, account for up to 95% of metal-plastic laminates known. They are made by adhesive bonding processes.

Particulate Reinforced Composites (PRC)

Microstructures of metal and ceramics composites, which show particles of one phase strewn in the other, are known as particle reinforced composites. Square, triangular and round shapes of reinforcement are known, but the dimensions of all their sides are observed to be more or less equal. The size and volume concentration of the dispersoid distinguishes it from dispersion hardened materials.

The dispersed size in particulate composites is of the order of a few microns and volume concentration is greater than 28%. The difference between particulate composite and dispersion strengthened ones is, thus, oblivious. The mechanism used to strengthen each of them is also different. The dispersed in the dispersion-strengthen materials reinforces the matrix alloy by arresting motion of dislocations and needs large forces to fracture the restriction created by dispersion.

In particulate composites, the particles strengthen the system by the hydrostatic coercion of fillers in matrices and by their hardness relative to the matrix.

Three-dimensional reinforcement in composites offers isotropic properties, because of the three systematical orthogonal planes. Since it is not homogeneous, the material properties acquire sensitivity to the constituent properties, as well as the interfacial properties and geometric shapes of the array. The composite's strength usually depends on the diameter of the particles, the inter-particle spacing, and the volume fraction of the reinforcement. The matrix properties influence the behavior of particulate composite too.

Advantages of Composites

- Light Weight: Composites are light in weight, compared to most woods and metals. Their lightness is important in automobiles and aircraft, for example, where less weight means better fuel efficiency (more miles to the gallon). People who design airplanes are greatly concerned with weight, since reducing a craft's weight reduces the amount of fuel it needs and increases the speeds it can reach. Some modern airplanes are built with more composites than metal including the new Boeing 787, Dreamliner.
- High Strength: Composites can be designed to be far stronger than aluminum or steel. Metals are equally strong in all directions. But composites can be engineered and designed to be strong in a specific direction.
- Strength Related to Weight: Strength-to-weight ratio is a material's strength in relation to how much it weighs. Some materials are very strong and heavy, such as steel. Other materials can be strong and light, such as bamboo poles. Composite materials can be designed to be both strong and light. This property is why composites are used to build airplanes—which need a very high strength material at the lowest possible weight. A composite can be made to resist bending in one direction, for example. When something is built with metal, and greater strength is needed in one direction, the material usually must be made thicker, which adds weight. Composites can be strong without being heavy. Composites have the highest strength-to-weight ratios in structures today.
- Corrosion Resistance: Composites resist damage from the weather and from harsh chemicals that can eat away at other materials. Composites are good choices where chemicals are handled or stored. Outdoors, they stand up to severe weather and wide changes in temperature.
- High-Impact Strength: Composites can be made to absorb impacts—the sudden force of a bullet, for instance, or the blast from an explosion. Because of this property, composites are used in bulletproof vests and panels, and to shield airplanes, buildings, and military vehicles from explosions.
- Design Flexibility: Composites can be molded into complicated shapes more easily than most other materials. This gives designers the freedom to create almost any shape or form. Most recreational boats today, for example, are built from fiberglass composites because these materials can easily be molded into complex shapes, which improve boat design while lowering costs. The surface of composites can also be molded to mimic any surface finish or texture, from smooth to pebbly.

- Part Consolidation: A single piece made of composite materials can replace an entire assembly of metal parts. Reducing the number of parts in a machine or a structure saves time and cuts down on the maintenance needed over the life of the item.
- Dimensional Stability: Composites retain their shape and size when they are hot or cool, wet or dry. Wood, on the other hand, swells and shrinks as the humidity changes. Composites can be a better choice in situations demanding tight fits that do not vary. They are used in aircraft wings, for example, so that the wing shape and size do not change as the plane gains or loses altitude.
- Nonconductive: Composites are nonconductive, meaning they do not conduct electricity. This property makes them suitable for such items as electrical utility poles and the circuit boards in electronics. If electrical conductivity is needed, it is possible to make some composites conductive.
- Nonmagnetic: Composites contain no metals; therefore, they are not magnetic. They can be used around sensitive electronic equipment. The lack of magnetic interference allows large magnets used in MRI (magnetic resonance imaging) equipment to perform better. Composites are used in both the equipment housing and table. In addition, the construction of the room uses composites rebar to reinforced the concrete walls and floors in the hospital.
- Radar Transparent: Radar signals pass right through composites, a property that makes composites ideal materials for use anywhere radar equipment is operating, whether on the ground or in the air. Composites play a key role in stealth aircraft, such as the U.S. Air Force's B-2 stealth bomber, which is nearly invisible to radar.
- Low Thermal Conductivity: Composites are good insulators—they do not easily conduct heat or cold. They are used in buildings for doors, panels, and windows where extra protection is needed from severe weather.
- Durable: Structures made of composites have a long life and need little maintenance. We do not know how long composites last, because we have not come to the end of the life of many original composites. Many composites have been in service for half a century.

Fabrication of Composite Materials

There are numerous methods for fabricating composite components. Some methods have been borrowed (injection molding, for example), but many were developed to meet specific design or manufacturing challenges. Selection of a method for a particular part, therefore, will depend on the materials, the part design and end-use or application. Composite fabrication processes involve some form of molding, to shape the resin and reinforcement. A mold tool is required to give the unformed resin /fiber combination its shape prior to and during cure.

The most basic fabrication method for thermoset composites is hand layup, which typically consists of laying dry fabric layers, or "plies," or prepreg plies, by hand onto a tool to form a laminate stack. Resin is applied to the dry plies after layup is complete (e.g., by means of resin infusion).

Several curing methods are available. The most basic is simply to allow cure to occur at room temperature. Cure can be accelerated, however, by applying heat, typically with an oven, and pressure, by means of a vacuum. Many high-performance thermoset parts require heat and high consolidation pressure to cure — conditions that require the use of an autoclave. Autoclaves, generally, are expensive to buy and operate. Manufacturers that are equipped with autoclaves usually cure a number of parts simultaneously. Computer systems monitor and control autoclave temperature, pressure, vacuum and inert atmosphere, which allows unattended and/or remote supervision of the cure process and maximizes efficient use of the technique. Electron-beam (E-beam) curing has been explored as an efficient curing method for thin laminates. In E-beam curing, the composite layup is exposed to a stream of electrons that provide ionizing radiation, causing polymerization and crosslinking in radiation sensitive resins. X-ray and microwave curing technologies work in a similar manner. A fourth alternative, ultraviolet (UV) curing, involves the use of UV radiation to activate a photoinitiator added to a thermoset resin, which, when activated, sets off a crosslinking reaction. UV curing requires light-permeable resin and reinforcements.

Open Molding

Open contact molding in one-sided molds is a low-cost, common process for making fiberglass composite products. Typically used for boat hulls and decks, RV components, truck cabs and fenders, spas, bathtubs, shower stalls and other relatively large, noncomplex shapes, open molding involves either hand layup or a semi-automated alternative, spray-up. In an open-mold spray-up application, the mold is first treated with mold release. If a gel coat is used, it is typically sprayed into the mold after the mold release has been applied. The gel coat then is cured and the mold is ready for fabrication to begin. In the spray-up process, catalyzed resin (viscosity from 500 to 1,000 cps) and glass fiber are sprayed into the mold using a chopper gun, which chops continuous fiber into short lengths, then blows the short fibers directly into the sprayed resin stream so that both materials are applied simultaneously. To reduce VOCs, piston pump-activated, non-atomizing spray guns and fluid impingement spray heads dispense gel coats and resins in larger droplets at low pressure. Another option is a roller impregnator, which pumps resin into a roller similar to a paint roller.

In the final steps of the spray-up process, workers compact the laminate by hand with rollers. Wood, foam or other core material may then be added, and a second spray-up layer imbeds the core between the laminate skins. The part is then cured, cooled and removed from the reusable mold. Hand layup and spray-up methods are often used in tandem to reduce labor.

Resin Infusion Processes

Ever-increasing demand for faster production rates has pressed the industry to replace hand layup with alternative fabrication processes and has encouraged fabricators to automate those processes wherever possible.

A common alternative is resin transfer molding (RTM), sometimes referred to as liquid molding. The benefits of RTM are impressive. Generally, the dry preforms and resins used in RTM are less expensive than prepreg material and can be stored at room temperature. The process can produce thick, near-net shape parts, eliminating most post-fabrication work. It also yields dimensionally accurate complex parts with good surface detail and delivers a smooth finish on all exposed surfaces. It is possible to place inserts inside the preform before the mold is closed, allowing the RTM process to accommodate core materials and integrate "molded in" fittings and other hardware into the part structure. Finally, RTM significantly cuts cycle times and can be adapted for use as one stage in an automated, repeatable manufacturing process for even greater efficiency, reducing cycle time from what can be several days, typical of hand layup, to just hours — or even minutes.

In contrast to RTM, where resin and catalyst are premixed prior to injection under pressure into the mold, reaction injection molding (RIM) injects a rapid-cure resin and a catalyst into the mold in two separate streams. Mixing and the resulting chemical reaction occur in the mold instead of in a dispensing head. Automotive industry suppliers combine structural RIM (SRIM) with rapid preforming methods to fabricate structural parts that don't require a Class A finish. Programmable robots have become a common means to spray a chopped fiberglass/binder combination onto a vacuum equipped preform screen or mold. Robotic spray-up can be directed to control fiber orientation. A related technology, dry fiber placement, combines stitched preforms and RTM. Fiber volumes of up to 68 percent are possible, and automated controls ensure low voids and consistent preform reproduction, without the need for trimming. Vacuum-assisted resin transfer molding (VARTM) refers to a variety of related processes that represent the fastest-growing new molding technology. The salient difference between VARTM-type processes and RTM is that in VARTM, resin is drawn into a preform through use of a vacuum only, rather than pumped in under pressure. VARTM does not require high heat or pressure. For that reason, VARTM operates with low-cost tooling, making it possible to inexpensively produce large, complex parts in one shot.

In the VARTM process, fiber reinforcements are placed in a one-sided mold, and a cover (typically a plastic bagging film) is placed over the top to form a vacuum-tight seal. The resin typically enters the structure through strategically placed ports and feed lines, termed a "manifold." It is drawn by vacuum through the reinforcements by means of a series of designed-in channels that facilitate wetout of the fibers. Fiber content in the finished part can run as high as 70 percent. Current applications include marine, ground transportation and infrastructure parts. This method has been employed by The Boeing Co. (Chicago, Ill.) and NASA, as well as small fabricating firms, to produce aerospace-quality laminates without an autoclave.

High-volume Molding Methods

Compression molding is a high-volume thermoset molding process that employs expensive but very durable metal dies. It is an appropriate choice when production quantities exceed 10,000 parts. As many as 200,000 parts can be turned out on a set of forged steel dies, using sheet molding compound (SMC), a composite sheet material made by sandwiching chopped fiberglass between two layers of thick resin paste. Low-pressure SMC formulations that are now on the market offer open molders low-capital investment entry into closed-mold processing with near-zero VOC emissions and the potential for very high-quality surface finish.

Automakers are exploring carbon fiber-reinforced SMC, hoping to take advantage of carbon's high strength- and stiffness-to-weight ratios in exterior body panels and other parts. Newer, toughened SMC formulations help prevent micro cracking, a phenomenon that previously caused paint "pops" during the painting process (surface craters caused by outgassing, the release of gasses trapped in the micro cracks during oven cure). Composites manufacturers in industrial markets are formulating their own resins and compounding SMC in-house to meet needs in specific applications that require UV, impact and moisture resistance and have surface-quality demands that drive the need for customized material development.

Injection molding is a fast, high-volume, low pressure, closed process using, most commonly, filled thermoplastics, such as nylon with chopped glass fiber. In the past 20 years, however, automated injection molding of BMC has taken over some markets previously held by thermoplastic and metal casting manufacturers. For example, the first-ever BMC-based electronic throttle control (ETC) valves (previously molded only from die-cast aluminum) debuted on engines in the BMW Mini and the Peugeot 207, taking advantage of dimensional stability offered by a specially-formulated BMC supplied by TetraDUR GmbH (Hamburg, Germany), a subsidiary of Bulk Molding Compounds Inc. (BMCI, West Chicago, Ill.,).Injection speeds are typically one to five seconds, and as many as 2,000 small parts can be produced per hour in some multiple-cavity molds.

Parts with thick cross-sections can be compression molded or transfer molded with BMC. Transfer molding is a closed-mold process wherein a measured charge of BMC is placed in a pot with runners that lead to the mold cavities. A plunger forces the material into the cavities, where the product cures under heat and pressure.

Filament winding is a continuous fabrication method that can be highly automated and repeatable, with relatively low material costs. Filament winding yields parts with exceptional circumferential or "hoop" strength. The highest-volume single application of filament winding is golf club shafts. Fishing rods, pipe, pressure vessels and other cylindrical parts comprise most of the remaining business.

Pultrusion, like RTM, has been used for decades with glass fiber and polyester resins, but in the last 10 years the process also has found application in advanced composites applications. In this relatively simple, low-cost, continuous process, the reinforcing fiber (usually roving, tow or continuous mat) is typically pulled through a heated resin bath and then formed into specific shapes as it passes through one or more forming guides or bushings. The material then moves through a heated die, where it takes its net shape and cures. Further downstream, after cooling, the resulting profile is cut to desired length. Pultrusion yields smooth finished parts that typically do not require post processing. A wide range of continuous, consistent, solid and hollow profiles are pultruded, and the process can be custom-tailored to fit specific applications

Tube rolling is a longstanding composites manufacturing process that can produce finite-length tubes and rods. It is particularly applicable to small diameter cylindrical or tapered tubes in lengths as great as 20 ft/6.2m. Tubing diameters up to 6 inches/152 mm can be rolled efficiently. Typically, a tacky prepreg fabric or unidirectional tape is used, depending on the part. The material is precut in patterns that have been designed to achieve the requisite ply schedule and fiber architecture for the application. The pattern pieces are laid out on a flat surface and a mandrel is rolled over each one under applied pressure, which compacts and debulks the material. When rolling a tapered mandrel — e.g., for a fishing rod or golf shaft — only the first row of longitudinal fibers falls on the true 0° axis. To impart bending strength to the tube, therefore, the fibers must be continuously reoriented by repositioning the pattern pieces at regular intervals.

Automated fiber placement (AFP) The fiber placement process automatically places multiple individual prepreg tows onto a mandrel at high speed, using a numerically controlled, articulating robotic placement head to dispense, clamp, cut and restart as many as 32 tows simultaneously. Minimum cut length (the shortest tow length a machine can lay down) is the essential ply-shape determinant. The fiber placement heads can be attached to a 5-axis gantry, retrofitted to a filament winder or delivered as a turnkey custom system. Machines are available with dual mandrel stations to increase productivity. Advantages of fiber placement include processing speed, reduced material scrap and labor costs, parts consolidation and improved part-to-part uniformity. Often, the process is used to produce large thermoset parts with complex shapes.

Automated tape laying (ATL) is an even speedier automated process in which prepreg tape, rather than single tows, is laid down continuously to form parts. It is often used for parts with highly complex contours or angles. Tape layup is versatile, allowing breaks in the process and easy direction changes, and it can be adapted for both thermoset and thermoplastic materials. The head includes a spool or spools of tape, a winder, winder guides, a compaction shoe, a position sensor and a tape cutter or slitter. In either case, the head may be located on the end of a multi axis articulating robot that moves around the tool or mandrel to which material is being applied, or the head may be located on a gantry suspended above the tool.

Although ATL generally is faster than AFP and can place more material over longer distances, AFP is better suited to shorter courses and can place material more effectively over contoured surfaces. These technologies grew out of the machine tool industry and have seen extensive use in the manufacture of the fuselage, wing skin panels, wing box, tail and other structures on the forthcoming Boeing 787 Dreamliner and the Airbus A350 XWB. ATL and AFP also are used extensively to produce parts for the F-35 Lightning II fighter jet the V-22 Osprey tiltrotor troop transport and a variety of other aircraft.

Centrifugal casting of pipe from 1 inch/25 mm to 14 inches/356 mm in diameter is an alternative to filament winding for high-performance, corrosion resistant service. In cast pipe, 0°/90° woven fiberglass provides both longitudinal and hoop strength throughout the pipe wall and brings greater strength at equal wall thickness compared to multiaxial fiberglass wound pipe. In the casting process, epoxy or vinyl ester resin is injected into a 150G centrifugally spinning mold, permeating the woven fabric wrapped around the mold's interior surface. The centrifugal force pushes the resin through the layers of fabric, creating a smooth finish on the outside of the pipe, and excess resin pumped into the mold creates a resin-rich, corrosion- and abrasion resistant interior liner. Fiber-reinforced thermoplastic components now can be produced by extrusion, as well. A huge market has emerged in the past decade for extruded thermoplastic/wood flour (or other additives, such as bast fibers or fly ash) composites. These wood plastic composites, or WPCs, used to simulate wood decking, siding, window and door frames, and fencing.

References

- Ceramic, entry: newworldencyclopedia.org, Retrieved 11 June 2018
- Materials, Engineering: dl4a.org, Retrieved 10 July 2018
- Electrical-properties, Ceramic: newworldencyclopedia.org, Retrieved 19 April 2018
- Composite-Materials: scienceclarified.com, Retrieved 09 July 2018
- Adv-composites, why-composites: premix.com, Retrieved 27 May 2018

8

Phase Diagrams

A phase diagram is a representation of the physical conditions of pressure, volume or temperature at which distinct thermodynamic phases coexist and occur at equilibrium. This is an important chapter, which analyzes in detail about phase diagrams, its various types, phase equilibria, etc.

Many of the engineering materials possess mixtures of phases, e.g. steel, paints, and composites. The mixture of two or more phases may permit interaction between different phases, and results in properties usually are different from the properties of individual phases. Different components can be combined into a single material by means of solutions or mixtures. A solution (liquid or solid) is phase with more than one component; a mixture is a material with more than one phase. Solute does not change the structural pattern of the solvent, and the composition of any solution can be varied. In mixtures, there are different phases, each with its own atomic arrangement. It is possible to have a mixture of two different solutions.

A pure substance, under equilibrium conditions, may exist as either of a phase namely vapor, liquid or solid, depending upon the conditions of temperature and pressure. A phase can be defined as a homogeneous portion of a system that has uniform physical and chemical characteristics i.e. it is a physically distinct from other phases, chemically homogeneous and mechanically separable portion of a system. In other words, a phase is a structurally homogeneous portion of matter. When two phases are present in a system, it is not necessary that there be a difference in both physical and chemical properties; a disparity in one or the other set of properties is sufficient.

There is only one vapor phase no matter how many constituents make it up. For pure substance there is only one liquid phase, however there may be more than one solid phase because of differences in crystal structure. A liquid solution is also a single phase, even as a liquid mixture (e.g. oil and water) forms two phases as there is no mixing at the molecular level. In the solid state, different chemical compositions and/or crystal structures are possible so a solid may consist of several phases. For the same composition, different crystal structures represent different phases. A solid solution has atoms mixed at atomic level thus it represents a single phase. A single-phase system is termed as homogeneous, and systems composed of two or more phases are termed as mixtures or heterogeneous. Most of the alloy systems and composites are heterogeneous.

It is important to understand the existence of phases under various practical conditions which may dictate the microstructure of an alloy, thus the mechanical properties and usefulness of it. Phase diagrams provide a convenient way of representing which state of aggregation (phase or phases) is stable for a particular set of conditions. In addition, phase diagrams provide valuable information about melting, casting, crystallization, and other phenomena.

Equilibrium Phase Diagrams, Particle Strengthening by Precipitation and Precipitation Reactions

Equilibrium Phase Diagrams

A diagram that depicts existence of different phases of a system under equilibrium is termed as phase diagram. It is also known as equilibrium or constitutional diagram. Equilibrium phase diagrams represent the relationships between temperature and the compositions and the quantities of phases at equilibrium. In general practice it is sufficient to consider only solid and liquid phases, thus pressure is assumed to be constant (1 atm.) in most applications. These diagrams do not indicate the dynamics when one phase transforms into another. However, it depicts information related to microstructure and phase structure of a particular system in a convenient and concise manner. Important information, useful for the scientists and engineers who are involved with materials development, selection, and application in product design, obtainable from a phase diagram can be summarized as follows:

- To show phases are present at different compositions and temperatures under slow cooling (equilibrium) conditions.
- To indicate equilibrium solid solubility of one element/compound in another.
- To indicate temperature at which an alloy starts to solidify and the range of solidification.
- To indicate the temperature at which different phases start to melt.
- Amount of each phase in a two-phase mixture can be obtained.

A phase diagram is actually a collection of solubility limit curves. The phase fields in equilibrium diagrams depend on the particular systems being depicted. Set of solubility curves that represents locus of temperatures above which all compositions are liquid are called liquidus, while solidus represents set of solubility curves that denotes the locus of temperatures below which all compositions are solid. Every phase diagram for two or more components must show a liquidus and a solidus, and an intervening freezing range, except for pure system, as melting of a phase occurs over a range of temperature. Whether the components are metals or nonmetals, there are certain locations on the phase diagram where the liquidus and solidus meet. For a pure component, a contact point lies at the edge of the diagram. The liquidus and solidus also meet at the other invariant positions on the diagram. Each invariant point represents an invariant reaction that can occur only under a particular set of conditions between particular phases, so is the name for it.

Phase diagrams are classified based on the number of components in the system. Single component systems have unary diagrams, two-component systems have binary diagrams, and three-component systems are represented by ternary diagrams, and so on. When more than two components are present, phase diagrams become extremely complicated and difficult to represent.

Particle Strengthening by Precipitation and Precipitation Reactions

By obstructing dislocation motion in different means, material's strength can be increased. One of the methods that are applicable to multi-phase material is particle strengthening in which second phase particles are introduced into the matrix by either mixing-and-consolidation (dispersion strengthening) or precipitated in solid state (precipitation hardening).

The object of the precipitation strengthening is to create in a heat-treated alloy a dense and fine dispersion of precipitated particles in a matrix of deformable metal. The particles act as obstacles to dislocation motion. In order for an alloy system to be able to precipitation-strengthened for certain alloy compositions; there must be a terminal solid solution which has a decreasing solid solubility as the temperature decreases. For example: Au-Cu in which maximum solid solubility of Cu in Al is 5.65% at 548 °C that decreases with decreasing temperature.

The precipitation strengthening process involves the following three basic steps:

- Solutionizing (solution heat treatment), where the alloy is heated to a temperature between solvus and solidus temperatures and kept there till a uniform solidsolution structure is produced.
- Quenching, where the sample is rapidly cooled to a lower temperature (room temperature) and the cooling medium is usually water. Alloy structure in this stage consists of supersaturated solid solution.
- Aging is the last but critical step. During this heat treatment step finely dispersed precipitate particle will form. Aging the alloy at room temperature is called natural aging, whereas at elevated temperatures is called artificial aging. Most alloys require artificial aging, and aging temperature is usually between 15-25% of temperature difference between room temperature and solution heat treatment temperature.

Precipitation strengthening and reactions that occur during precipitation can be best illustrated using the Al-4%Cu (duralumin) system. Figure depicts the Al-rich end of the Al-Cu phase diagram. It can be observed that the alloy with 4%Cu exists as a single phase α-solid solution at around 550 °C, and at room temperature as a mixture of α (with less than 0.5%Cu) and an inter-metallic compound, CuAl2 (θ) with 52%Cu. On slow cooling α rejects excess Cu as precipitate particles of θ. These particles relatively coarse in size and can cause only moderate strengthening effect.

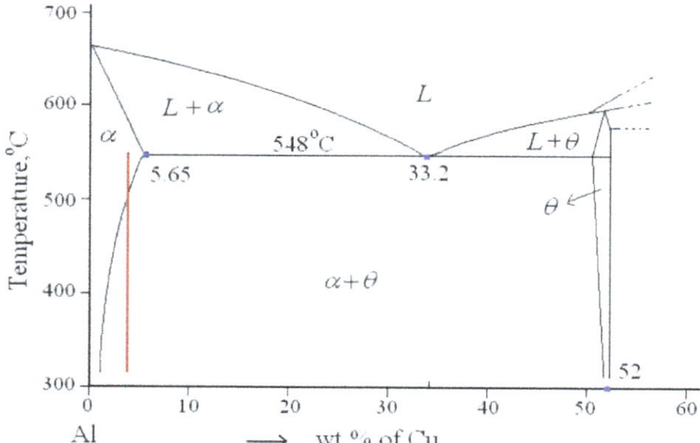

Figure: Aluminium rich end of Al-Cu phase diagram.

By rapidly cooling the alloy, a supersaturated solution can be obtained at room temperature. As a function of time at room temperature, and at higher temperatures up to 200 ° C, the diffusion of Cu atoms may take place and the precipitate particles can form. For this particular alloy, Al-4%Cu, five sequential structures can be identified: (a) supersaturated solid solution α, (b) GP1 zones, (c) GP2 zones (θ" phase), (d) θ' phase and (e) θ phase, $CuAl_2$. Not all these phases can be produced at all aging temperatures. GP1 and GP2 zones are produced at lower temperatures, and θ' and θ phases occur at higher temperatures. The initial stages of precipitation are the most difficult to analyze because of the extremely small size of the particles and their relatively uniform distribution. GP zones meant for Guinier-Preston zones which have a definite composition and structure that is not the same as that of the final stable precipitate. Evidently these particles are easier to nucleate than the final precipitate, as a result, form first. Eventually they disappear as later more stable phases appear. θ" and θ' are metastable transition precipitates with distinct crystal structure of their own, while θ is the equilibrium stable precipitate of $CuAl_2$

- GP1 zones:- These zones are created by Cu atoms segregating in α, and the segregated regions are of disk shape with thickness of 0.4-0.6 nm, and 8-10 nm in diameter and form on the {100} cubic planes of the matrix. As Cu atoms which replace Al atoms are smaller in diameter, matrix lattice strains tetragonally. These zones are said to be coherent with the matrix lattice.
- GP2 zones / θ" phase:- With additional aging, ordering of larger clumps of Cu atoms on {100} occurs. These zones have tetragonal structure which therefore introduces coherency in the lattice with {100} planes of the matrix, accompanied by further hardening. However, their size ranges from 1-4 nm thick and 10-100 nm in diameter as aging proceeds.
- θ' phase:- This phase nucleates heterogeneously especially on dislocations. It has tetragonal structure but is partially coherent with the matrix. This phase forms platelets with thickness 10-150 nm.
- θ phase:- With still further aging the equilibrium phase $CuAl_2$ or θ is formed from the transition lattice θ' or directly from the matrix accompanied by a reduction in hardness. It has a BCT (body-centered-tetragonal) structure, and is incoherent with the matrix. As these particles are no longer coherent with the matrix, hardness is lower than at the stage when coherent was present. Over-aging continues with the growth of these particles controlled by diffusion. Variation of hardness with aging time is shown in figure below.

The general sequence of precipitation in binary Al-Cu alloys can represented as: Supersaturated α → GP1 zones → GP2 zones (θ" phase) → θ' phase → θ phase ($CuAl_2$)

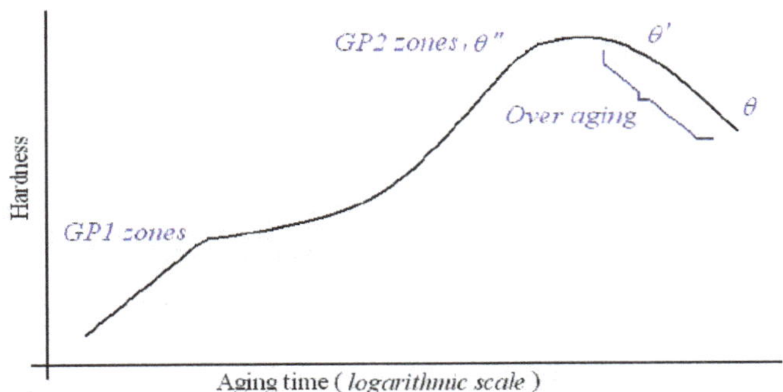

Figure: Correlation of structures and hardness for Al-4%Cu alloy.

Most precipitation-hardening systems operate in a similar way, peak hardness usually being attained in the later stages of coherency or at the onset of incoherency. It is quite common for a coherent precipitate to form and then lose coherency when the particle grows to a critical size. However, in some systems there is no evidence of coherency strains, and the fine particles appear to act alone as impediments to dislocation movements, for example – systems with dispersion strengthening.

Kinetics of Nucleation and Growth

Structural changes in metallic systems usually take place by nucleation and growth whether it is just a phase change within one of the three states, or a simple structural rearrangement within a single phase, or a phase transformation. An equilibrium phase diagram presents the phases and phase changes expected under equilibrium conditions, but it provides no information about the rates of transformation. Although changes in pressure, composition, or temperature can cause phase transformations, it is temperature changes that are more important. From a micro structural standpoint, the first process to accompany a phase transformation is nucleation (i.e. the formation of very small particles or nuclei of the product phase from the parent phase) of the new phase particles which are capable of growing. The second stage is growth, in which the nucleated particles increase their size. The transformation reaches completion if growth of these new phase particles is allowed to proceed until the equilibrium fraction is attained.

Both nucleation and growth require that the accompanying free energy change be negative. Consequently, the super-heating or super-cooling that is necessary for a phase change is to be expected. That is a transformation cannot tale place precisely at the equilibrium transformation temperature because at that temperature free energies of phases are equal. In addition to temperature, two other factors that affect transformation rate – first, diffusion controlled rearrangement of atoms because of compositional and/or crystal structural differences; second, difficulty encountered in nucleating small particles via change in surface energy associated with the interface. Diffusion limits both the nucleation and growth rates in many cases.

With the nucleation of new particle, new interface is created between the particle and liquid. This interface will have positive energy that must be supplied during the transformation process. A tiny particle has a large surface area to volume ratio and therefore be unstable. Thus energy of the surface can effectively prevent the initial formation of a tiny particle. A particle said to have nucleated when it becomes stable and will not disappear due to thermal fluctuations. After a particle attained a critical size, it can grow further with a continuous decrease in energy. The surface energy is no longer a dominant factor in the growth process.

Nucleation Kinetics

In homogeneous nucleation, the probability of nucleation occurring at any given site is identical to that at any other site within the volume of the parent phase. When a pure liquid metal is cooled below its equilibrium freezing temperature to a sufficient degree, numerous homogeneous nuclei are created by slow-moving atoms bonding together. Homogeneous nucleation usually requires a considerable amount of undercooling (cooling a material below the equilibrium temperature for a given transformation without the transformation occurring). Undercooling enhances the formation of nuclei that eventually grow. If Δf is the free energy change accompanying the formation of a spherical new phase particle,

$$\Delta f = \frac{4}{3}\pi r^3 \Delta g + 4\pi r^2 \gamma$$

where r is the radius of the particle, Δg is the Gibbs free energy change per unit volume and γ is the surface energy of the interface. As surface energy, γ, is always positive, and Δg is negative, passes through a maximum. From calculus, critical values can be found from the following:

$$r^* = \frac{2\gamma}{\Delta g}; \Delta f^*_{\text{hom}} = \frac{16}{3}\pi\gamma^3 / (\Delta g)^2$$

Particles which are smaller than the critical size are called embryos; those larger than the critical size are called nuclei. As Δg becomes more negative with a lowering of the temperature, the critical values of Δf and r becomes smaller as shown in figure below. At sufficiently low temperatures, nucleation can be triggered by a few atoms statistically clustering as a nucleus, so a small critical radius is exceeded. With added growth, the new phase attains stability. Of course atom movements are sluggish at low temperatures, so growth is generally slow.

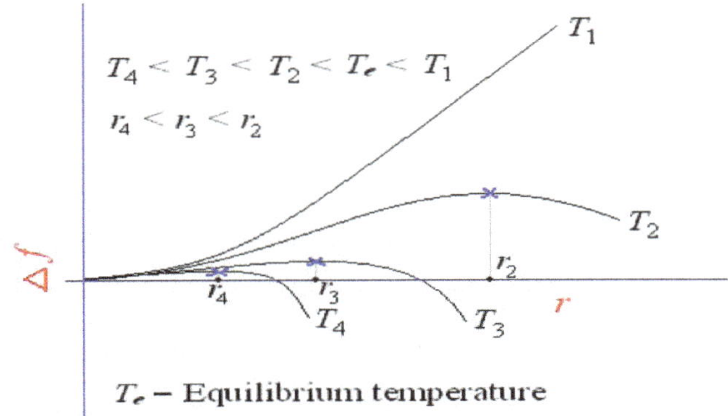

Figure: Effect of temperature on free energy change and particle radius.

The greater the degree of undercooling below the equilibrium melting temperature of the metal, the greater the change in volume free energy, however, the change in free energy due to surface energy does not change much with temperature. Thus, the critical size of nuclei is mainly determined by volume free energy. Near the freezing temperature, critical nucleus size must be infinite since ΔT approaches zero. As the amount of undercooling increases, critical size decreases, and are related as follows:

$$r^* = \frac{2\gamma T_m}{\Delta H_f \Delta T}$$

where T_m – freezing temperature (in K), ΔH_f – latent heat of fusion, ΔT – amount of undercooling at which nucleus is formed.

In heterogeneous nucleation, the probability of nucleation occurring at certain preferred sites is much greater than that at other sites. During solidification, inclusions of foreign particles (inoculants), walls of container holding the liquid provide preferred sites. Irregularities in crystal structure such as point defects and dislocations possess strain energy. In solid-solid transformation, foreign inclusions, grain boundaries, interfaces, stacking faults and dislocations can act as preferred sites for nucleation as the strain energy associated with them will be reduced. The released strain energy can reduce the energy requirements for free energy change, Δf. Therefore, nucleation proceeds with a smaller critical radius. A majority of reactions are initiated by some type of heterogeneous nucleation which is common among the two types.

For example, consider the nucleation of β from α occurring on a foreign inclusion, δ, as shown in figure below. Considering the force equilibrium in surface tension terms,

$$\gamma_{\alpha\delta} = \gamma_{\alpha\beta}\cos\theta + \gamma_{\beta\delta}$$

where θ is the contact angle. An expression for Δf can be written in terms of volume energy and surface energies as follows:

$$\Delta f^{*}_{het} = \frac{4\pi\gamma^{3}_{\alpha\beta}}{3(\Delta g)^{2}}(2-3\cos\theta+\cos^{3}\theta) = \Delta f^{*}_{\text{hom}}\frac{2-3\cos\theta+\cos^{3}\theta}{4}$$

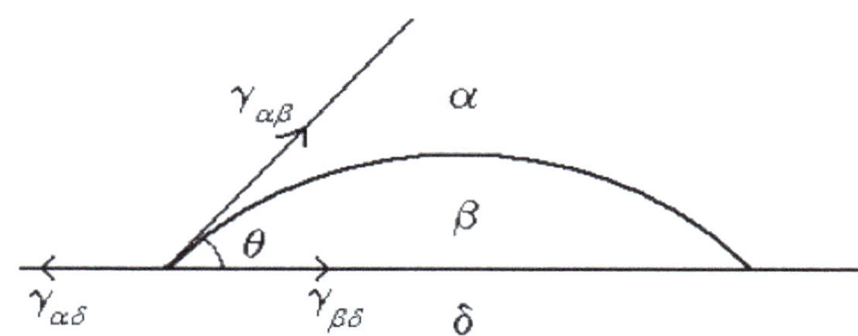

Figure: Schematic of heterogeneous nucleation.

By comparing the free energy terms for homogeneous and heterogeneous nucleation processes for various contact conditions:

- When product particle makes only a point contact with the foreign surface, i.e. θ = 180°, the foreign particle does not play any role in the nucleation process → $\Delta f^{*}_{het} = \Delta f^{*}_{\text{hom}}$
- If the product particle completely wets the foreign surface, i.e. θ = 0°, there is no barrier for heterogeneous nucleation → $\Delta f^{*}_{het} = 0$
- In intermediate conditions such as where the product particle attains hemispherical shape, θ = 0→ $\Delta f^{*}_{het} = \frac{1}{2}\Delta f^{*}_{\text{hom}}$

The above derivations are helpful in selecting a heterogeneous nucleation agent. It shows that a small contact angle is very helpful in heterogeneous nucleation. For a system of α – β interface, θ can be minimized by choosing δ such that energy of β- δ interface is kept to minimum. If the crystal structure two phases are similar and their lattice parameters are nearly equal, energy of the interface between those two phases will be minimum. This criterion is useful in selecting an agent for heterogeneous nucleation.

Growth Kinetics

Many transformations occur as a result of continuous formation of critical nuclei in the parent phase and the subsequent growth of the particles. Growth is the increase in size of the particle after it has nucleated i.e. growth kinetics become important once an embryo has exceeded the critical size and become a stable nucleus. Growth may proceed in two adically different manners. In one type of growth, individual atoms move independently from the parent to the product phase, thus it is diffusion controlled and is thermally activated. In the other type of growth that occurs in solid-solid transformations many atoms move cooperatively without thermal assistance. Growth that is diffusion controlled is more common the other.

Growth usually occurs by the thermally activated jump of atoms from the parent phase to the product phase. The unit step in the growth process thus consists of an atom leaving the parent phase and jumping across the interface to join the product phase. At the equilibrium temperature, both phases have the same free energy, hence the frequency of jumps from parent phase to product phase will be equal to that from product phase to parent phase i.e. the net growth rate is zero. At lower temperatures, product phase is expected to have lower free energy, and thus a net flow of atoms from parent phase to product phase. This net flux of atoms results in interface motion i.e. growth rate is taken as the rate of increase of a linear dimension of a growing particle. As a function of temperature, the growth rate first increases with increasing degree of supercooling, but eventually slows-down as thermal energy decreases. This is same as for nucleation; however the maximum in the growth rate usually occurs at a higher temperature than the maximum in the nucleation rate. Figure(a) below depicts the temperature dependence of nucleation rate (U), growth rate (I) and overall transformation rate (dX/dt) that is a function of both nucleation rate and growth rate i.e. dX/dt= fn (U, I). On the other-hand, the time required for a transformation to completion has a reciprocal relationship to the overall transformation rate. Temperature dependence of this time is shown in figure (b) below. The C-curve shown in figure (b) below is characteristic of all thermally activated nucleation and growth transformations involving the transformation of a high temperature phase to low-temperature phase. This curve is also known as time temperature-transformation (TTT) curve. The nose of the C-curve corresponds to the minimum time for a specified fraction of transformation. It is also the place where overall transformation rate is a maximum.

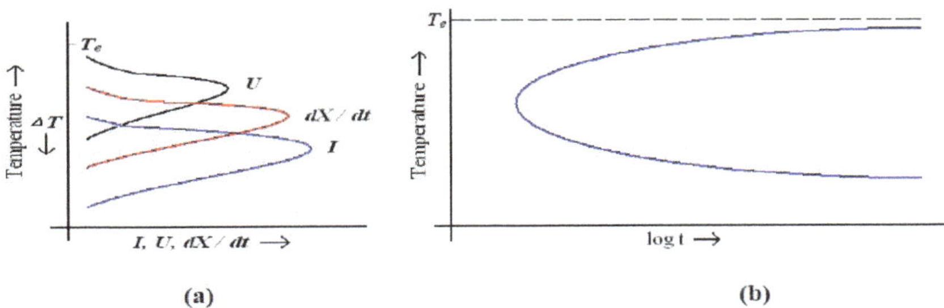

(a) (b)

Figure: (a) Temperature dependence of rates, (b) Time dependence of transformation as a function of temperature.

In many investigations, the fraction of transformation that has occurred is measured as a function of time, while the temperature is maintained constant. Transformation progress is usually ascertained by either microscopic examination or measurement of some physical property. Transformation data are plotted as the fraction of transformed material versus the logarithm of time, which results in characteristic S-curve, largely because diffusion plays such an important role in both nucleation and growth. Conditions under which transformation may take place greatly affect the nature of the resulting microstructure. At small degrees of supercooling, where slow nucleation and rapid growth prevail, relatively coarse particles appear; at larger degrees of supercooling, relatively fine particles result. The time dependence of the transformation rate is an important consideration in the heat treatment of materials.

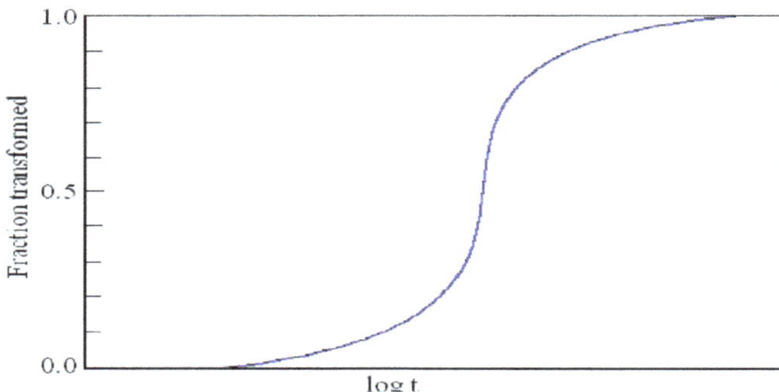

Figure: Fraction of transformation Vs the logarithm of time at constant temperature.

The other kind of growth involves congruent transformation which is considered as diffusion-less because it takes place at a rate approaching the speed of sound. It can be visualized as a cooperative type of process in which, without aid of thermal activation, atoms move into new locations because of the strain energy resulting from like movements of adjacent atoms. However the strains set-up in the parent phase may impede the further transformation, thus a lower temperature or mechanical deformation may be required to complete this martensitic transformation. The cooperative displacement of atoms here resembles a shear process during which, for example, FCC structure of Co transforms into HCP-Co or FCC-austenite into BCT-martensite. This merely requires that atoms in the FCC-phase move a fraction of an inter-atomic distance.

Because of its crystallographic nature, a martensitic transformation only occurs in the solid state. In addition, the crystal structure of the product phase must be easily generated from that of the parent phase without diffusive motion of atoms. This is true for most allotropic transformations in metals that occur at low temperatures or for high temperature transformations of metals brought about by a quench. Reasons for martensitic transformation: (a) the free energy difference between the high-temperature phase and low-temperature phases becomes increasingly negative with decreasing temperature (b) the crystal structures of allotropes of a metal are relatively simple and share similar features with each other. It is also said that diffusion-controlled nucleation and growth and a martensitic change are competitive processes in many cases.

The martensitic transformation starts at a temperature designated Ms, which is generally below the equilibrium temperature, Te. The transformation is completed at a lower temperature, Mf. The amount of parent phase transformed into product phase depends on temperature only, and is independent of time. Furthermore, in most cases, Ms temperature and the fractional amount of product phase as a function of temperature are independent of quenching rate. Consequently, Ms and Mf are presented as horizontal lines on a TTT diagram. Catalytic effect of cold working can be used to make Ms approach Te. Martensitic transformations in Fe-C alloys and Ti are of great technological importance.

The iron – Carbon System, Phase Transformations

A study of iron-carbon system is useful and important in many respects. This is because (1) steels constitute greatest amount of metallic materials used by man (2) solid state transformations that occur in steels are varied and interesting. These are similar to those occur in many other systems and helps explain the properties.

Iron-carbon phase diagram shown in figure below is not a complete diagram. Part of the diagram after 6.67 wt% C is ignored as it has little commercial significance. The 6.67%C represents the composition where an inter-metallic compound, cementite (Fe_3C), with solubility limits forms. In addition, phase diagram is not true equilibrium diagram because cementite is not an equilibrium phase. However, in ordinary steels decomposition of cementite into graphite never observed because nucleation of cementite is much easier than that of graphite. Thus cementite can be treated as an equilibrium phase for practical purposes.

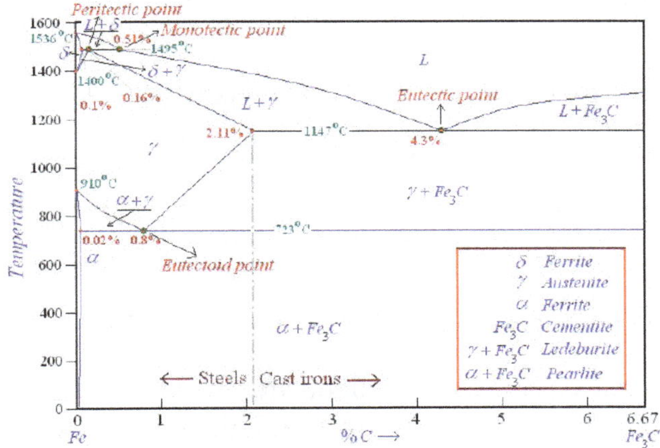

Figure: Iron – Iron carbide phase diagram.

The Fe-Fe_3C is characterized by five individual phases and four invariant reactions. Five phases that exist in the diagram are: α–ferrite (BCC) Fe-C solid solution, γ-austenite (FCC) Fe-C solid solution, δ-ferrite (BCC) Fe-C solid solution, Fe_3C (iron carbide) or cementite - an inter-metallic compound and liquid Fe-C solution. Four invariant reactions that cause transformations in the system are namely eutectoid, eutectic, monotectic and peritectic.

As depicted by left axes, pure iron upon heating exhibits two allotropic changes. One involves α–ferrite of BCC crystal structure transforming to FCC austenite, γ-iron, at 910 °C. At 1400 °C, austenite changes to BCC phase known as δ-ferrite, which finally melts at 1536 °C.

Carbon present in solid iron as interstitial impurity, and forms solid solution with ferrites / austenite as depicted by three single fields represented by α, γ and δ. Carbon dissolves least in α–ferrite in which maximum amount of carbon soluble is 0.02% at 723 °C. This limited solubility is attributed to shape and size of interstitial position in BCC α–ferrite. However, carbon present greatly influences the mechanical properties of α–ferrite. α– ferrite can be used as magnetic material below 768 °C. Solubility of carbon in γ-iron reaches its maximum, 2.11%, at a temperature of 1147 °C. Higher solubility of carbon in austenite is attributed to FCC structure and corresponding interstitial sites. Phase transformations involving austenite plays very significant role in heat treatment of different steels. Austenite itself is non-magnetic. Carbon solubility in δ-ferrite is maximum (0.1%) at 1495 °C. As this ferrite exists only at elevated temperatures, it is of no commercial importance. Cementite, Fe_3C an inter-metallic compound forms when amount of carbon present exceeds its solubility limit at respective temperatures. Out of these four solid phases, cementite is hardest and brittle that is used in different forms to increase the strength of steels. α–ferrite, on the other hand, is softest and act as matrix of a composite material. By combining these two phases in a solution, a material's properties can be varied over a large range.

For technological convenience, based on %C dissolved in it, a Fe-C solution is classified as: commercial pure irons with less than 0.008%C; steels having %C between 0.008- 2.11; while cast irons have carbon in the range of 2.11%-6.67%. Thus commercial pure iron is composed of exclusively α–ferrite at room temperature. Most of the steels and cast irons contain both α–ferrite and cementite. However, commercial cast irons are not simple alloys of iron and carbon as they contain large quantities of other elements such as silicon, thus better consider them as ternary alloys. The presence of Si promotes the formation of graphite instead of cementite. Thus cast irons may contain carbon in form of both graphite and cementite, while steels will have carbon only in combined from as cementite.

As shown in figure, and mentioned earlier, Fe-C system constitutes four invariant reactions:

- Peritectic reaction at 1495C and 0.16%C, δ-ferrite + L ↔ γ-iron (austenite).
- Monotectic reaction 1495°C and 0.51%C, L ↔ L + γ-iron (austenite).
- Eutectic reaction at 1147°C and 4.3 %C, L ↔ γ-iron + Fe3C (cementite) [ledeburite].
- Eutectoid reaction at 723°C and 0.8%C, γ-iron ↔ α–ferrite + Fe3C (cementite) [pearlite].

Product phase of eutectic reaction is called ledeburite, while product from eutectoid reaction is called pearlite. During cooling to room temperature, ledeburite transforms into pearlite and cementite. At room temperature, thus after equilibrium cooling, Fe-C diagram consists of either α–ferrite, pearlite and/or cementite. Pearlite is actually not a single phase, but a micro-constituent having alternate thin layers of α–ferrite (~88%) and Fe3C, cementite (~12%). Steels with less than 0.8%C (mild steels up to 0.3%C, medium carbon steels with C between 0.3%-0.8% i.e. hypo-eutectoid Fe-C alloys) i.e. consists pro-eutectoid α–ferrite in addition to pearlite, while steels with carbon higher than 0.8% (high-carbon steels i.e. hyper-eutectoid Fe-C alloys) consists of pearlite and pro-eutectoid cementite. Phase transformations involving austenite i.e. processes those involve eutectoid reaction are of great importance in heat treatment of steels.

In practice, steels are almost always cooled from the austenitic region to room temperature. During the cooling upon crossing the boundary of the single phase γ-iron, first pro-eutectoid phase (either α–ferrite or cementite) forms up to eutectoid temperature. With further cooling below the eutectoid temperature, remaining austenite decomposes to eutectoid product called pearlite, mixture of thin layers of α–ferrite and cementite. Though pearlite is not a phase, nevertheless, a constituent because it has a definite appearance under the microscope and can be clearly identified in a structure composed of several constituents. The decomposition of austenite to form pearlite occurs by nucleation and growth. Nucleation, usually, occurs heterogeneously and rarely homogeneously at grain boundaries. When it is not homogeneous, nucleation of pearlite occurs both at grain boundaries and in the grains of austenite. When austenite forms pearlite at a constant temperature, the spacing between adjacent lamellae of cementite is very nearly constant. For a given colony of pearlite, all cementite plates have a common orientation in space, and it is also true for the ferrite plates. Growth of pearlite colonies occurs not only by the nucleation of additional lamellae but also through an advance at the ends of the lamellae. Pearlite growth also involves the nucleation of new colonies at the interfaces between established colonies and the parent austenite. The thickness ratio of the ferrite and cementite layers in pearlite is approximately 8 to 1. However, the absolute layer thickness depends on the temperature at which the isothermal transformation is allowed to occur.

The temperature at which austenite is transformed has a strong effect on the interlamellar spacing of pearlite. The lower the reaction temperature, the smaller will be interlamellar spacing. For example, pearlite spacing is in order of 10-3 mm when it formed at 700 °C, while spacing is in order of 10-4 mm when formed at 600 °C. The spacing of the pearlite lamellae has a practical significance because the hardness of the resulting structure depends upon it; the smaller the spacing, the harder the metal. The growth rate of pearlite is also a strong function of temperature. At temperatures just below the eutectoid, the growth rate increases rapidly with decreasing temperature, reaching a maximum at 600 °C, and then decreases again at lower temperatures.

Additions of alloying elements to Fe-C system bring changes (alternations to positions of phase boundaries and shapes of fields) depends on that particular element and its concentration. Almost all alloying elements causes the eutectoid concentration to decrease, and most of the alloying elements (e.g.: Ti, Mo, Si, W, Cr) causes the eutectoid temperature to increase while some other (e.g.: Ni, Mn) reduces the eutectoid temperature. Thus alloying additions alters the relative amount of pearlite and proeutectoid phase that form.

Fe-C alloys with more than 2.11% C are called cast irons. Phase transformations in cast irons involve formation of pro-eutectic phase on crossing the liquidus. During the further cooling, liquid of eutectic composition decomposes in to mixture of austenite and cementite, known as ledeburite. On further cooling through eutectoid temperature, austenite decomposes to pearlite. The room temperature microstructure of cast irons thus consists of pearlite and cementite. Because of presence of cementite, which is hard, brittle and white in color, product is called white cast iron. However, depending on cooling rate and other alloying elements, carbon in cast iron may be present as graphite or cementite. Gray cast iron contains graphite in form of flakes. These flakes are sharp and act as stress risers. Brittleness arising because of flake shape can be avoided by producing graphite in spherical nodules, as in malleable cast iron and SG (spheroidal graphite) cast iron. Malleable cast iron is produced by heat treating white cast iron (Si < 1%) for prolonged periods at about 900 °C and then cooling it very slowly. The cementite decomposes and temper carbon appears approximately as spherical particles. SG iron is produced by adding inoculants to molten iron. In these Si content must be about 2.5%, and no subsequent heat treatment is required.

Transformation rate effects and TTT diagrams, Microstructure and Property Changes in Fe-C Alloys

Solid state transformations, which are very important in steels, are known to be dependent on time at a particular temperature. Isothermal transformation diagram, also known as TTT diagram, measures the rate of transformation at a constant temperature i.e. it shows time relationships for the phases during isothermal transformation. Information regarding the time to start the transformation and the time required to complete the transformation can be obtained from set of TTT diagrams. One such set of diagram for reaction of austenite to pearlite in steel is shown in figure below. The diagram is not complete in the sense that the transformations of austenite that occur at temperatures about 550 °C are not shown.

If the transformation took place at a temperature that is just below the eutectoid temperature, relatively thick layers of α– ferrite and cementite are produced in what is called coarse pearlite. This is because of high diffusion rates of carbon atoms. Thus with decreasing transformation temperature, sluggish movement of carbon results in thinner layers α–ferrite and cementite i.e. fine pearlite is produced.

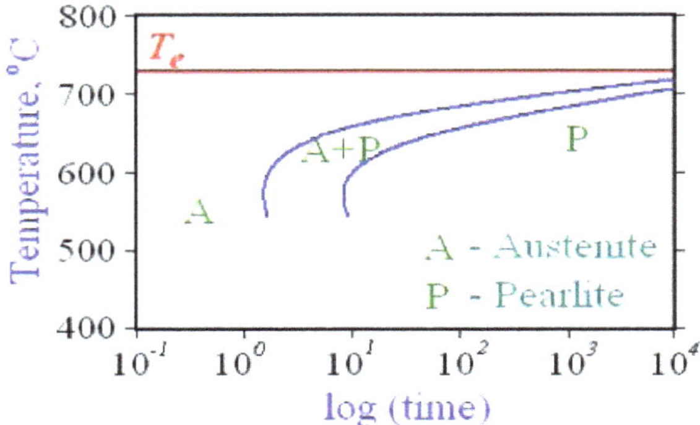

Figure: Partial TTT diagram for a eutectoid Fe-C alloy.

At transformation temperatures below 550 °C, austenite results in different product known as bainite. Bainite also consists of α–ferrite and cementite phases i.e. transformation is again diffusion controlled but morphologically it consists of very small particles of cementite within or between fine ferrite plates. Bainite forms needles or plates, depending on the temperature of the transformation; the microstructural details of bainite are so fine that their resolution is only possible using electron microscope. It differs from pearlite in the sense that different mechanism is involved in formation ob bainite which does not have alternating layers of α–ferrite and cementite. In addition, because of equal growth rates in all directions pearlite tends to form spherical colonies, whereas bainite grows as plates and has a characteristic acicular (needlelike) appearance. Upper bainite, formed at the upper end of the temperature range (550 °C-350 °C), is characterized by relatively coarse, irregular shaped cementite particles in α–ferrite plates. If the transformation is taking place at lower temperatures (350 °C-250 °C), the α– ferrite plates assume a more regular needlelike shape, and the transformation product is called lower bainite. At the same time carbide particles become smaller in size and appear as cross-striations making an angle of about 55 ° to the axis of the α–ferrite plate. Upper bainite has large rod-like cementite regions, whereas lower bainite has much finer cementite particles as a result of sluggish diffusion of carbon atoms at lower temperatures. Lower bainite is considerably harder than upper bainite. Another characteristic of bainite is that as it has crystallographic orientation that is similar to that found in simple ferrite nucleating from austenite, it is believed that bainite is nucleated by the formation of ferrite. This is in contrast to pearlite which is believed to be nucleated by formation of cementite.

Basically, bainite is a transformation product that is not as close to equilibrium as pearlite. The most puzzling feature of the bainite reaction is its dual nature. In a number of respects, it reveals properties that are typical of a nucleation and growth type of transformation such as occurs in the formation pearlite and also a mixture of α–ferrite and cementite though of quite different morphology (no alternate layers), but at the same time it differs from the Martensite as bainite formation is athermal and diffusion controlled though its microstructure is characterized by acicular (needlelike) appearance.

The time-temperature dependence of the bainite transformation can also be presented using TTT diagram. It occurs at temperatures below those at which pearlite forms i.e. it does not form until the transformation temperature falls below a definite temperature, designated as BB s. Above this temperature austenite does not form bainite except under external stresses. Below BsB, austenite does not transform completely to bainite. The amount of bainite formed increases as the isothermal reaction temperature is lowered. By reaching a lower limiting temperature, BB f, it is possible to transform austenite completely to bainite. The BsB and BB f temperatures are equivalent to the Ms and Mf temperatures for Martensite.

In simple eutectoid steels, pearlite and bainite transformations overlap, thus transition from the pearlite to bainite is smooth and continuous i.e. knees of individual pearlite and bainite curves are merged together. However each of the transformations has a characteristic C-curve, which can be distinguishable in presence of alloying elements. As shown in complete TTT diagram for eutectoid steel in figure, above approximately 550° C-600° C, austenite transforms completely to pearlite. Below this range up to 450° C, both pearlite and bainite are formed. Finally, between 450° C and 210° C, the reaction product is bainite only. Thus bainite transformation is favored at a high degree of supercooling, and the pearlite transformation at a low degree of supercooling. In middle region, pearlitic and bainitic transformations are competitive with each other.

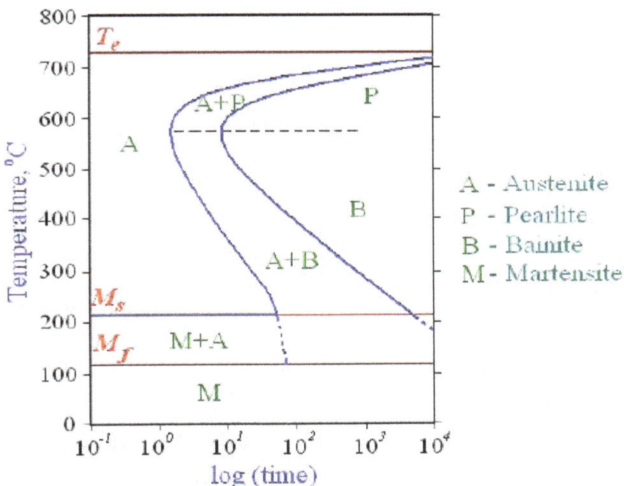

Figure: Complete TTT (isothermal transformation) diagram for eutectoid steel.

Martensitic transformation can dominate the proceedings if steel is cooled rapid enough so that diffusion of carbon can be arrested. Transformation of austenite to Martensite is diffusion-less, time independent and the extent of transformation depends on the transformation temperature. Martensite is a meta-stable phase and decomposes into ferrite and pearlite but this is extremely slow (and not noticeable) at room temperature. Alloying additions retard the formation rate of pearlite and bainite, thus rendering the martensitic transformation more competitive. Start of the transformation is designated by Ms, while the completion is designated by Mf in a transformation diagram. Martensite forms in steels possesses a body centered tetragonal crystal structure with carbon atoms occupying one of the three interstitial sites available. This is the reason for characteristic structure of steel Martensite instead of general BCC. Tetragonal distortion caused by carbon atoms increases with increasing carbon content and so is the hardness of Martensite. Austenite is slightly denser than Martensite, and therefore, during the phase transformation upon quenching, there is a net volume increase. If relatively large pieces are rapidly quenched, they may crack as a result of internal stresses, especially when carbon content is more than about 0.5%.

Mechanically, Martensite is extremely hard, thus its applicability is limited by brittleness associated with it. Characteristics of steel Martensite render it unusable for structural applications in the as-quenched form. However, structure and thus the properties can be altered by tempering, heat treatment observed below eutectoid temperature to permit diffusion of carbon atoms for a reasonable period of time. During tempering, carbide particles attain spherical shape and are distributed in ferrite phase – structure called spheroidite. Spheroidite is the softest yet toughest structure that steel may have. At lower tempering temperature, a structure called tempered Martensite forms with similar microstructure as that of spheroidite except that cementite particles are much, much smaller. The tempering heat treatment is also applicable to pearlitic and bainitic structures. This mainly results in improved machinability. The mechanism of tempering appears to be first the precipitation of fine particles of hexagonal ε-carbide of composition about $Fe_{2\cdot4}C$ from Martensite, decreasing its tetragonality. At higher temperatures or with increasing tempering times, precipitation of cementite begins and is accompanied by dissolution of the unstable ε-carbide. Eventually the Martensite loses its tetragonality and becomes BCC ferrite, the cementite coalesces into spheres. A schematic of possible transformations involving austenite decomposition are shown in figure below.

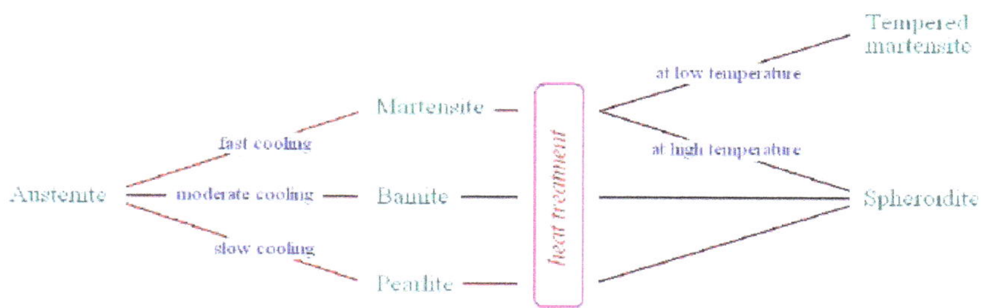

Figure: Possible transformation involving austenite decomposition.

Tempering of some steels may result in a reduction of toughness what is known as temper embrittlement. This may be avoided by (1) compositional control, and/or (2) tempering above 575 or below 375, followed by quenching to room temperature. The effect is greatest in Martensite structures, less severe in bainitic structures and least severe in pearlite structures. It appears to be associated with the segregation of solute atoms to the grain boundaries lowering the boundary strength. Impurities responsible for temper brittleness are: P, Sn, Sb and As. Si reduces the risk of embrittlement by carbide formation. Mo has a stabilizing effect on carbides and is also used to minimize the risk of temper brittleness in low alloy steels.

TTT diagrams are less of practical importance since an alloy has to be cooled rapidly and then kept at a temperature to allow for respective transformation to take place. However, most industrial heat treatments involve continuous cooling of a specimen to room temperature. Hence, Continuous Cooling Transformation (CCT) diagrams are generally more appropriate for engineering applications as components are cooled (air cooled, furnace cooled, quenched etc.) from a processing temperature as this is more economic than transferring to a separate furnace for an isothermal treatment. CCT diagrams measure the extent of transformation as a function of time for a continuously decreasing temperature. For continuous cooling, the time required for a reaction to begin and end is delayed, thus the isothermal curves are shifted to longer times and lower temperatures.

Both TTT and CCT diagrams are, in a sense, phase diagrams with added parameter in form of time. Each is experimentally determined for an alloy of specified composition. These diagrams allow prediction of the microstructure after some time period for constant temperature and continuous cooling heat treatments, respectively. Normally, bainite will not form during continuous cooling because all the austenite will have transformed to pearlite by the time the bainite transformation has become possible. Thus, as shown in figure below, region representing austenite-pearlite transformation terminates just below the nose.

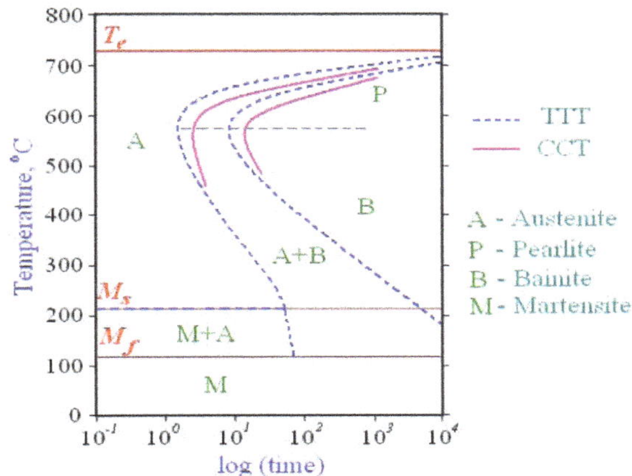

Figure: Superimposition of TTT and CCT diagrams for a eutectoid steel.

Industrial Applications of Phase Diagrams

The following are but a few of the many instances where phase diagrams and phase relationships have proved invaluable in the efficient solving of practical metallurgical problems. The areas covered include alloy design, processing, and performance.

Alloy Design

Four examples of the application of phase diagrams to alloy design are given: the development of a basis for age-hardening aluminum alloys, material substitution in two types of wrought stainless steel alloys to reduce costs, and an improvement in the manufacturing process for FeNd-B-base magnets.

- Age-Hardening Alloys: The age hardening of certain aluminum-copper alloys (then called "Duralumin" alloys) had been accidentally discovered in 1904, but this process was thought to be a unique and curious phenomenon. This work led to the development of several families of commercial "agehardening" alloys covering different base metals.
- Austenitic Stainless Steel: In connection with a research project aimed at the conservation of always expensive, sometimes scarce, materials, the question arose: Can manganese and aluminum be substituted for nickel and chromium in stainless steels? In other words, can standard chromium nickel stainless steels be replaced with an austenitic alloy system? The answer came in two stages—in both instances with the help of phase diagrams. It was first determined that manganese should be capable of replacing nickel because it stabilizes the γ-iron phase (austenite), and aluminum may substitute for chromium because it stabilizes the α-iron phase (ferrite), leaving only a small γ loop.

Aluminum is known to impart good high-temperature oxidation resistance to iron. Next, A nonmagnetic alloy with austenitic structure containing 44% Fe, 45% Mn, and 11% Al was prepared.

However, it proved to be very brittle, presumably because of the precipitation of a phase based on β-manganese. By examining the phase diagram for C-Fe-Mn, as well as the diagram for Al-CFe, the researcher determined that the problem could be solved through the addition of carbon to the Al-Fe-Mn system, which would move the composition away from the β-manganese phase field. The carbon addition also would further stabilize the austenite phase, permitting reduced manganese content. With this information, the composition of the alloy was modified to 7 to 10% Al, 30 to 35% Mn, and 0.75 to 1% C, with the balance iron. It had good mechanical properties, oxidation resistance, and moderate stainlessness.

- Permanent Magnets: A problem with permanent magnets based on Fe-Nd-B is that they show high magnetization and coercivity at room temperature but unfavorable properties at higher temperatures. Because hard magnetic properties are limited by nucleation of severed magnetic domains, the surface and interfaces of grains in the sintered and heat treated material is the controlling factor. Therefore, the effects of alloying additives on the phase diagrams and microstructural development of the Fe-Nd-B alloy system plus additives were studied. These studies showed that the phase relationships and domain-nucleation difficulties were very unfavorable for the production of a magnet with good magnetic properties at elevated temperatures by the sintering method. However, such a magnet might be produced from Fe-Nd-C material by some other process, such as melt spinning or bonding.

Processing

Two examples of the application of phase diagrams to alloy design are discussed: alloy additions to a hacksaw blade steel to allow the production of more cost-effective blades, and alloy additions to a hard facing alloy that produced superior properties.

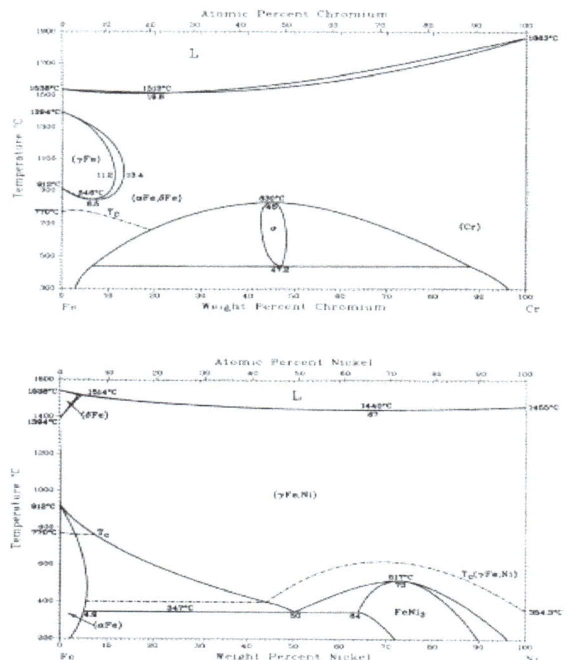

Figure: Two binary iron phase diagrams, showing ferrite stabilization (ironchromium) and austenite stabilization (iron-nickel).

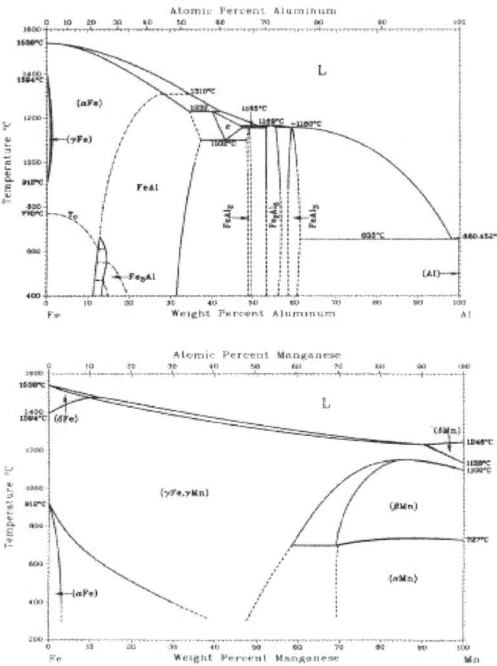

Figure: The aluminum-iron and iron-manganese phase diagrams.

- Hacksaw Blades: In the production of hacksaw blades, a strip of highspeed steel for the cutting edges is joined to a backing strip of low-alloy steel by laser or electron beam welding. As a result, a very hard martensitic structure forms in the weld area that must be softened by heat treatment before the composite strip can be further rolled or set.

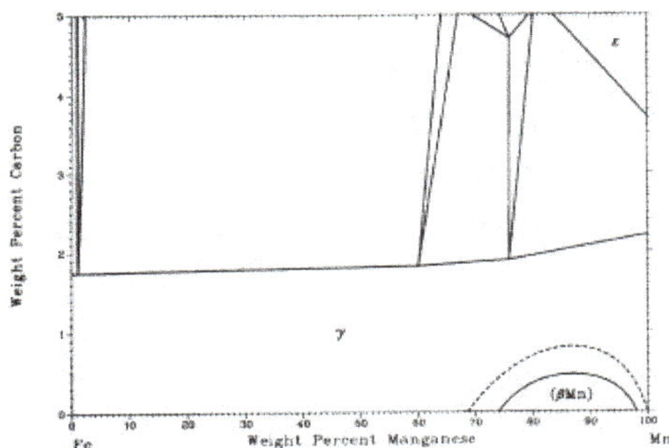

Figure: The isothermal section at 1100 °C (2012 °F) of the Fe-Mn-C phase diagram.

To avoid the cost of the heat treatment, an alternative technique was investigated. This technique involved alloy additions during welding to create a microstructure that would not require subsequent heat treatment. Instead of expensive experiments, several mathematical simulations were made based on additions of various steels or pure metals. In these simulations, the hardness of the weld was determined by combining calculations of the equilibrium phase diagrams and available information to calculate (assuming the average composition of the weld) the martensite transformation temperatures and amounts of retained austenite, untransformed ferrite, and carbides formed in the postweld microstructure. Of those alloy additions considered, chromium was found to be the most efficient.

- Hardfacing: A phase diagram was used to design a nickel-base hardfacing alloy for corrosion and wear resistance. For corrosion resistance, a matrix of at least 15% Cr was desired; for abrasion resistance, a minimum amount of primary chromium-boride particles was desired. After consulting the B-Cr-Ni phase diagram, a series of samples having acceptable amounts of total chromium borides and chromium matrix were made and tested. Subsequent fine tuning of the composition to ensure fabricability of welding rods, weldability, and the desired combination of corrosion, abrasion, and impact resistance led to a patented alloy.

Performanc

Four examples of the application of phase diagrams to performance are listed: the elimination of sulfur contamination from Nichrome heating elements, the elimination of lead and bismuth contaminants from extruded aluminum electric motor housings, a deficiency in the amount of carbon in sintered tungsten-carbide cutting tools, and a problem in which components were failing where the gold lead wires were fused to aluminized transistor and integrated circuits.

- Heating elements: made of Nichrome (a Ni-Cr-Fe alloy registered by Driver-Harris Company, Inc., Harrison, NJ) in a heat-treating furnace were failing prematurely. Reference to nickel-base phase diagrams suggested that low-melting eutectics can be produced by very small quantities of the chalcogens (sulfur, selenium, or tellurium), and it was thought that one of these eutectics could be causing the problem. Investigation of the furnace system resulted in the discovery that the tubes conveying protective atmosphere to the furnace were made of sulfur-cured rubber, which could result in liquid metal being formed at temperatures as low as 637 °C (1179 °F), as shown in figure below. With this information, a metallurgist solved the problem by substituting neoprene for the rubber.
- Electric Motor Housings: At moderately high service temperatures, cracks developed in electric motor housings that had been extruded from aluminum produced from a combination of recycled and virgin metal. Extensive studies revealed that the cracking was caused by small amounts of lead and bismuth in the recycled metal reacting to form bismuth-lead eutectic at the grain boundaries at 327 and ~270 °C (621 and ~518 °F), respectively, much below the melting point of pure aluminum (660.45 °C, or 1220.81 °F).

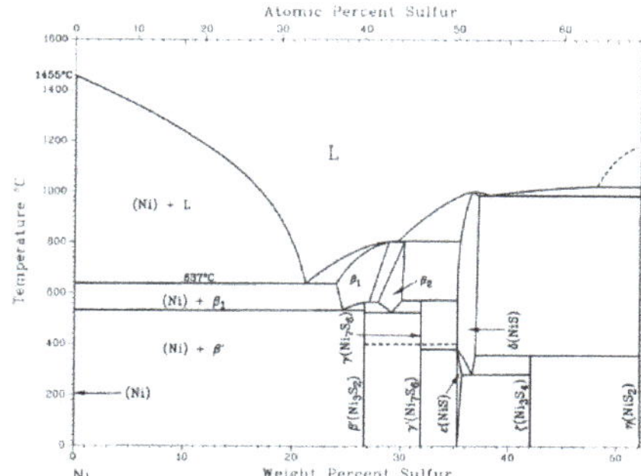

Figure: The nickel-sulfur phase diagram.

The question became: How much lead and bismuth can be tolerated in this instance? The phase diagrams showed that aluminum alloys containing either lead or bismuth in amounts exceeding their respective solubility limits (< 0.05%*and :* 0.2%) can lead to hot cracking of the aluminum.

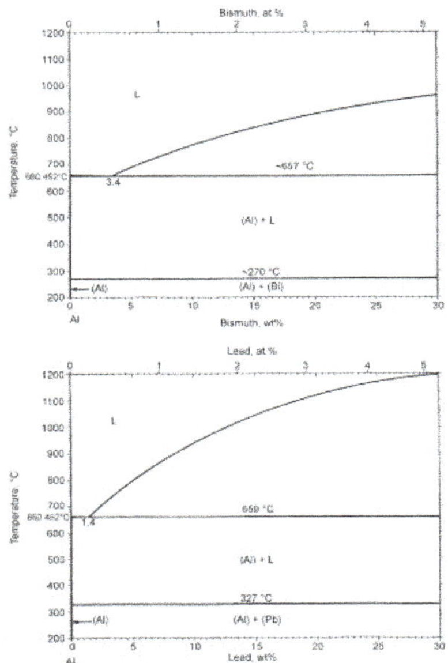

Figure: The aluminum-bismuth and aluminum-lead phase diagrams.

- Carbide Cutting Tools: A manufacturer of carbide cutting tools once experienced serious trouble with brittleness of the sintered carbide. No impurities were found. The range of compositions for cobalt-bonded sintered carbides is shown in the shaded area of the ternary phase diagram in figure below, along the dashed line connecting pure tungsten carbide (WC) on the right and pure cobalt at the lower left. At 1400 °C (2552 °F), materials with these compositions consist of particles of tungsten carbide suspended in liquid metal. However, when there is a deficiency of carbon, compositions drop into the region labeled WC + η + liquid, or the region labeled WC + η where tungsten carbide particles are surrounded by a matrix of η phase. The η phase is known to be brittle. The upward adjustment of the carbon content by only a few hundredths of a weight percent eliminated this problem.
- Solid-State Electronics: In the early stages of the solid-state industry, a phenomenon known as the "purple plague" nearly destroyed the fledgling industry. Components were failing where the gold lead wires were fused to aluminized transistor and integrated circuits. A purple residue was formed, which was thought to be a product of corrosion. Actually, what was happening was the formation of an intermetallic compound, an aluminum-gold precipitate (Al2Au) that is purple in color and very brittle.

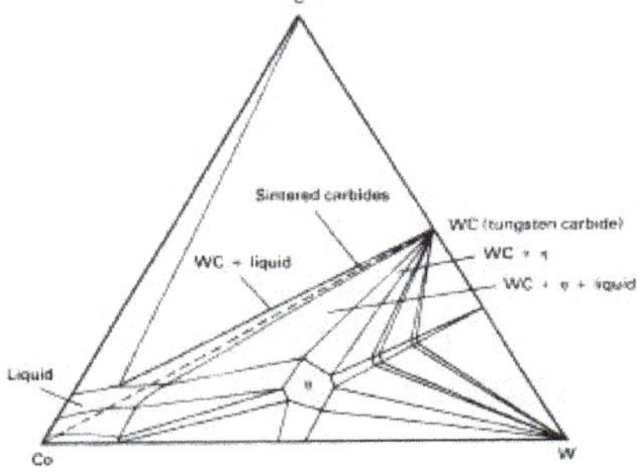

Figure: The isothermal section at 1400 °C (2552 °F) of the Co-W-C phase diagram.

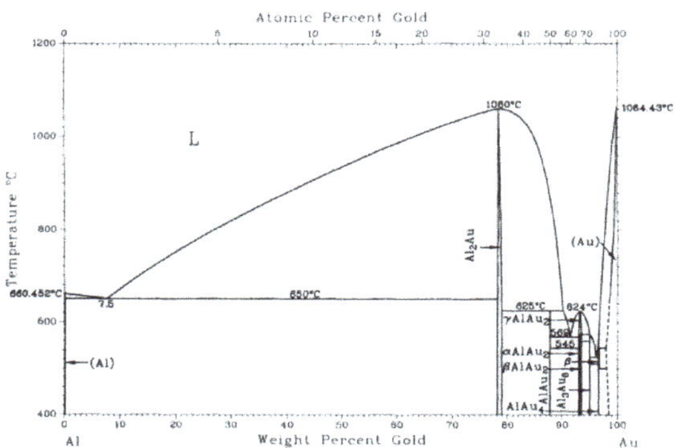

Figure: The aluminum-gold phase diagram.

Millions of actual and opportunity dollars were lost in identifying the problem and its solution, which could have been avoided had the proper phase diagram been examined. A question concerning purple plague problems, however, has remained unresolved: whether or not the presence of silicon near the gold-aluminum interface has an influence on the stability and rate of formation of the damaging intermetallic phase. An examination of the phase relationships in the Al-Al_2Au-Si subternary system showed no stable ternary Al-Au-Si phases. It was suggested instead that the reported effect of silicon may be due to a reaction between silicon and alumina (Al_2O_3) at the aluminum-gold interface that becomes thermodynamically feasible in the presence of gold.

Limitations of Phase Diagrams

Phase diagrams play an extremely useful role in the interpretation of the microstructures developed in alloys, but they have several limitations:

- Phase diagrams only show the equilibrium state of alloys (i.e., under very slow cooling rates); however, in normal industrial processes, alloys are rarely cooled slowly enough to approach equilibrium.
- Phase diagrams do not indicate whether a high-temperature phase can be retained at room temperature by rapid cooling. Phase diagrams do not indicate whether a particular transformation (e.g., a eutectoid transformation) can be suppressed, and what should be the rate of cooling of the alloy to avoid the transformation.
- Phase diagrams do not indicate the phases produced by fast cooling rates. For example, the formation of martensite is not shown in the Fe-Fe3C phase diagram. Thus, they do not indicate the temperature of the start of such transformations (e.g., the Ms Temperature) and their kinetics of formation.
- Even under equilibrium conditions, phase diagrams do not indicate the character of the transformations. They do not indicate the rate at which the equilibrium will be attained.
- Phase diagrams only give information on the constitution of alloys, such as the number of phases present at a point, but do not give information about the structural distribution of the phases; that is, they do not indicate the size, shape, or distribution of the phases, which affects final mechanical properties. The structural distribution of phases is affected by the surface energy between phases and the strain energy produced by the transformation. For

example, if the β phase, in a mixture of α and β, is present in small amounts and is totally distributed with the α grains, the mechanical properties will largely be governed by the α phase. However, if β is present around the grain boundaries of α, then the strength and ductility of the alloy is largely dictated by properties of the β phase.

Binary Phase Diagram

Binary phase diagrams describe the co-existence of two phases at a range of pressures for a given temperature.

A binary system has two components; C equals 2, and the number of degrees of freedom is F=4−PF=4−P. There must be at least one phase, so the maximum possible value of F is 3. Since F cannot be negative, the equilibrium system can have no more than four phases.

We can independently vary the temperature, pressure, and composition of the system as a whole. Instead of using these variables as the coordinates of a three-dimensional phase diagram, we usually draw a two-dimensional phase diagram that is either a temperature–composition diagram at a fixed pressure or a pressure–composition diagram at a fixed temperature. The position of the system point on one of these diagrams then corresponds to a definite temperature, pressure, and overall composition. The composition variable usually varies along the horizontal axis and can be the mole fraction, mass fraction, or mass percent of one of the components, as will presently be illustrated by various examples.

The way in which we interpret a two-dimensional phase diagram to obtain the compositions of individual phases depends on the number of phases present in the system.

- If the system point falls within a one-phase area of the phase diagram, the composition variable is the composition of that single phase. There are three degrees of freedom. On the phase diagram, the value of either T or p has been fixed, so there are two other independent intensive variables. For example, on a temperature–composition phase diagram, the pressure is fixed and the temperature and composition can be changed independently within the boundaries of the one-phase area of the diagram.
- If the system point is in a two-phase area of the phase diagram, we draw a horizontal tie line of constant temperature (on a temperature–composition phase diagram) or constant pressure (on a pressure–composition phase diagram). The lever rule applies. The position of the point at each end of the tie line, at the boundary of the two-phase area, gives the value of the composition variable of one of the phases and also the physical state of this phase: either the state of an adjacent one-phase area, or the state of a phase of fixed composition when the boundary is a vertical line. Thus, a boundary that separates a two-phase area for phases α and β from a one-phase area for phase α is a curve that describes the composition of phase αα as a function of T or p when it is in equilibrium with phase β. The curve is called a solidus, liquidus, or vaporus depending on whether phase α is a solid, liquid, or gas.
- A binary system with three phases has only one degree of freedom and cannot be represented by an area on a two-dimensional phase diagram. Instead, there is a horizontal boundary

line between areas, with a special point along the line at the junction of several areas. The compositions of the three phases are given by the positions of this point and the points at the two ends of the line. The position of the system point on this line does not uniquely specify the relative amounts in the three phases.

The examples that follow show some of the simpler kinds of phase diagrams known for binary systems.

Solid–liquid Systems

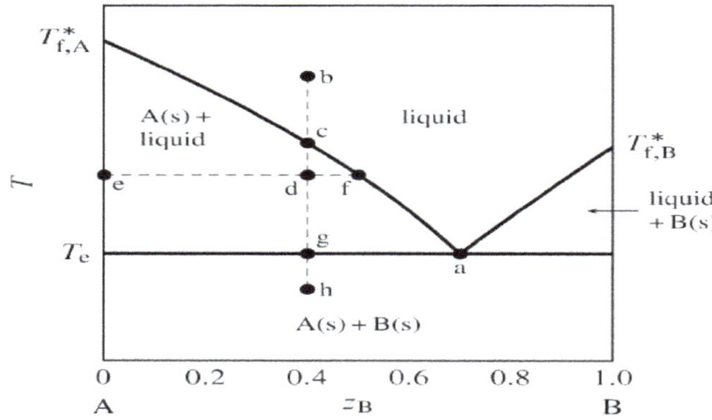

Figure: Temperature–composition phase diagram for a binary system exhibiting a eutectic point.

Figure above is a temperature–composition phase diagram at a fixed pressure. The composition variable z_B is the mole fraction of component B in the system as a whole. The phases shown are a binary liquid mixture of A and B, pure solid A, and pure solid B.

The one-phase liquid area is bounded by two curves, which we can think of either as freezing-point curves for the liquid or as solubility curves for the solids. These curves comprise the liquidus. As the mole fraction of either component in the liquid phase decreases from unity, the freezing point decreases. The curves meet at point a, which is a eutectic point. At this point, both solid A and solid B can coexist in equilibrium with a binary liquid mixture. The composition at this point is the *eutectic composition*, and the temperature here (denoted T_e is the *eutectic temperature*. ("Eutectic" comes from the Greek for *easy melting*.) T_e is the lowest temperature for the given pressure at which the liquid phase is stable.

Suppose we combine 0.60mol0.60mol A and 0.40mol0.40mol B (z_B=0.40z_B=0.40) and adjust the temperature so as to put the system point at b. This point is in the one-phase liquid area, so the equilibrium system at this temperature has a single liquid phase. If we now place the system in thermal contact with a cold reservoir, heat is transferred out of the system and the system point moves down along the *isopleth* (path of constant overall composition) b–h. The cooling rate depends on the temperature gradient at the system boundary and the system's heat capacity.

At point c on the isopleth, the system point reaches the boundary of the one-phase area and is about to enter the two-phase area labeled A(s) + liquid. At this point in the cooling process, the liquid is saturated with respect to solid A, and solid A is about to freeze out from the liquid. There is an abrupt decrease (break) in the cooling rate at this point, because the freezing process involves an extra enthalpy decrease.

At the still lower temperature at point d, the system point is within the two-phase solid–liquid area. The tie line through this point is line e–f. The compositions of the two phases are given by the values of z_B at the ends of the tie line: $x^S{}_B = 0$ for the solid and $x^1{}_B = 0.50$ for the liquid. From the general lever rule, the ratio of the amounts in these phases is,

$$\frac{n^1}{n^s} = \frac{z_B - x_B^s}{x_B^1 - Z_B} = \frac{0.40 - 0}{0.50 - 0.40} = 4.0$$

Since the total amount is $n^s + n^l = 1.00$ mol, the amounts of the two phases must be $n^s = 0.20$ mol and $n^l = 0.80$ mol.

When the system point reaches the eutectic temperature at point g, cooling halts until all of the liquid freezes. Solid B freezes out as well as solid A. During this *eutectic halt*, there are at first three phases: liquid with the eutectic composition, solid A, and solid B. As heat continues to be withdrawn from the system, the amount of liquid decreases and the amounts of the solids increase until finally only 0.60mol0.60mol of solid A and 0.40mol0.40mol of solid B are present. The temperature then begins to decrease again and the system point enters the two-phase area for solid A and solid B; tie lines in this area extend from $z_B=0$ to $z_B=1$.

Temperature–composition phase diagrams such as this are often mapped out experimentally by observing the cooling curve (temperature as a function of time) along isopleths of various compositions. This procedure is *thermal analysis*. A break in the slope of a cooling curve at a particular temperature indicates the system point has moved from a one-phase liquid area to a two-phase area of liquid and solid. A temperature halt indicates the temperature is either the freezing point of the liquid to form a solid of the same composition, or else a eutectic temperature.

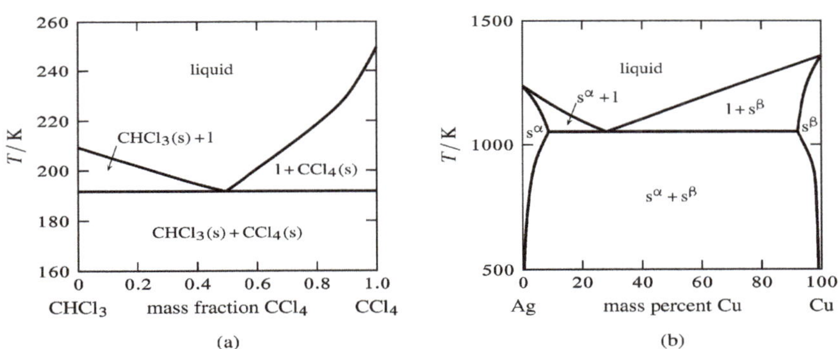

Figure: Temperature–composition phase diagrams with single eutectics. (a) Two pure solids and a liquid mixture (b) Two solid solutions and a liquid mixture.

Figure above shows two temperature–composition phase diagrams with single eutectic points. The left-hand diagram is for the binary system of chloroform and carbon tetrachloride, two liquids that form nearly ideal mixtures. The solid phases are pure crystals. The right-hand diagram is for the silver–copper system and involves solid phases that are solid solutions (substitutional alloys of variable composition). The area labeled s^α is a solid solution that is mostly silver, and s^β is a solid solution that is mostly copper. Tie lines in the two-phase areas do not end at a vertical line for a pure solid component as they do in the system shown in the left-hand diagram. The three phases that can coexist at the eutectic temperature of 1,052K are the melt of the eutectic composition and the two solid solutions.

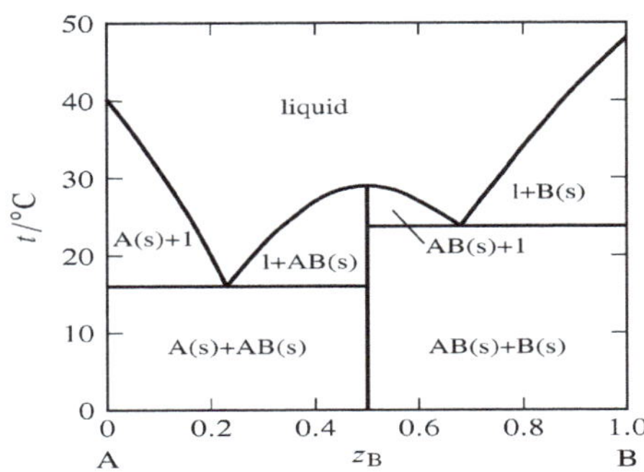

Figure: Temperature–composition phase diagram for the binary system of αα-naphthylamine (A) and phenol (B) at 1bar1bar.

An example of this behavior is shown in figure above, in which the solid compound contains equal amounts of the two components αα-naphthylamine and phenol. The possible solid phases are pure A, pure B, and the solid compound AB. Only one or two of these solids can be present simultaneously in an equilibrium state. The vertical line in the figure at z_B=0.5 represents the solid compound. The temperature at the upper end of this line is the melting point of the solid compound,29°C. The solid melts *congruently* to give a liquid of the same composition. A melting process with this behavior is called a *dystectic reaction*. The cooling curve for liquid of this composition would display a halt at the melting point

The phase diagram in figure above has two eutectic points. It resembles two simple phase diagrams placed side by side. There is one important difference: the slope of the freezing-point curve (liquidus curve) is nonzero at the composition of a pure component, but is zero at the composition of a solid compound that is completely dissociated in the liquid. Thus, the curve in figure above has a relative maximum at the composition of the solid compound (zB=0.5zB=0.5) and is rounded there, instead of having a cusp—like a Romanesque arch rather than a Gothic arch.

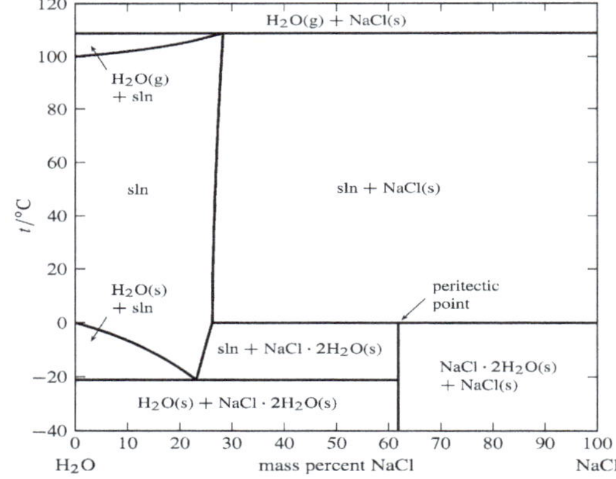

Figure: Temperature–composition phase diagram for the binary system of H_2O and NaCl at 1bar1bar.

An example of a solid compound that does not melt congruently is shown in figure above. The solid hydrate NaCl·^2H_2O is 61.9% NaCl by mass. It decomposes at 0° C to form an aqueous solution of composition 26.3%26.3% NaCl by mass and a solid phase of anhydrous NaCl. These three phases can coexist at equilibrium at 0° C. A phase transition like this, in which a solid compound changes into a liquid and a different solid, is called *incongruent* or *peritectic* melting, and the point on the phase diagram at this temperature at the composition of the liquid is a *peritectic point*.

Figure above shows there are two other temperatures at which three phases can be present simultaneously: −21° C,where the phases are ice, the solution at its eutectic point, and the solid hydrate; and 109°C,where the phases are gaseous H_2O, a solution of composition 28.3%28.3% NaCl by mass, and solid NaCl. Note that both segments of the right-hand boundary of the one-phase solution area have positive slopes, meaning that the solubilities of the solid hydrate and the anhydrous salt both increase with increasing temperature.

Partially-miscible Liquids

When two liquids that are partially miscible are combined in certain proportions, phase separation occurs. Two liquid phases in equilibrium with one another are called *conjugate phases*. Obviously the two phases must have different compositions or they would be identical; the difference is called a *miscibility gap*. A binary system with two phases has two degrees of freedom, so that at a given temperature and pressure each conjugate phase has a fixed composition.

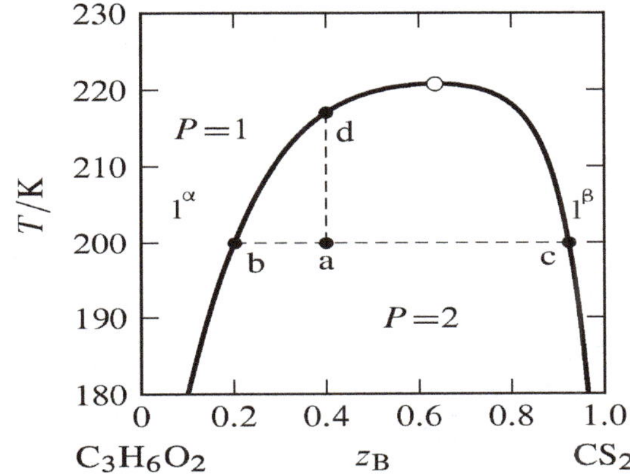

Figure: Temperature–composition phase diagram for the binary system of methyl acetate (A) and carbon disulfide (B) at 1bar1bar. All phases are liquids. The open circle indicates the critical point.

The typical dependence of a miscibility gap on temperature is shown in figure above. The miscibility gap (the difference in compositions at the left and right boundaries of the two-phase area) decreases as the temperature increases until at the *upper consolute temperature*, also called the *upper critical solution temperature*, the gap vanishes. The point at the maximum of the boundary curve of the two-phase area, where the temperature is the upper consolute temperature, is the *consolute point* or *critical point*. At this point, the two liquid phases become identical, just as the liquid and gas phases become identical at the critical point of a pure substance. Critical opalescence is observed in the vicinity of this point, caused by large local composition fluctuations. At temperatures at and above the critical point, the system is a single binary liquid mixture.

Suppose we combine 6.0 mol of component A (methyl acetate) and 4.0mol4.0mol of component B (carbon disulfide) in a cylindrical vessel and adjust the temperature to 200K. The overall mole fraction of B is z_B=0.40. The system point is at point a in the two-phase region. From the positions of points b and c at the ends of the tie line through point a, we find the two liquid layers have compositions $x\frac{\alpha}{B} = 0.20$ and $x\frac{\beta}{B} = 0.92$. Since carbon disulfide is the more dense of the two pure liquids, the bottom layer is phase β, the layer that is richer in carbon disulfide. According to the lever rule, the ratio of the amounts in the two phases is given by

$$\frac{n^{\beta}}{n^{\alpha}}=\frac{z_B-x_B{}^{\alpha}}{x_B{}^{\beta}-Z_B}=\frac{0.40-0.20}{0.92-0.40}=0.38$$

Combining this value with $n^{\alpha}+n^{\beta}=10.0 mol\, gives\, us\, n^{\alpha}=7.2 mol\, and\, n^{\alpha}=2.8 mol.$

If we gradually add more carbon disulfide to the vessel while gently stirring and keeping the temperature constant, the system point moves to the right along the tie line. Since the ends of this tie line have fixed positions, neither phase changes its composition, but the amount of phase β increases at the expense of phase α. The liquid–liquid interface moves up in the vessel toward the top of the liquid column until, at overall composition z_B=0.92 (point c), there is only one liquid phase.

Now suppose the system point is back at point a and we raise the temperature while keeping the overall composition constant at z_B=0.40. The system point moves up the isopleth a–d. The phase diagram shows that the ratio $\left(z_B=x^{\alpha}{}_B\right)/(x^{\alpha}{}_B-Z_B)$ decreases during this change. As a result, the amount of phase αα increases, the amount of phase ββ decreases, and the liquid–liquid interface moves down toward the bottom of the vessel until at 217K (point d) there again is only one liquid phase.

Liquid–gas Systems with Ideal Liquid Mixtures

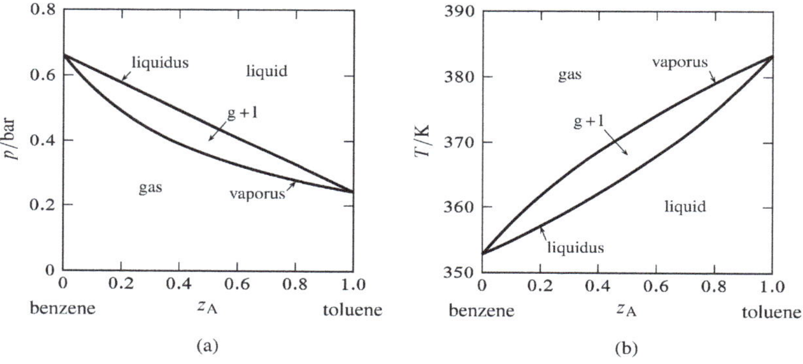

Figure: Phase diagrams for the binary system of toluene (A) and benzene (B). The curves are calculated from Eqs. $x_A=\frac{P-P_B^*}{P_A^*-P_B^*}$ and $y_A=\frac{p_A}{p}=\frac{xAp_A^*}{p}=\left(\frac{p-p_B^*}{p_A^*-p_B^*}\right)\frac{p_A^*}{p}$ and the saturation vapor pressures of the pure liquids. (a)Pressure–compositiondiagramat T=340K. (b) Temperature–composition diagram at p=1bar.

Toluene and benzene form liquid mixtures that are practically ideal and closely obey Raoult's law for partial pressure. For the binary system of these components, we can use the vapor pressures of the pure liquids to generate the liquidus and vaporus curves of the pressure–composition and temperature–composition phase diagram. The results are shown in figure above. The composition variable z_A is the overall mole fraction of component A (toluene).

The equations needed to generate the curves can be derived as follows. Consider a binary liquid mixture of components A and B and mole fraction composition xA that obeys Raoult's law for partial pressure:

$$PA=x_A P_A^* \qquad pB=(1-x_A)P_B^*$$

Strictly speaking, Raoult's law applies to a liquid–gas system maintained at a constant pressure by means of a third gaseous component, and p∗A and p∗B are the vapor pressures of the pure liquid components at this pressure and the temperature of the system. However, when a liquid phase is equilibrated with a gas phase, the partial pressure of a constituent of the liquid is practically independent of the total pressure,so that it is a good approximation to apply the equations to a *binary*liquid–gas system and treat p^*_A and p^*_B as functions only of T.

When the binary system contains a liquid phase and a gas phase in equilibrium, the pressure is the sum of $p^{*A}$and p^*_B, which from equation $PA = x_A P_A^*$ $\quad p_B = (1-x_A)P_B^*$ is given by,

$$\begin{aligned} p &= x_A p_A^* + (1-x_A)p_B^* \\ &= p_B^* + (p_A^* - p_B^*)x_A \end{aligned} \qquad \text{(C=2, ideal liquid mixture)}$$

Where x_A is the mole fraction of A in the liquid phase. Equation above shows that in the two-phase system, p has a value between p^*A and p^*B, and that if T is constant, p is a linear function of x_A. The mole fraction composition of the gas in the two-phase system is given by,

$$y_A = \frac{p_A}{p} = \frac{x_A p_A^*}{p_B^* + (p_A^* - p_B^*)x_A}$$

A binary two-phase system has two degrees of freedom. At a given T and ρ, each phase must have a fixed composition. We can calculate the liquid composition by rearranging equation

$$\begin{aligned} p &= x_A p_A^* + (1-x_A)p_B^* \\ &= p_B^* + (p_A^* - p_B^*)x_A \end{aligned} \quad ,$$

$$x_A = \frac{p - p_B^*}{p_A^* - p_B^*} \qquad \text{(C=2, ideal liquid mixture)}$$

The gas composition is then given by,

$$\begin{aligned} y_A &= \frac{p_A}{p} = \frac{x_A p_A^*}{p} \\ &= \left(\frac{p - p_B^*}{p_A^* - p_B^*}\right)\frac{p_A^*}{p} \end{aligned} \qquad \text{(C=2, ideal liquid mixture)}$$

If we know p^*_A and p^*_B as functions of T, we can use equation above to calculate the compositions for any combination of T and *p* at which the liquid and gas phases can coexist, and thus construct a pressure–composition or temperature–composition phase diagram.

Figure, the liquidus curve shows the relation between *p* and x_A for equilibrated liquid and gas phases at constant *T*, and the vaporus curve shows the relation between p and y_A under these conditions. We see that p is a linear function of xA but not of yA.

In a similar fashion, the liquidus curve in figure (b) shows the relation between *T* and x_A, and the vaporus curve shows the relation between T and y_A, for equilibrated liquid and gas phases at constant *p*. Neither curve is linear.

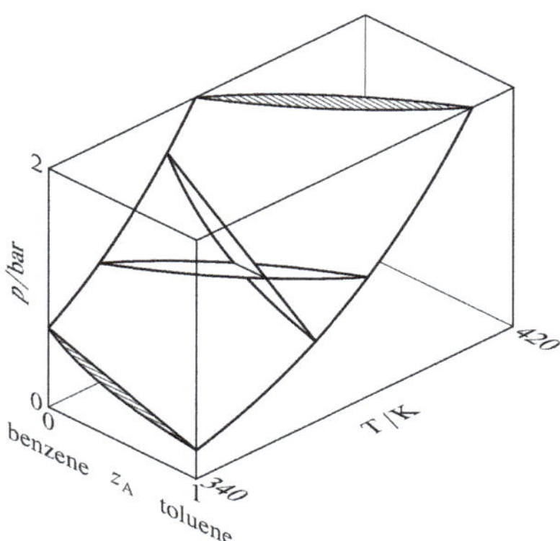

Figure: Liquidus and vaporus surfaces for the binary system of toluene (A) and benzene.

Cross-sections through the two-phase region are drawn at constant temperatures of 340K and 370K and at constant pressures of 1bar and 2bar. Two of the cross-sections intersect at a tie line at T=370 *K* and p=1bar, and the other cross-sections are hatched in the direction of the tie lines.

A liquidus curve is also called a bubble-point curve or a boiling-point curve. Other names for a vaporus curve are dew-point curve and condensation curve. These curves are actually cross-sections of liquidus and vaporus surfaces in a three-dimensional $T - p - z_A$ phase diagram, as shown in figure above. In this figure, the liquidus surface is in view at the front and the vaporus surface is hidden behind it.

Liquid–gas Systems with Nonideal Liquid Mixtures

Most binary liquid mixtures do not behave ideally. The most common situation is positive deviations from Raoult's law. Some mixtures, however, have specific A–B interactions, such as solvation or molecular association that prevents random mixing of the molecules of A and B, and the result is then negative deviations from Raoult's law. If the deviations from Raoult's law, either positive or negative, are large enough, the constant-temperature liquidus curve exhibits a maximum or minimum and azeotropic behavior results.

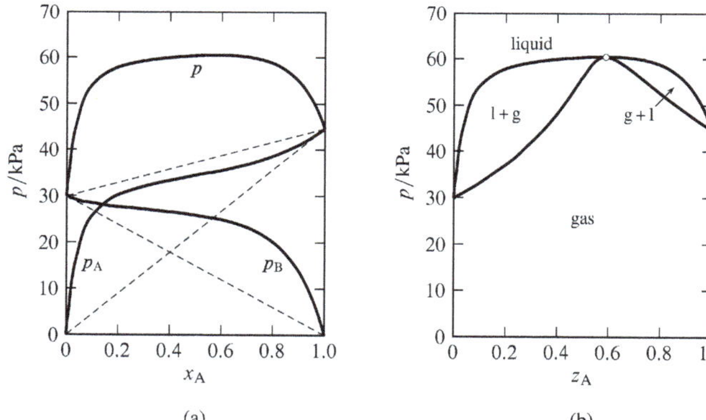

Figure: Binary system of methanol (A) and benzene at 45°C ((a) Partial pressures and total pressure in the gas phase equilibrated with liquid mixtures. The dashed lines indicate Raoult's law behavior. (b) Pressure–composition phase diagram at 45°C. Open circle: azeotropic point at zA=0.59 and p=60.5kPa.

Figure shows the azeotropic behavior of the binary methanol-benzene system at constant temperature. In figure (a) above, the experimental partial pressures in a gas phase equilibrated with the nonideal liquid mixture are plotted as a function of the liquid composition. The partial pressures of both components exhibit positive deviations from Raoult's law, that if one constituent of a binary liquid mixture exhibits positive deviations from Raoult's law, with only one inflection point in the curve of fugacity versus mole fraction, the other constituent also has positive deviations from Raoult's law. The total pressure (equal to the sum of the partial pressures) has a maximum value greater than the vapor pressure of either pure component. The curve of p versus xA becomes the liquidus curve of the pressure–composition phase diagram shown in figure (b). Points on the vaporus curve are calculated from $p=p_A/y_A$.

In practice, the data needed to generate the liquidus and vaporus curves of a nonideal binary system are usually obtained by allowing liquid mixtures of various compositions to boil in equilibrium still at a fixed temperature or pressure. When the liquid and gas phases have become equilibrated, samples of each are withdrawn for analysis. The partial pressures shown in figure (a) above were calculated from the experimental gas-phase compositions with the relations $p_A=y_Ap$ and $p_B=p-p_A$.

If the constant-temperature liquidus curve has a maximum pressure at a liquid composition not corresponding to one of the pure components, which is the case for the methanol–benzene system, then the liquid and gas phases are mixtures of identical compositions at this pressure. On the pressure–composition phase diagram, the liquidus and vaporus curves both have maxima at this pressure, and the two curves coincide at an azeotropic point. A binary system with negative deviations from Raoult's law can have an isothermal liquidus curve with a minimum pressure at a particular mixture composition, in which case the liquidus and vaporus curves coincide at an azeotropic point at this minimum. The general phenomenon in which equilibrated liquid and gas mixtures have identical compositions is called azeotropy, and the liquid with this composition is an azeotropic mixture or azeotrope. An azeotropic mixture vaporizes as if it were a pure substance, undergoing an equilibrium phase transition to a gas of the same composition.

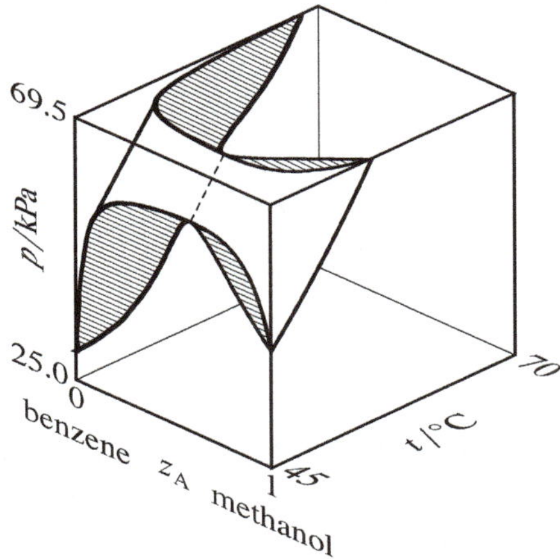

Figure: Liquidus and vaporus surfaces for the binary system of methanol (A) and benzene. Cross-sections are hatched in the direction of the tie lines. The dashed curve is the azeotrope vapor-pressure curve.

If the liquidus and vaporus curves exhibit a maximum on a pressure–composition phase diagram, then they exhibit a minimum on a temperature–composition phase diagram. This relation is explained for the methanol–benzene system by the three-dimensional liquidus and vaporus surfaces drawn in figure. In this diagram, the vaporus surface is hidden behind the liquidus surface. The hatched cross-section at the front of the figure is the same as the pressure–composition diagram of figure (b), and the hatched cross-section at the top of the figure is a temperature–composition phase diagram in which the system exhibits a minimum-boiling azeotrope.

A binary system containing an azeotropic mixture in equilibrium with its vapor has two species, two phases, and one relation among intensive variables: $x_A=y_A$. The number of degrees of freedom is then $F=2+s-r-P=2+2-1-2=1$; the system is univariant. At a given temperature, the azeotrope can exist at only one pressure and have only one composition. As T changes, so do p and z_A along an azeotrope vapor-pressure curve as illustrated by the dashed curve in figure above.

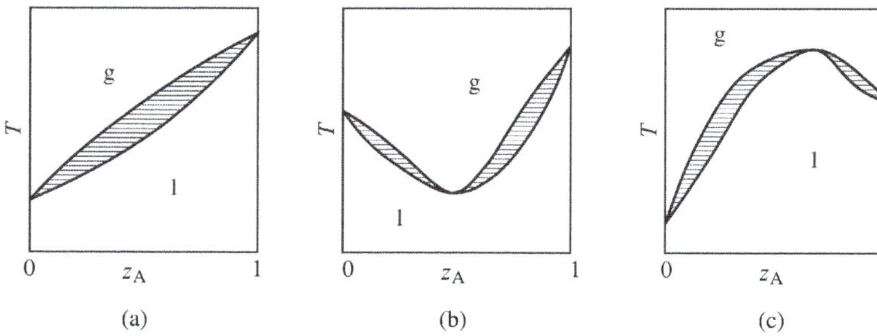

Figure: Temperature–composition phase diagrams of binary systems exhibiting (a) no azeotropy, (b) a minimum-boiling azeotrope, and (c) a maximum-boiling azeotrope. Only the one-phase areas are labeled; two-phase areas are hatched in the direction of the tie lines.

Figure above summarizes the general appearance of some relatively simple temperature–composition phase diagrams of binary systems. If the system does not form an azeotrope (*zeotropic* behavior), the equilibrated gas phase is richer in one component than the liquid phase at all liquid compositions, and the liquid mixture can be separated into its two components by fractional distillation. The gas in equilibrium with an azeotropic mixture, however, is not enriched in either component. Fractional distillation of a system with an azeotrope leads to separation into one pure component and the azeotropic mixture.

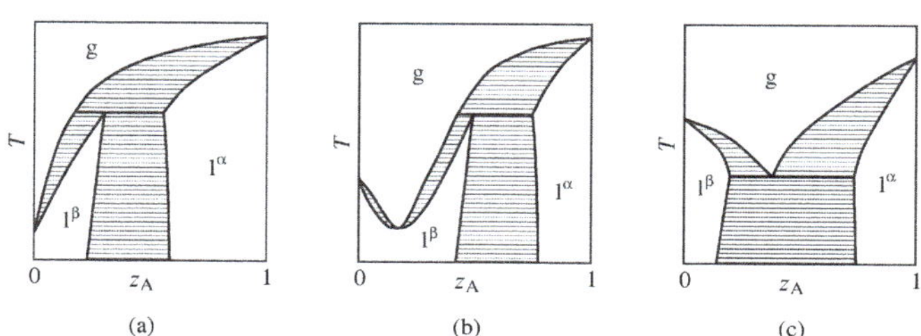

Above figure show temperature–composition phase diagrams of binary systems with partially-miscible liquids exhibiting (a) the ability to be separated into pure components by fractional distillation, (b) a minimum-boiling azeotrope, and (c) boiling at a lower temperature than the boiling point of either pure component. Only the one-phase areas are labeled; two-phase areas are hatched in the direction of the tie lines.

More complicated behavior is shown in the phase diagrams of figure above. These are binary systems with partially-miscible liquids in which the boiling point is reached before an upper consolute temperature can be observed.

Solid–gas Systems

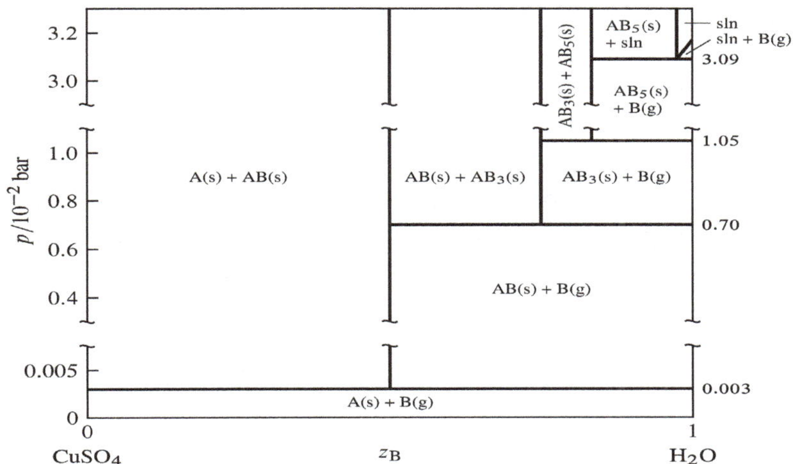

Figure: Pressure–composition phase diagram for the binary system of CuSO44 (A) and H22O (B) at 25°C25°C .

As an example of a two-component system with equilibrated solid and gas phases, consider the components $CuSO_4$ and H_2O , denoted A and B respectively. In the pressure–composition phase diagram shown in figure above, the composition variable zB is as usual the mole fraction of component B in the system as a whole.

The anhydrous salt and its hydrates (solid compounds) form the series of solids $CuSO_4$, CuSO4·H_2O, CuSO4·3H_2O, and CuSO4·5H_2O. In the phase diagram these formulas are abbreviated A, AB, AB 3, and AB 5. The following dissociation equilibria (dehydration equilibria) are possible:

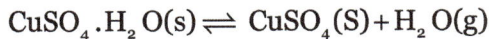

$$CuSO_4.H_2O(s) \rightleftharpoons CuSO_4(S) + H_2O(g)$$

$$\frac{1}{2}CuSO_4.3H_2O(s) \rightleftharpoons \frac{1}{2}CuSO_4.H_2O(s) + H_2O(g)$$

$$\frac{1}{2}CuSO_4.5H_2O(s) \rightleftharpoons \frac{1}{2}CuSO_4.3H_2O(s) + H_2O(g)$$

The equilibria are written above with coefficients that make the coefficient of H_2O(g) unity. When one of these equilibria is established in the system, there are two components and three phases; the phase rule then tells us the system is univariant and the pressure has only one possible value at a given temperature. This pressure is called the *dissociation pressure* of the higher hydrate.

The dissociation pressures of the three hydrates are indicated by horizontal lines in figure above. For instance, the dissociation pressure of CuSO4·5H_2O is 1.05×10^{-2}bar. At the pressure of each horizontal line, the equilibrium system can have one, two, or three phases, with compositions given by the intersections of the line with vertical lines. A fourth three-phase equilibrium is shown at $p=3.09\times10^{-2}$bar; this is the equilibrium between solid CuSO4·5H_2O, the saturated aqueous solution of this hydrate, and water vapor.

Consider the thermodynamic equilibrium constant of one of the dissociation reactions. At the low pressures shown in the phase diagram, the activities of the solids are practically unity and the fugacity of the water vapor is practically the same as the pressure, so the equilibrium constant is almost exactly equal to pd/p°, where pd is the dissociation pressure of the higher hydrate in the reaction. Thus, a hydrate cannot exist in equilibrium with water vapor at a pressure below the dissociation pressure of the hydrate because dissociation would be spontaneous under these conditions. Conversely, the salt formed by the dissociation of a hydrate cannot exist in equilibrium with water vapor at a pressure above the dissociation pressure because hydration would be spontaneous.

If the system contains dry air as an additional gaseous component and one of the dissociation equilibria is established, the partial pressure pH_2O of H_2O is equal (approximately) to the dissociation pressure pd of the higher hydrate. The prior statements regarding dissociation and hydration now depend on the value of pH_2O. If a hydrate is placed in air in which pH_2O is less than pd, dehydration is spontaneous; this phenomenon is called efflorescence. If pH_2O is greater than the vapor pressure of the saturated solution of the highest hydrate that can form in the system, the anhydrous salt and any of its hydrates will spontaneously absorb water and form the saturated solution; this is deliquescence.

If the two-component equilibrium system contains only two phases, it is bivariant corresponding to one of the areas in figure. Here both the temperature and the pressure can be varied. In the case of areas labeled with two solid phases, the pressure has to be applied to the solids by a fluid (other than H_2O) that is not considered part of the system.

Systems at High Pressure

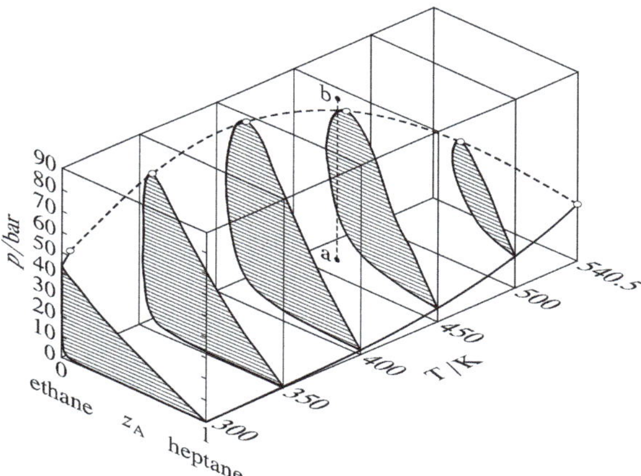

Figure: Pressure–temperature–composition behavior in the binary heptane–ethane system. The open circles are critical points; the dashed curve is the critical curve. The dashed line a–b illustrates retrograde condensation at 450K.

Binary phase diagrams begin to look different when the pressure is greater than the critical pressure of either of the pure components. Various types of behavior have been observed in this region. One common type, that found in the binary system of heptane and ethane, is shown in figure above. This figure shows sections of a three-dimensional phase diagram at five temperatures. Each section is a pressure–composition phase diagram at constant T. The two-phase areas are hatched in the direction of the tie lines. At the left end of each tie line (at low z_A) is a vaporus curve, and at the right end is a liquidus curve. The vapor pressure curve of pure ethane (z_A=0) ends at the critical point of ethane at 305.4K; between this point and the critical point of heptane at 540.5K, there is a continuous critical curve, which is the locus of critical points at which gas and liquid mixtures become identical in composition and density.

Consider what happens when the system point is at point a in figure and the pressure is then increased by isothermal compression along line a–b. The system point moves from the area for a gas phase into the two-phase gas–liquid area and then out into the gas-phase area again. This curious phenomenon, condensation followed by vaporization, is called retrograde condensation.

Under some conditions, an isobaric increase of T can result in vaporization followed by condensation; this is retrograde vaporization.

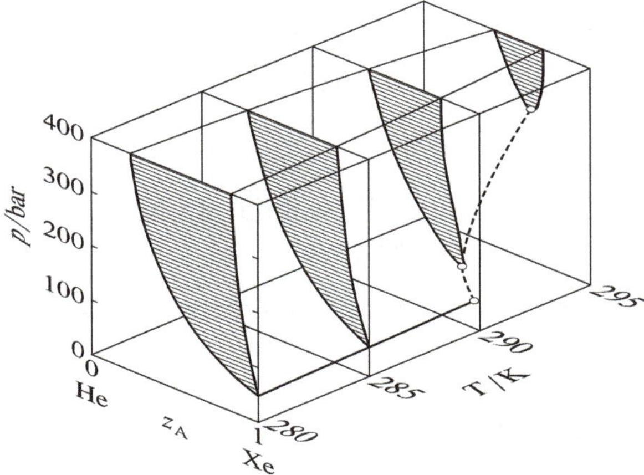

Figure: Pressure–temperature–composition behavior in the binary xenon–helium system. The open circles are critical points; the dashed curve is the critical curve.

A different type of high-pressure behavior, that found in the xenon–helium system, is shown in figure above. Here, the critical curve begins at the critical point of the less volatile component (xenon) and continues to higher temperatures and pressures than the critical temperature and pressure of either pure component. The two-phase region at pressures above this critical curve is sometimes said to represent gas–gas equilibrium, or gas–gas immiscibility, because we would not usually consider a liquid to exist beyond the critical points of the pure components. Of course, the coexisting phases in this two-phase region are not gases in the ordinary sense of being tenuous fluids, but are instead high-pressure fluids of liquid-like densities. If we want to call both phases gases, then we have to say that pure gaseous substances at high pressure do not necessarily mix spontaneously in all proportions as they do at ordinary pressures.

If the pressure of a system is increased isothermally, eventually solid phases will appear; these are not shown in figures above.

Binary Eutectic Phase Diagram

The binary eutectic phase diagram explains the chemical behavior of two immiscible (unmixable) crystals from a completely miscible (mixable) melt, such as olivine and pyroxene, or pyroxene and Ca plagioclase. Here we are going to generalize to two minerals, A and B, or P and Q. We want to observe the behavior of this system under two conditions, one of complete equilibrium during crystallization when all chemical reactions can run to completion, and the second of disequilibrium when fractionation occurs and igneous rocks evolve.

The conventions for the phase diagram include the following:

- Two immiscible components: such as CaAl2Si2O8 (calcic plagioclase) and FeMgSiO4 (olivine) plotted along the horizontal axis, OR olivine (isolated tetrahedra) and pyroxene (single chain tetrahedra). They are immiscible because they have different crystal structures.
- One variable, temperature, plotted along the vertical axis. Pressure is held constant at 1 atmosphere.
- Three phases, crystal A, crystal B, and melt.
- Complete miscibility of the melt (magma).

The assumptions are:

- The system remains in equilibrium throughout its history so that all reactions can take place and everything can come to stability.
- Everything in the original melt remains in communication throughout the crystallization process.

Eutectic or Syntectic Reactions

A eutectic reaction can occur in a phase diagram if there is (i) a miscibility gap in the solid phase, (ii) a minimum in the iso-G curve and (iii) the composition corresponding to the minimum lies within the composition range of the miscibility gap. The limiting case thus corresponds to the intersection of the miscibility gap and the iso-G curve at the minimum in the latter. In case this minimum lies outside the miscibility gap, a peritectic instead of a eutectic reaction occurs as mentioned earlier. Thus,

$$T_o^m = T_{mg}^a.$$

Substituting from $T_{mg}^a = \dfrac{(1-2x^a)}{R\,In\left[\{1-x^a\}/x^a\right]}W^a$, $x^m = \dfrac{1}{2} + \dfrac{\Delta T\,\Delta S}{\Delta W}$. and

$T_0^m = \dfrac{1}{2}(T_B + T_A) + \dfrac{\Delta T^2\,\Delta S}{4\Delta W} + \dfrac{\Delta W}{4\Delta S}$. in the above, we have, for the special case of $\Delta S_A = \Delta S_B = \Delta S$,

$$\mathrm{W}^a = -R\omega\left[\frac{T_B + T_A}{2} + \frac{\Delta T}{4}\left(\omega + \frac{1}{\omega}\right)\right]\mathrm{In}\left(\frac{\omega - 1}{\omega + 1}\right),$$

where $\omega = \Delta W(\Delta T\ \Delta S)$. A more general relation valid for the case of $\Delta S_B \neq \Delta S_A$ can be found by substituting from $x^m = -\dfrac{\Delta S_A \Delta W \pm \sqrt{\Delta S_A \Delta S_B}\sqrt{\Delta W^2 + \Delta T(\Delta S_B - \Delta S_A)\Delta W}}{(\Delta S_B - \Delta S_A)\Delta W}$ in

$T_0 = \dfrac{(1-x)T_A\,\Delta S_A + xT_B\,\Delta S_B + x(1-x)\Delta W}{(1-x)\Delta S_A + x\Delta S_B}$ to find T_o^m and in $T_{mg}^a = \dfrac{(1-2x^a)}{R\,In\left[\{1-x^a\}/x^a\right]}W^a$,

to find T_{mg}^a. These can then be utilized in to yield W^a in terms of ΔW. W^a has been evaluated for a series of values of DW. Subsequently, W b has been found from $\mathrm{W}^\beta = \mathrm{W}^a + \Delta W$. The computed boundary denoted by 5 is displayed in figure. Hence the eutectic reaction occurs in the regions P and I lying to the right of this boundary. As in the case of the peritectic boundary, a limiting value for ΔW can be found by observing that $W^{a} \geq 0$. At $W^{a} = 0$,

$$W^{\beta} = \Delta W = -(\sqrt{T_B} + \sqrt{T_A})^2 \Delta S$$

Thus, this eutectic boundary also terminates at $(W^{a}, W^{\beta}) = (0, -39595{\cdot}9)$. The occurrence of syntectic reaction can be analysed in an analogous manner. Thus, $T_o^{m} = T^{\beta}_{mg}$

Substituting from an equation analogous to $T^{a}_{mg} = \frac{(1-2x^{a})}{R\,In\left[\{1-x^{a}\}/x^{a}\right]} W^{a}$, for the b-phase and

$x^{m} = \frac{1}{2} + \frac{\Delta T\,\Delta S}{\Delta W}$. and $T_0^{m} = \frac{1}{2}(T_B + T_A) + \frac{\Delta T^2\,\Delta S}{4\Delta W} + \frac{\Delta W}{4\Delta S}$., we obtain, for the special case of $\Delta S_A = \Delta S_B = \Delta S$,

$$W^{\beta} = -R\omega\left[\frac{T_B + T_A}{2} + \frac{\Delta T}{4}\left(\omega + \frac{1}{\omega}\right)\right] \mathrm{In}\left(\frac{\omega - 1}{\omega + 1}\right)$$

An equation for W^{β} for the general case of $\Delta S_B \neq \Delta S_A$ can be found easily. The above expression has been utilized for evaluating W^{β} for various values of ΔW. Subsequently, W^{a} has been found from $W^{a} = W^{\beta} - \Delta W$. The boundary between the regions of occurrence of syntectic and monotectic reactions is denoted as 6 in figure 1. Thus, the syntectic reaction occurs in the regions B, C and D, which lie above this boundary.

Non-eutectic Compositions

Compositions of eutectic systems that are not at the eutectic composition can be classified as *hypoeutectic* or *hypereutectic*. Hypoeutectic compositions are those with a smaller percent composition of species β and a greater composition of species α than the eutectic composition (E) while hypereutectic solutions are characterized as those with a higher composition of species β and a lower composition of species α than the eutectic composition. As the temperature of a non-eutectic composition is lowered the liquid mixture will precipitate one component of the mixture before the other. In a hypereutectic solution, there will be a proeutectoid phase of species β whereas a hypoeutectic solution will have a proeutectic α phase.

Types

Alloys

Eutectic alloys have two or more materials and have a eutectic composition. When a non-eutectic alloy solidifies, its components solidify at different temperatures, exhibiting a plastic melting range. Conversely, when a well-mixed, eutectic alloy melts, it does so at a single, sharp temperature. The various phase transformations that occur during the solidification of a particular alloy composition can be understood by drawing a vertical line from the liquid phase to the solid phase on the phase diagram for that alloy.

Some uses include:

- NEMA Eutectic Alloy Overload Relays for electrical protection of 3-phase motors for pumps, fans, conveyors, and other factory process equipment.
- Eutectic alloys for soldering, composed of tin (Sn), lead (Pb) and sometimes silver (Ag) or gold (Au) — especially $Sn_{63}Pb_{37}$ alloy formula for electronics.
- Casting alloys, such as aluminium-silicon and cast iron (at the composition of 4.3% carbon in iron producing an austenite-cementite eutectic).
- Silicon chips are bonded to gold-plated substrates through a silicon-gold eutectic by the application of ultrasonic energy to the chip.
- Brazing, where diffusion can remove alloying elements from the joint, so that eutectic melting is only possible early in the brazing process.
- Temperature response, e.g., Wood's metal and Field's metal for fire sprinklers.
- Non-toxic mercury replacements, such as galinstan.
- Experimental glassy metals, with extremely high strength and corrosion resistance.
- Eutectic alloys of sodium and potassium (NaK) that are liquid at room temperature and used as coolant in experimental fast neutron nuclear reactors.

Others

- Sodium chloride and water form a eutectic mixture whose eutectic point is −21.2°C and 23.3% salt by mass. The eutectic nature of salt and water is exploited when salt is spread on roads to aid snow removal, or mixed with ice to produce low temperatures (for example, in traditional ice cream making).
- Ethanol–water has an unusually biased eutectic point, *i.e.* it is close to pure ethanol, which sets the maximum proof obtainable by fractional freezing.

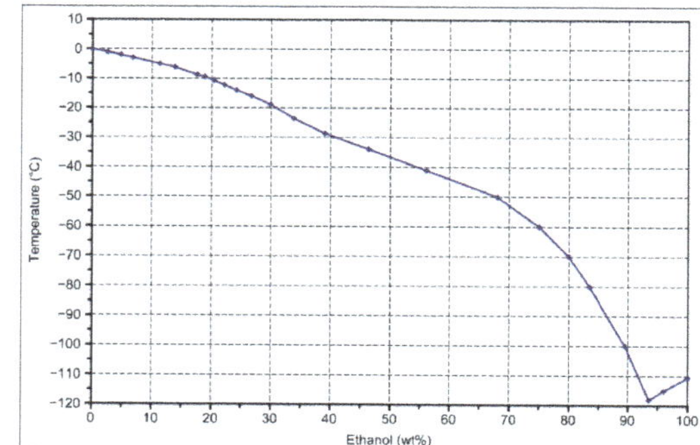

Solid-liquid phase change of ethanol water mixtures

- "Solar salt", 60% $NaNO_3$ and 40% KNO_3, forms a eutectic molten salt mixture which is used for thermal energy storage in concentrated solar power plants. To reduce the eutectic melting point in the solar molten salts calcium nitrate is used in the following proportion: 42% $Ca(NO_3)_2$, 43% KNO_3, and 15% $NaNO_3$.

- Lidocaine and prilocaine—both are solids at room temperature—form a eutectic that is an oil with a 16 °C (61 °F) melting point that is used in eutectic mixture of local anesthetic (EMLA) preparations.
- Menthol and camphor, both solids at room temperature, form a eutectic that is a liquid at room temperature in the following proportions: 8:2, 7:3, 6:4, and 5:5. Both substances are common ingredients in pharmacy extemporaneous preparations.
- Minerals may form eutectic mixtures in igneous rocks, giving rise to characteristic intergrowth textures exhibited, for example, by granophyre.
- Some inks are eutectic mixtures, allowing inkjet printers to operate at lower temperatures.

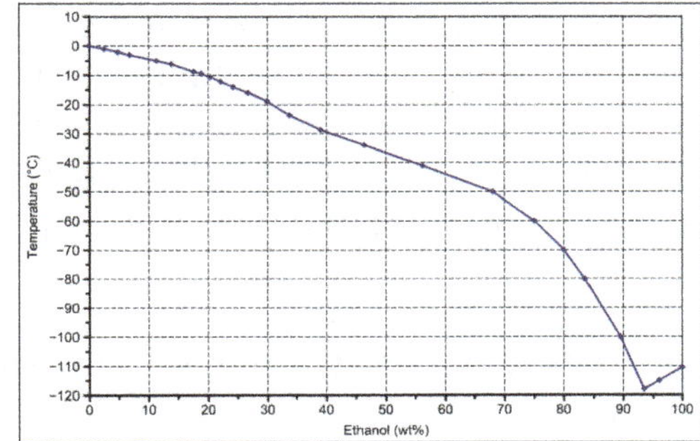

Solid-liquid phase change of ethanol water mixtures

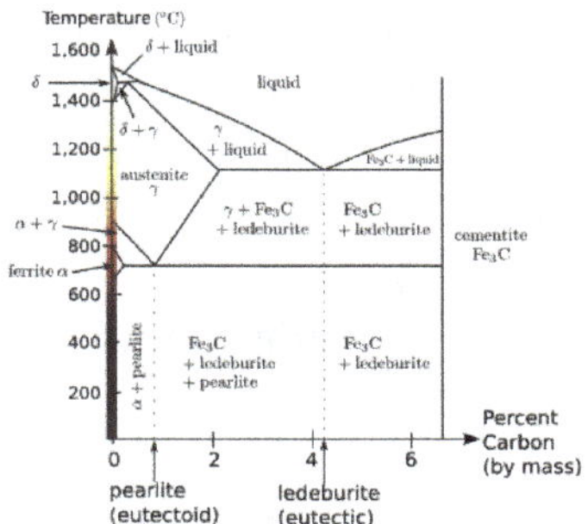

Iron-carbon phase diagram, showing the eutectoid transformation between austenite (γ) and pearlite

Other Critical Points

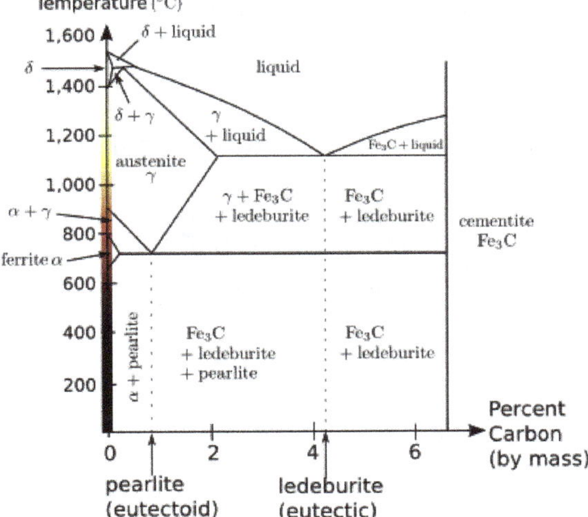

Iron-carbon phase diagram, showing the eutectoid transformation between austenite (γ) and pearlite.

Eutectoid

When the solution above the transformation point is solid, rather than liquid, an analogous eutectoid transformation can occur. For instance, in the iron-carbon system, the austenite phase can undergo a eutectoid transformation to produce ferrite and cementite, often in lamellar structures such as pearlite and bainite. This eutectoid point occurs at 727 °C (1,341 °F) and about 0.76% carbon.

Peritectoid

A *peritectoid* transformation is a type of isothermal reversible reaction that has two solid phases reacting with each other upon cooling of a binary, ternary, ary alloy to create a completely different and single solid phase. The reaction plays a key role in the order and decomposition of quasicrystalline phases in several alloy types.

Peritectic

Peritectic transformations are also similar to eutectic reactions. Here, a liquid and solid phase of fixed proportions reacts at a fixed temperature to yield a single solid phase. Since the solid product forms at the interface between the two reactants, it can form a diffusion barrier and generally causes such reactions to proceed much more slowly than eutectic or eutectoid transformations. Because of this, when a peritectic composition solidifies it does not show the lamellar structure that is found with eutectic solidification.

Such a transformation exists in the iron-carbon system, as seen near the upper-left corner of the figure. It resembles an inverted eutectic, with the δ phase combining with the liquid to produce pure austenite at 1,495 °C (2,723 °F) and 0.17% carbon.

At the peritectic decomposition temperature the compound, rather than melting, decomposes into another solid compound and a liquid. The proportion of each is determined by the lever rule. In the Al-Au phase diagram, for example, it can be seen that only two of the phases melt congruently, $AuAl_2$ and Au_2Al , while the rest peritectically decompose.

Eutectic Calculation

The composition and temperature of a eutectic can be calculated from enthalpy and entropy of fusion of each components.

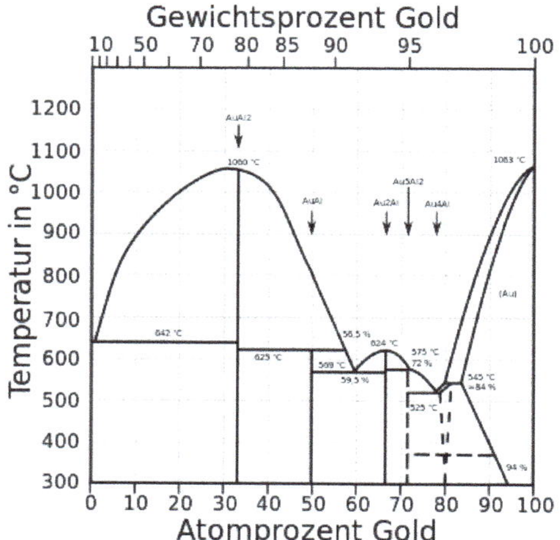

Gold-aluminium phase diagram (German). Top axis title reads "Weight-percent Gold", lower axis title reads "Atomic-percent Gold"

The Gibbs free energy, G, depends on its own differential,

$$G = H - TS \Rightarrow \left\{ \begin{matrix} H = G + TS \\ \left(\frac{\partial G}{\partial T}\right)_P = -S \end{matrix} \Rightarrow H = G - T\left(\frac{\partial G}{\partial T}\right)_P \right\}$$

Thus, the G/T derivative at constant pressure is calculated by the following equation,

$$\left(\frac{\partial G/T}{\partial T}\right)_p = \frac{1}{T}\left(\frac{\partial G}{\partial T}\right)_p - \frac{1}{T^2}G = -\frac{1}{T^2}\left(G - T\left(\frac{\partial G}{\partial T}\right)_p\right) = -\frac{H}{T^2}$$

The chemical potential μ_i is calculated if we assume the activity is equal to the concentration.

$$\mu_i = \mu_i^o + RTIn\frac{a_i}{a} \approx \mu_i^o + RTInx_i$$

At the equilibrium, μ_i^o =0, thus μ_i^o is obtained by:

$$\mu_i = \mu_i^o + RTInx_i = 0 \Rightarrow \mu_i^o = -RTInx_i.$$

Using and integrating gives,

$$\left(\frac{\partial u_i/T}{\partial T}\right)_p = \frac{\partial}{\partial T}(RInx_i) \Rightarrow RInx_i = -\frac{H_i^o}{T} + K$$

The integration constant K may be determined for a pure component with a melting temperature T^o and an enthalpy of fusion H^o Eq.

$$x_i = 1 \Rightarrow T = T_i^o \Rightarrow K = \frac{H_i^o}{T_i^o}$$

We obtain a relation that determines the molar fraction as a function of the temperature for each component.

$$R\ln x_i = -\frac{H_i^\circ}{T} + \frac{H_i^\circ}{T_i^\circ}$$

The mixture of n components is described by the system,

$$\left\{ \begin{matrix} \ln x_i + \frac{H_i^\circ}{RT} - \frac{H_i^\circ}{RT_i^\circ} = 0 \\ \sum_{i=1}^{n} x_i = 1 \end{matrix} \right. \quad \left\{ \begin{matrix} \forall i < n \Rightarrow \ln x_i + \frac{H_i^\circ}{RT} - \frac{H_i^\circ}{RT_i^\circ} = 0 \\ \ln\left(1 - \sum_{i=1}^{n-1} x_i\right) + \frac{H_n^\circ}{RT} - \frac{H_n^\circ}{RT_n^\circ} = 0 \end{matrix} \right.$$

that can be solved by,

$$\begin{bmatrix} \Delta x_1 \\ \Delta x_2 \\ \Delta x_3 \\ \vdots \\ \Delta x_{n-1} \\ \Delta T \end{bmatrix} = \begin{bmatrix} 1/x_1 & 0 & 0 & 0 & 0 & -\frac{H_1^\circ}{RT^2} \\ 0 & 1/x_2 & 0 & 0 & 0 & -\frac{H_2^\circ}{RT^2} \\ 0 & 0 & 1/x_3 & 0 & 0 & -\frac{H_3^\circ}{RT^2} \\ \vdots & \ddots & \ddots & \ddots & \ddots & \vdots \\ 0 & 0 & 0 & 0 & 1/x_{n-1} & -\frac{H_{n-1}^\circ}{RT^2} \\ \frac{-1}{1-\sum_{i=1}^{n-1} x_i} & \frac{-1}{1-\sum_{i=1}^{n-1} x_i} & \frac{-1}{1-\sum_{i=1}^{n-1} x_i} & \frac{-1}{1-\sum_{i=1}^{n-1} x_i} & \frac{-1}{1-\sum_{i=1}^{n-1} x_i} & -\frac{H_n^\circ}{RT^2} \end{bmatrix}^{-1} \cdot \begin{bmatrix} \ln x_1 + \frac{H_1^\circ}{RT} - \frac{H_1^\circ}{RT_1^\circ} \\ \ln x_2 + \frac{H_2^\circ}{RT} - \frac{H_2^\circ}{RT_2^\circ} \\ \ln x_3 + \frac{H_3^\circ}{RT} - \frac{H_3^\circ}{RT_3^\circ} \\ \vdots \\ \ln x_{n-1} + \frac{H_{n-1}^\circ}{RT} - \frac{H_{n-1}^\circ}{RT_{n-1}^\circ} \\ \ln\left(1-\sum_{i=1}^{n-1} x_i\right) + \frac{H_n^\circ}{RT} - \frac{H_n^\circ}{RT_n^\circ} \end{bmatrix}$$

Peritectic Phase Diagram

Sometimes a solid solution phase, which has already been formed, and the residual liquid phase react and form another solid solution phase or intermetallic compound, having a composition between the compositions of the liquid and the first solid. This is peritectic transformation (peritectic reaction).

An example of a phase diagram with peritectic transformation is shown in the figure:

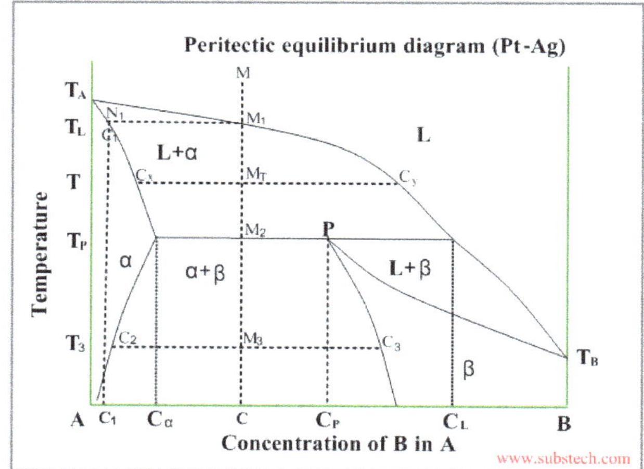

Consider solidification of an alloy with concentration C. When the alloy temperature is higher than T_L, single liquid phase exists (point M on the diagram).

When the temperature reaches the value T_L (point M_1 on the liquidus curve) solidification starts. According to solidus curve (point N_1) the first solid crystals (primary crystals) of the α-phase have composition C_1.

Further cooling of the alloy causes changing of the liquid phase composition according to the liquidus curve and when the alloy temperature reaches a certain intermediate value T (position M_T), liquid phase of composition C_y and solid α-phase of composition C_x are in equilibrium.

At the temperature equal to T_P (peritectic temperature) formation of the α-phase crystals stops and the remainding liquid phase, having composition C_L (peritectic point) reacts with α-phase crystals, forming β-phase of composition C_P (peritectic phase transformation).

Peritectic point is the point on a phase diagram where a previously precipitated α-phase reacts with the liquid phase producing a new solid β-phase. At this point two liquidus curves intersect at the peritectic temperature.

At peritectic temperature remaining α-phase crystals have composition C_a and all crystals of β-phase have composition C_P.

Relative amounts of the α-phase crystals and the liquid phase just above the peritectic transformation may be calculated by the "lever rule":

$$W_a / W_L = (C_L - C) / (C - C_a)$$

Where:

W_a – weight of the α-phase crystals;

W_L – weight of the liquid phase;

Just below the peritectic temperature T_P the alloy consists of two solid phase: a-phase and β-phase, a relative amount of which is determined by the "lever rule":

$$W_a / W_\beta = (C_P - C) / (C - C_a)$$

Where:

W_a – weight of the α-phase;

W_β – weight of theβ-phase;

During further cooling solid solution phases (a-phase andβ-phase) change their compositions according to the corresponding solvus curves.

At the temperature T_3 a-phase crystals have composition C_a and all crystals of β-phase have composition C_β.

If the alloy composition is exactly equal to C_P, a-phase and liquid phase are consumed completely in the peritectic reaction.

Alloys with composition C lower than C_P, some quantity of α-phase remains after the peritectic reaction (it may be calculated by the "lever rule").

If the alloy composition C is higher than C_P, some liquid phase remains after the peritectic reaction. This remaining liquid transforms to β-phase during the further cooling.

Phase Equilibria

Phase equilibrium conditions of a system comprising v components (where v ≥ 1) and ϕ phases (where ϕ ≥ 1) results from the Second Law of Thermodynamics and may be expressed by following equalities: temperatures $T^{(\alpha)}$ (thermal equilibrium, in which there are no heat flows); pressures $p^{(\alpha)}$ (mechanical equilibrium, i.e., the phases are not separated); chemical potentials $p_i^{(\alpha)}$ of each of the components (no mass transfer):

$$T^{(1)} = T^{(2)} = \ldots = T^{(\varphi)} = T;$$
$$p^{(1)} = p^{(2)} = \ldots = p^{(\varphi)} = p;$$
$$p_i^1 = p_i^2 = \ldots = p_i^{(\varphi)} = p_{(i)}; 1 \le i \le v.$$

In the presence of force fields, and/or surface tension (which occurs at the phase interfaces) condition (2) needs to be modified.

The chemical potential of a pure substance is equal to the molar (specific) Gibbs energy at the given aggregate state $p^{(\alpha)} = g^{(\alpha)}(T,p)$. The chemical potential of a component in a mixture is equal to the partial Gibbs energy, which depends on the phase composition of the v-component heterogeneous system:

$$p_i^{(\alpha)} = \overline{g}_i^{(\alpha)}\left(T,p,\overline{x}_1^{(\alpha)},\ldots,\overline{x}_v^{(\alpha)}\right),$$

where $\overline{x}_1^{(\alpha)} = \overline{N}_1^{(\alpha)} / \overline{N}^{(\alpha)}$ is the mole fraction of the ith component in phase α, where $\overline{N}_1^{(\alpha)}$ is the number of moles of the component in phase α and $\overline{N}^{(\alpha)}$ is the total number of moles of all components in the given phase (α).

$$\sum_{i-1}^{v} \overline{x}_1^{(\alpha)} = 1 \; 1 \le \alpha \le \varphi.$$

In the v-component, φ phase system there are vφ component fractions $\overline{x}_1^{(k)}$. Together with temperature T and pressure p, this gives a total of (vφ + 2) variables defining the (equilibrium) system. However, the equilibria and summation expression imply that the variables are not independent and only f variables were.

$$f = v - \varphi + 2$$

are independent. The f value is called the number of the degrees of freedom (or the variants) of the system. Equation above expresses the Gibbs Phase Rule. As soon as f ≥ 0, the number of the phases which are in equilibrium cannot exceed (v + 2) in a v-component system (if a pure substance, then f ≤ 3); a system is called nonvariant if f = v + 2.

For a pure substance, equilibrium of two phases (α and β, say) obeys the condition,

$$g^{(\alpha)}(T,p) = g^{(\beta)}(T,p)$$

A differential equation of the equilibrium curve follows from (6) if the thermodynamic correlations,

$$(\partial g/\partial T)_p = -s, (\partial g/\partial p)_T = v$$

are taken into account; the equation is the well-known Clapeyron-Clausius Equation:

$$dp1^{(\alpha\beta)} / dt = \left(s_1^{(\beta)} - s_1^{(\alpha)}\right) / \left(v_1^{(\beta)} - v_1^{(\alpha)}\right),$$

where $v_1^{(\alpha)}, v_1^{(\beta)}, s_1^{(\alpha)}, s_1^{(\beta)}$ are the molar (specific) volumes and entropies of the coexisting phases. The enthalpy change accompanying the isothermal-isobaric phase transformation α to β is determined by the difference in entropy or enthalpy of the coexisting phases, which are participating in the process transformation:

$$r^{(\alpha\beta)} = T\left(s_1^{(\beta)} - s_1^{(\alpha)}\right) = h_1^{(\beta)} - h_1^{(\alpha)}.$$

Figures below present the Gibbs energy dependency on phase temperature gaseous (g), liquid (l), solid (s) phases at three different pressure values. The correlations explain a shape of the curves considering that the entropy should be less for those phases which have a more regular molecular structure; on the contrary, a density for such a phase should be greater, as a rule. Consistent with equation $g^{(\alpha)}(T,p) = g^{(\beta)}(T,p)$, the intersection points of the curves define the parameters of the phase transformations at equilibrium: sublimation (s → g), melting (s → 1), evaporation (1 → g). The curves formed by such points are given in figure below often referred to as a phase diagram. If all the lines intersect at the same point, then they give a triple point, with parameters p_{tr} , T_{tr}, at which three phases (solid, liquid, and gaseous) exist simultaneously. Triple points of the type s-l-g are inherent to all pure substances except He; because of quantum effects, the liquid phase of He forms crystals at higher pressures. As a result, the phase diagram shape essentially differs in comparison with figure below. Also, this figure does not present the equilibrium curves and the triple points corresponding to phase transformations which occur between solid states due to allotropic modifications. The slope of the melting curve given in figure below is typical for the majority of the normal substances which obey the correlation $v^{(l)} > v^{(s)}$. However, for the case where $v^{(l)} < v^{(s)}$ in the vicinity of the triple point of some anomolous substances (for example, water), the slope of the curve is negative ($dp_l^{(sl)}/dT < 0$). The vapor-liquid equilibrium curve starts at the triple point and finishes at the critical point T_c, p_c at which liquid and gaseous phases become identical. At supercritical parameters $T > T_c$, $p > p_c$. There is no phase transformation of type l → g and the substance is in high density state, which has sometimes been called the fluid state (f).

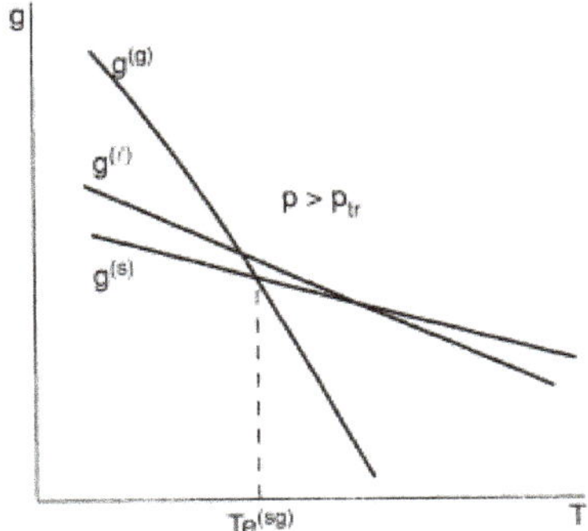

Figure: Phasic Gibbs energy as functions of temperature at pressures above the triple point.

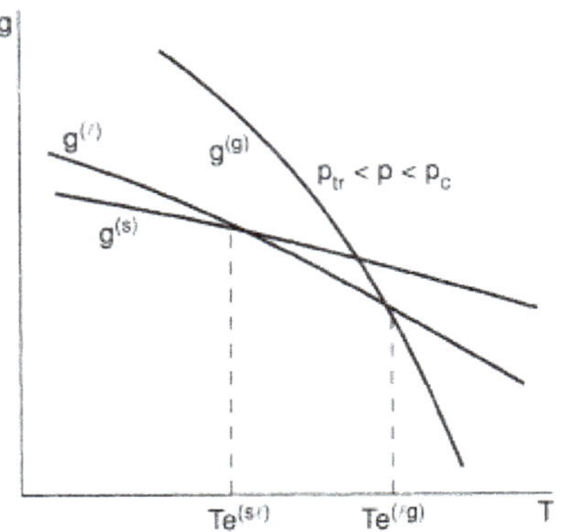

Figure: Phasic Gibbs energy as a function of temperature at a pressure between the triple and critical points.

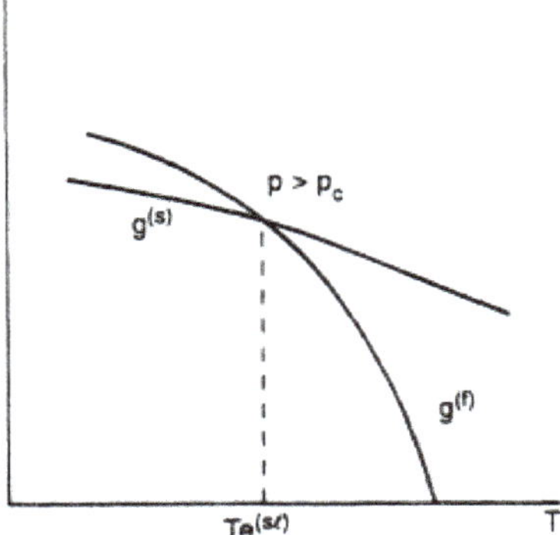

Figure: Phasic Gibbs energy as a function of temperature at pressure.

According to the Second Law of Thermodynamics, the stable states of substances at given T and p relate to minimal values of the Gibbs energy g (T,p). An unstable state can only exist transiently. Metastable states may exist indefinitely until some sufficiently strong external influence causes the system to pass into the stable state. Stability of the metastable state diminishes as the parameters of the system depart more and more from the equilibrium state. The metastable states of a superheated liquid (which doesn't boil) and a supercooled liquid (which doesn't form crystals) as well as a supercooled vapor (which doesn't condense) are frequent in practice; these states are represented by lines above the lines which have minimum Gibbs energy.

The thermodynamic properties ψ of the heterogeneous system— which may be defined as a first derivative of the Gibbs energy with respect to the parameters in Eq. $(\partial g/\partial T)_p = -s, (\partial g/\partial p)_T = v$ —specific (molar) volume, entropy, enthalpy, etc.— may be calculated by summing the same property $\psi^{(\alpha)}$ related to the phase under equilibrium

$$\psi = \sum_{\alpha-1}^{\varphi} z^{(\alpha)} \psi^{(\alpha)},$$

where $z^{(\alpha)} = N^{(\alpha)}/N$ is a mass (or mole) fraction of phase α, N is the total number of moles in the system. Of course,

$$\sum_{\alpha-1}^{\varphi} z^{(\alpha)} = 1.$$

Under equilibrium, for a pure substance, the fraction zα may have any value between 0 and 1 at given p and T subject to these values obeying equation above. An example is the moisture fraction $z^{(1)}$ in the case of vapor-liquid equilibrium. Here, $z^{(1)}$ is referred to as Quality. This type of state of the heterogeneous systems is called indifferent. In such states, as well as at the critical point of pure substances—thermal expansion $(\partial s/\partial T)_p$, isothermal compressibility $(\partial v/\partial p)_T$, as well as an isobaric heat capacity $c_p = T(\partial v/\partial T)_p$ —are infinitely large. Also, constant volume values of specific heat capacity $c_v = T(\partial s/\partial T)_v$, as well as the derivative $(\partial p/\partial T)_v$ vary with a jump, when the substance passes from a one-phase into two-phase state; they stay defined in the two-phase region. Figures below illustrate the described situation for pure substances with dependencies of specific volume and entropy on temperature and pressure. The heavy lines relate to borders of the coexisting phases regions, the weak lines are isotherms, and the broken ones are isochores.

If the total composition of the v-component, φ-phase system is given, then it follows that:

$$\bar{x}_i = N_i / N; N_i = \sum_{\alpha-1}^{\varphi} N^{(\alpha)};$$

$$N = \sum_{i-1}^{v} N_i = \sum_{\alpha-1}^{\varphi} N^{(\alpha)}; \sum_{i-1}^{v} \bar{x}_i = 1,$$

and hence:

$$\sum_{\alpha-1}^{\varphi} z^{(\alpha)} \bar{x}_i^{(\alpha)} = \bar{x}_i; 1 \le i \le v.$$

The numbers of the equilibrium equations together with the material balance equations are exactly equal to the number of the values $x_i^{\alpha}, z^{(\alpha)}$, which need to be defined.

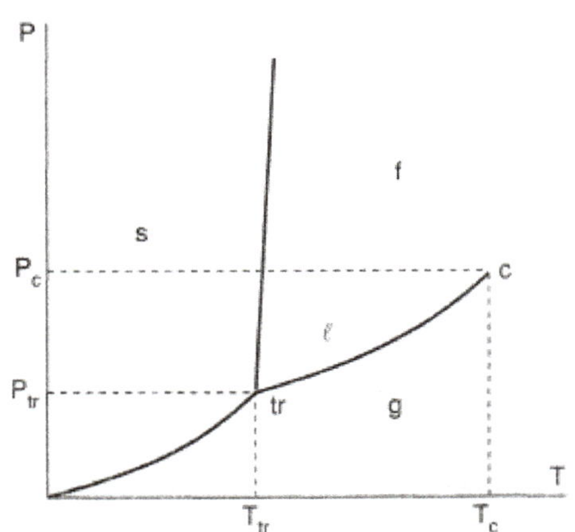

Figure: Phase diagram in terms of pressure and temperature.

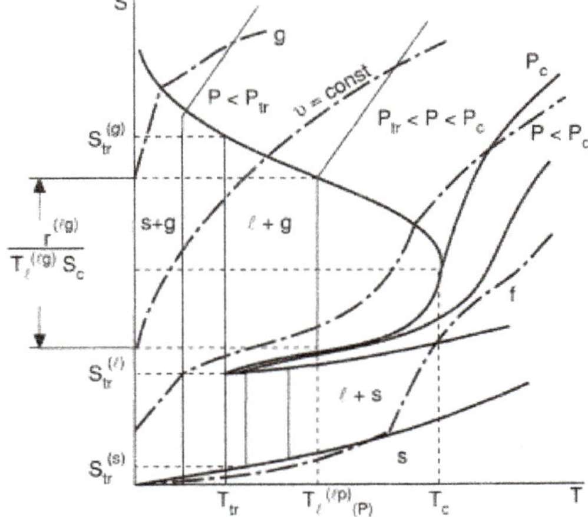

Figure: Phase diagram in terms of entropy and temperature.

The thermodynamic behavior of the multicomponent heterogeneous systems is identical to that which has been described above for pure substances. In particular, appearance (or disappearance) of the phase causes a discontinuity in the temperature and pressure dependencies of those thermodynamic functions, which may be defined as the first derivatives of the Gibbs energy (v, s, h, etc.). As to the second derivatives ($(\partial v/\partial T)_p$, $(\partial v/\partial p)_T$, c_p, c_v, etc.), they change the values in a jumpwise manner. Indifferent thermodynamic states are inherent to azeotropic mixtures and non- or mono-variant multicomponent mixtures.

The number of types of the phase equilibrium is far greater in multicomponent systems in comparison with pure substances. In particular, at certain conditions, some liquids or high-density gases may split into two phases, which have the same aggregate states but different compositions (liquid-liquid or gas-gas equilibria).

The types of phase equilibrium described above relate to the first order of phase change. By contrast, the second order of phase changes (ferromagnetic-paramagnetic, normal conductor— superconductor, normal—superfluid helium, etc.) are not accompanied by volume and heat effects; that is why, in such situations, there are no phase equilibrium states.

Lever Rule

The lever rule is one of the cornerstones of understanding and interpreting phase diagrams. A portion of a binary phase diagram is shown in figure In this diagram, all phases present are solid phases. There are two single-phase fields labeled α and β , separated by a two-phase field labeled $\alpha+\beta$. It indicates that, at a temperature such as b, pure metal A can dissolve metal B in any proportion up to the limit of the single-phase α field at composition a.

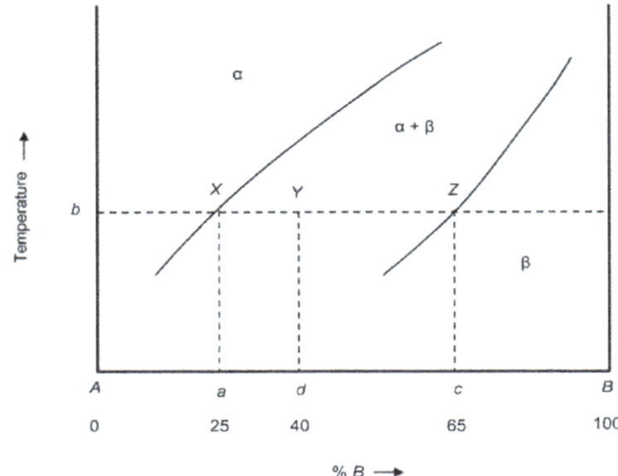

Figure: Portion of hypothetical phase diagram.

At the same temperature, metal B can dissolve metal A in any proportion up to composition c, which, at this temperature, is the boundary of the single-phase β field. Therefore, at temperature b, any alloy that contains less than a% of metal B will exist at equilibrium as the homogeneous α solid solution; and any alloy containing more than c% of metal B will exist as the β solid solution. However, any alloy whose overall composition is between a and c (e.g., at d) will, at the same temperature, contain more metal B than can be dissolved by the α and more metal A than can be dissolved by the b. It will therefore exist as a mixture of α and β solid solutions. At equilibrium, both solid solutions will be saturated. The composition of the α phase is therefore a% of metal B and that of the α phase is c% of metal B.

When two phases are present, as at composition Y in Fig. 9, their relative amounts are determined by the relation of their chemical compositions to the composition of the alloy. This is true because the total weight of one of the metals, for example metal A, present in the alloy must be divided between the two phases. This division can be represented by:

$$W_o\left(\frac{\%A_o}{100}\right)=W_\alpha\left(\frac{\%A_\alpha}{100}\right)+W_\beta\left(\frac{\%A_\beta}{100}\right)$$

$$\begin{pmatrix}\text{weight of metal}\\ \text{A in alloy}\end{pmatrix}=$$

$$\begin{pmatrix}\text{weight of metal}\\ \text{A in } \alpha \text{ phase}\end{pmatrix}+\begin{pmatrix}\text{weight of metal}\\ \text{A in } \beta \text{ phase}\end{pmatrix}$$

where W$_o$, W_α, and W_β are the weights of the alloy, the α phase, and the b phase, respectively, and %A$_o$, $\% A_\alpha$, and $\% A_\beta$ are the respective chemical compositions in terms of metal A. Because the weight of the alloy is the sum of the weight of the α phase and the weight of the b phase, the following relationship exists:

$$W_o=W_\alpha+W_\beta$$

This equation can be used to eliminate W_α from equation,

$$W_o\left(\frac{\%A_o}{100}\right)=W_\alpha\left(\frac{\%A_\alpha}{100}\right)+W_\beta\left(\frac{\%A_\beta}{100}\right)$$

$$\begin{pmatrix}\text{weight of metal}\\ \text{A in alloy}\end{pmatrix}=$$

$$\begin{pmatrix}\text{weight of metal}\\ \text{A in } \alpha \text{ phase}\end{pmatrix}+\begin{pmatrix}\text{weight of metal}\\ \text{A in } \beta \text{ phase}\end{pmatrix}$$

,

and the resulting equation can be solved for W_β to give the expression:

$$W_\beta=W_o\left(\frac{\%A_o-\%A_\alpha}{\%A_\beta-\%A_\alpha}\right)$$

Although a similar expression can be obtained for the weight of the α phase W_α, the weight of the a phase is more easily obtained by means of equation $W_o=W_\alpha+W_\beta$.

Because the weight of each phase is determined by chemical composition values according to Eq above, the tie-line ac shown in figure above can be used to obtain the weights of the phases. In terms of the lengths in the tie-line, equation above can be written as:

$$W \quad W\left(\frac{\text{length of line } da}{\text{length of line } ca}\right)$$

where the lengths are expressed in terms of the numbers used for the concentration axis of the diagram. The lever rule, or inverse lever rule, can be stated: The relative amount of a given phase is proportional to the length of the tieline on the opposite side of the alloy point of the tie-line. Thus, the weights of the two phases are such that they would balance.

Using equation above, the weight of the β phase at composition Y:

$$W\beta = W\alpha\left(\frac{40-25}{65-25}\right) = W\alpha(0.375)$$

The percentage of β b phase can be determined by use of:

$$\text{Percentage of } \beta \text{ phase} = \left(\frac{W_\beta}{W_o}\right)100$$

$$= \left(\frac{\%A_o - \%A_\alpha}{\%A_\beta - \%A_\alpha}\right)100$$

At composition Y the percentage of β phase is:

$$\%\beta\text{phase} = \left(\frac{40-25}{65-25}\right)100 = 37.5\%$$

The percentage of α phase is the difference between 100% and 37.5%, or 62.5%.

www.ingramcontent.com/pod-product-compliance
Lightning Source LLC
Chambersburg PA
CBHW061300190326
41458CB00011B/3729
9781632386977